CDMA Cellular Mobile Communications and Network Security

Dr. Man Young Rhee

Professor Emeritus
Hanyang University

President
Korea Institute of Information Security and Cryptology

Chairman, Board of Directors
Korea Information Security Agency
Seoul, Korea

Prentice Hall PTR
Upper Saddle River, NJ 07458

ISBN 0-13-598418-1

9 780135 984185

90000

Library of Congress Cataloging-in-Publication Data

Rhee, Man Young.
    CDMA cellular mobile communications & network security / by Man
  Young Rhee.
      p.   cm.
    ISBN 0-13-598418-1
    1. Code division multiple access. 2. Telecommunication systems-
  -Security measures. I. Title
  TK5103.45.R49 1998
  621.3845'6--dc21   97-26391
                              CIP

Editorial/Production Supervision: *Eileen Clark*
Acquisitions Editor: *Bernard Goodwin*
Marketing Manager: *Betsy Carey*
Buyer: *Julia Meehan*
Cover Design: *Design Source*
Cover Design Direction: *Jerry Votta*
Editorial Assistant: *Diane Spina*

© 1998 Prentice Hall PTR
Prentice-Hall, Inc.
A Simon & Schuster Company
Upper Saddle River, NJ 07458

Prentice Hall books are widely used by corporations and government agencies for training, marketing, and resale. The publisher offers discounts on this book when ordered in bulk quantities.

For more information, contact
        Corporate Sales Department,
        Phone: 800-382-3419; FAX: 201-236-7141
        E-mail (Internet): corpsales@prenhall.com
Or write: Prentice Hall PTR
        Corp. Sales Department
        One Lake Street
        Upper Saddle River, NJ 07458

Printed in the United States of America

10 9 8 7 6 5 4 3 2 1

ISBN 0-13-598418-1

Prentice-Hall International (UK) Limited, *London*
Prentice-Hall of Australia Pty. Limited, *Sydney*
Prentice-Hall Canada Inc., *Toronto*
Prentice-Hall Hispanoamericana, S.A., *Mexico*
Prentice-Hall of India Private Limited, *New Delhi*
Prentice-Hall of Japan, Inc., *Tokyo*
Simon & Schuster Asia Pte. Ltd., *Singapore*
Editora Prentice-Hall do Brasil, Ltda., *Rio de Janeiro*

Table of Contents

Chapter 7
Brief Survey of One-Way Hash Functions and Messge Digest *349*

Chapter 8
Authentication, Secrecy, and Identification *355*

Chapter 10
Forward W-CDMA Channel   463

Preface

This book mainly covers topics that can be applied to Code Division Multiple Access (CDMA) which is receiving a great deal of attention as a promising technology for future generations of mobile communications systems.

For the cellular industry, selection of the most appropriate access method is a challenging task. CDMA is an attractive technique for wireless access to broadband services. To meet this challenge, we have to be familiar with the technologies and system architectures on the CDMA digital cellular system.

In the last six years, the wireless (or radio) communications field has changed very rapidly. This book is intended to motivate the reader to further explore this challenging area. In early 1990, QUALCOMM Incorporated of San Diego, California, pioneered to introduce the intensive system concepts and the innovative implementation approaches on CDMA spread spectrum digital cellular systems. This CDMA system was standardized and is known as the IS-95 standard of the Telecommunications Industry Association and the Electronic Industries Association (TIA/EIA/IS-95).

A comprehensive analytical treatment is given for helping practicing engineers in planning and designing an efficient CDMA cellular network. It is also designed for graduate students to study the principles underlying spread spectrum cellular systems. Most materials, particularly CDMA channel structures, presented in this text are the embodiment of many of the principles and system architectures from IS-95.

Numerous worked-out examples contained within the book are presented by quantitative values in order for the beginner to better understand the CDMA cellular system.

The following is a summary of the contents of each chapter.

Chapter 1 presents a general overview of the CDMA digital cellular system which is simply explained in terms of modulation and multiple access based on spread spectrum communications.

Chapter 2 introduces fundamental and practical elements essentially required for CDMA channel operation.

Chapters 3 and 4 cover the overall structures of CDMA channel in detail and discuss their characteristics and functions. The forward CDMA channel consists of the pilot, sync, paging, and forward traffic channels. The pilot channel is an unmodulated, direct-sequence spread spectrum signal transmitted at all times by each CDMA base station. The mobile station monitors the pilot channel to acquire the timing of the forward CDMA channel and provides a phase reference for coherent demodulation. The sync channel transports the synchronization message to the mobile station in order to acquire initial time synchronization. The paging channel is also an encoded, interleaved, spread and modulated spread spectrum signal used for transmission of control information and pages from a base station to a mobile station. The forward traffic channel is used for the transmission of user and signaling traffic from the base station to a specific mobile station during a call. Each of these code channels except the pilot channel is convolutionally encoded, block interleaved, orthogonally spread by the appropriate Walsh function and is then spread by a quadrature pair of pilot PN sequences at a fixed chip rate of 1.2288 Mcps.

Data scrambling applies to the paging channel and the forward traffic channel as well. Data scrambling is performed on the block interleaver output at the modulation symbol rate 19.2 ksps. Data scrambling is accomplished by performing the modulo-2 addition of the interleaver output with decimated binary value of the long code. The long code is a PN sequence with period $2^{42}-1$ that is used for *scrambling* on the forward CDMA channel (i.e., paging and forward traffic channels) and *spreading* on the reverse CDMA channel (i.e., access and reverse traffic channels). The reverse CDMA channel is composed of the access channels and reverse traffic channels. All data transmitted on the reverse CDMA channel are convolutionally encoded, block interleaved, orthogonally modulated by the 64-ary Walsh functions and direct-sequence spread by the long code chips prior to transmission. The data burst randomizer is used in the reverse traffic channel which generates a masking pattern of 0s and 1s that randomly masks out the redundant data generated by the code repetition.

Chapters 5 and 6 describe call processing of CDMA code channels, based on the Qualcomm system, including handoff procedures.

Chapter 7 presents a brief survey of one-way hash functions and message digest. One-way functions are a fundamental building block for most of the protocols for either conventional symmetric algorithm or public-key cryptography. A one-way hash function is relatively easy to compute but significantly harder to reverse. Hash code algorithms for authentication data are listed systematically.

Chapter 8 presents authentication and message privacy. The scope of analysis deals with numerous techniques for computation of 18-bit hash codes from the 152-bit message block for CDMA cellular systems. The mobile station operates in conjunction with the base station to authenticate the identity of the mobile station. Authentication is the process by which information is exchanged between a mobile station and base station for the purpose of confirming the identity of the mobile station. A successful outcome of the authentication process occurs only when it can be demonstrated that the mobile station and base station process identical sets of shared secret data (SSD). SSD is a 128-bit shared secret data to be stored in semi-permanent memory in the mobile station. SSD is divided into two distinct subsets: SSD-A and SSD-B. SSD-A is used to support the authentication procedure; and SSD-B is used for supporting voice privacy and message confidentiality. The SSD update procedure is completely explained to compute SSD-A-NEW and SSD-B-NEW as the SSD-generated output. It also includes signaling message encryption and network security.

Chapters 9 and 10 deal with Wideband CDMA links based on JTC (AIR)/95. Data rates of the reverse and forward information channels are 64, 32, or 16 kbps; PN chip rate is 4.096 Mcps; and the symbol rate is 64 ksps.

Chapter 9 presents the reverse W-CDMA channel which is the communication link from the personal station to the base station. The reverse W-CDMA channel is composed of the access channel and reverse traffic channels. The access channel consists of the reverse pilot channel and reverse access channel. The reverse traffic channel consists of three different channels: the reverse pilot, information, and signaling channels. All data transmitted on the reverse traffic channel are convolutionally encoded, interleaved, and modulated by direct-sequence spreading prior to transmission. In Chapter 10, the forward W-CDMA channel consists of one pilot channel, one sync channel, up to eight paging channels, and a number of forward traffic channels (i.e., forward information and signaling channels). Each of these code channels is orthogonally spread by the appropriate Walsh code and is then spread by a pilot PN sequence at a fixed chip rate of 4.096 Mcps. The forward signaling channel of the forward traffic channel is convolutionally encoded, block interleaved, orthogonally spread with a Walsh function, quadrature modulated by a pilot PN sequence at a fixed chip rate of 4.096 Mcps, filtered, and finally transmitted by QPSK waveform.

This book may be considered a fundamental text book on the technical aspects of digital cellular systems for helping further research and development. I hope that the inclusion of a total of 148 problems with complete solutions makes the book more profitable for independent study. It is a real challenge to write this book in the early stage of this very rapidly evolving field. I would like to hear from readers who may find any serious mistakes in this text. Your feedback is sincerely welcome and shall help in improving future editions of this book.

I have benefited from the Qualcomm system which proved a great influence on my selection of CDMA systems covered in this book. I wish to express my special appreciation to Ms. Karen J. Gettman, (Executive Editor–Acquisitions), and Ms. Eileen Clark (Production Editor), Professional, Technical and Reference Division at Prentice Hall, who provided the guidance and support throughout the publishing stages from acquisition to production. I am indebted to Dr. Robert K. Morrow Jr. who provided helpful comments and criticism to evaluate the manuscript at the early stages and Mr. Michael Schiaparelli (Copy Editor) who reviewed the manuscript for general sense and organization, ensuring consistency of usage. Special thanks are due to Dr. Ji Hong Kim for his computer programming of various examples contained in this book. I owe a special thanks to Ms. Shin Jean for word processing the entire manuscript and the revised versions.

Man Young Rhee

CHAPTER 1

Introduction to
Cellular CDMA

T echniques involving spread spectrum (SS) modulation have been evolving over the last 40 years. Spread spectrum techniques were well established for anti-jam and multipath applications as well as for accurate ranging and tracking. These SS techniques are also proposed for CDMA to support simultaneous services for digital communication among a large community of users. CDMA is an attractive technique for wireless access to broadband services. The CDMA concept is explained simply in terms of modulation and multiple access schemes based on spread spectrum communication.

This chapter provides a general overview of the CDMA digital cellular system pioneered by QUALCOMM, Inc., of San Diego. Digital cellular applications based on a multiple access scheme were also developed in cooperation with a number of participating carrier and equipment manufacturers (AT&T, Motorola, Northern Telecom, and others). The CDMA system is in full compliance with the Cellular Telecommunications Industry Association (CTIA) requirements as a candidate for standardization (IS-95).

Typical digital cellular systems can be listed such as GSM (European scheme, 1990), NA-TDMA (North American IS-54 scheme, 1990), PDC (Japanese standard scheme, 1991), and CDMA (US IS-95 scheme, 1993).

The Global System for Mobile Communications (GSM) TDMA system was developed in Western Europe starting in June 1982. GSM offers the capability of extending through diverse telecommunication networks (i.e., ISDN) and compatibility throughout the European continent. In 1992, the first commercial GSM system was devised in Germany. GSM is based on a combination of Frequency Division Multiple Access (FDMA) and Time Division Multiple Access (TDMA).

The NA-TDMA system is similar to the GSM scheme. The only difference is that there is only one common radio interface in this system. Personal Digital Cellular (PDC) is the Japanese

TDMA cellular system operating at 800 MHz and 1.5 GHz. This system provides nine interfaces among the digital cellular networks and 1.5 GHz PDC was publicly put in service in 1994.

Besides digital multiple-access systems, there are TDD cordless phone systems like PHP, CT-2, DCT-900 (or CT-3), and DECT. Time-division duplexing (TDD) systems are digital systems and use only one carrier to transmit and receive information. Personal handy phone (PHP) is a wireless communication TDD system which supports personal communication services (PCS). PHP can be used for residential cordless phones, private wireless PBXs, public telepoints, and walkie-talkie communications. The cordless telecommunications system 2 (CT-2) is a digital second generation cordless telephone system which was developed by GPT Ltd. in the United Kingdom and was the first TDD system for mobile radio communications. The CT-2 system is one of the simplest PCS systems having a simple control structure with no multi-channel multiplexing. This CT-2 system has no channel coding, and handoff and paging are eliminated. Therefore, only outgoing calls are allowed. Call sizes are typically less than 200 meters in radius, allowing a single user to occupy a large bandwidth. Digital cordless telephone at 900 MHz (DCT-900, or CT-3) was developed by Ericsson, Sweden in 1988 as an upgraded CT-2 version and it is sometimes refered to as CT-3. CT-2 and DCT-900 were allowed to exist in the U.K. and Sweden until the availability of DECT. Digital European Cordless Telecommunication (DECT) is a European standard system for consideration as the second generation PCS system. DECT has been accepted as a European standard for cordless telephony much like CT-2 or DCT-900 but with improved resources for handling data transfer as well as voice. CDMA development started in early 1989 after the NA-TDMA standard (IS-54) was established. A CDMA feasibility test was held in November 1989. The CDMA IS-95 intermediate standard of the Electronic Industries Association was issued in December 1992. The CDMA system can employ dual-mode subscriber units to provide compatibility with the analog system. However, in this book we confine ourselves to discussing only the CDMA digital mode.

1.1 CDMA CELL COVERAGE

For CDMA cellular systems, the service area is divided into hexagonal cells as shown in Fig. 1.1. Each cell contains a base station which is connected to the Mobile Telephone Switching Office (MTSO) prior to voice encoding and decoding. In each cell there are two links consisting of the forward and reverse CDMA channels between the base station and each mobile station in the cell. The forward CDMA channel interprets the forward link from a base station to a mobile station in the cell. The reverse CDMA channel denotes the reverse link from the mobile station to the base station.

CDMA reuses the cellular ratio frequency and controls system capacity effectively because it is inherently an excellent anti-interference mode.

The forward CDMA channel consists of one or more code channels that are transmitted on a CDMA frequency assignment using a particular pilot PN offset. Each base station uses a time offset of the pilot PN sequence (called a spreading pseudonoise sequence) to identify a forward CDMA channel. Time offset can be reused within a CDMA cellular system.

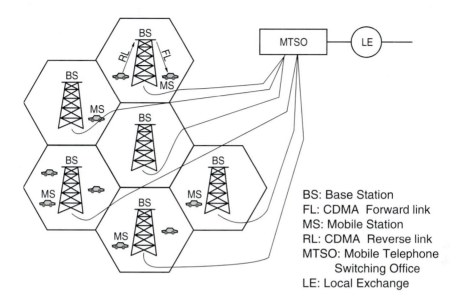

BS: Base Station
FL: CDMA Forward link
MS: Mobile Station
RL: CDMA Reverse link
MTSO: Mobile Telephone
Switching Office
LE: Local Exchange

Figure 1.1 CDMA forward/reverse cellular link geometry
in hexagonal cell coverage area

Each code channel transmitted on the forward CDMA channel is orthogonally spread by the appropriate Walsh function to provide orthogonal channelization among all code channels and is then spread by a quadrature pair (i.e., in-phase and quadrature-phase) of pilot PN sequences in order to transmit them by Quadrature Phase Shift Keying (QPSK) waveform.

The reverse CDMA channel is composed of access channels and reverse traffic channels. The access channel is used for short signaling message exchanges by providing for call originations, response to pages, orders, and registrations. A reverse traffic channel is used to transport user and signaling traffic from a single mobile station to one or more base stations. All data to be transmitted on the reverse CDMA channel are convolutionally encoded for error correction, block interleaved to avoid burst errors and to improve the system performance by the access redundancy, modulated by the 64-ary Walsh function to provide orthogonal channelization, and direct-sequence spread by the long code to achieve limited privacy prior to transmission.

1.2 STRUCTURAL LAYOUT OF CDMA CHANNELS

The overall structures of forward/reverse CDMA channels are shown in Figs. 1.2 and 1.3, respectively. The forward CDMA link consists of the pilot channel, sync channel, paging channels, and a number of forward traffic channels. A typical example of a forward CDMA channel consists of 64 code channels available for use. Out of the 64 code channels, the forward CDMA link comprises the pilot channel, one sync channel, seven paging channels, and 55 forward traffic channels.

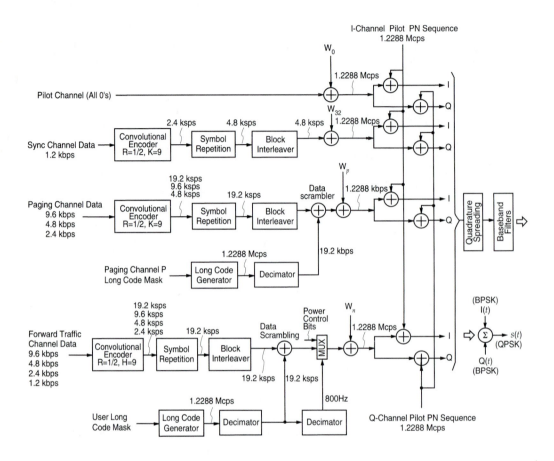

Figure 1.2 Forward CDMA code channel structure

The pilot channel is an unmodulated, direct-sequence spread spectrum signal transmitted at all times by each CDMA base station. The mobile station monitors the pilot channel to acquire the timing of the forward CDMA channel and provides a phase reference for coherent demodulation. Code channel number zero (W_0) is always assigned to the pilot channel.

The sync channel is assigned to the code channel number 32 (W_{32}) which transports the synchronization message to the mobile station. More importantly, the sync channel is an encoded, interleaved, spread, and modulated spread spectrum signal that is used by mobile stations to acquire initial time synchronization.

The paging channel is also an encoded, interleaved, spread and modulated spread spectrum signal used for transmission of control information and pages from a base station to a mobile station. Paging channels are assigned to code channel numbers one through seven (W_1–W_7) in sequence.

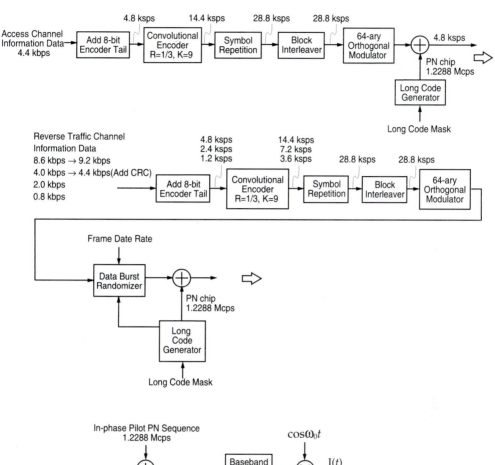

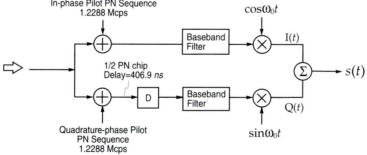

Figure 1.3 Reverse CDMA code channel structure

The forward traffic channel is used for the transmission of user and signaling traffic from the base station to a specific mobile station during a call. The maximum number of forward traffic channels is equal to 63 minus the number of sync and paging channels operating on the same forward CDMA channel.

Data rates at the channel input are as follows:

(1) the pilot channel sends all 0's at a 19.2 kbps rate, (2) the sync channel operates at a fixed rate of 1200 bps, (3) the paging channel supports the fixed data rate operation at 9600, 4800, or 2400 bps, and (4) the forward traffic channel supports variable data rate at 9600, 4800, 2400, or 1200 bps.

The sync channel, paging channel, and forward traffic channel are convolutionally encoded for error correction prior to transmission, but the pilot channel does not use convolutional encoding. Code symbols are generally defined as the output of an error-correcting encoder. Information bits are input to the encoder and code symbols are output from the encoder. For all code channels except the pilot channel, each encoded symbol is repeated prior to block interleaving whenever the information rate is lower than 9600 bps.

For paging and forward traffic channels, repetition depends on the data rate of each channel. Each code symbol at the 4.8 kbps data rate is repeated once (2 of each symbol). Each code symbol at the 2.4 kbps data rate is repeated 3 times (4 of each symbol). Each code symbol at the 1.2 kbps data rate is repeated 7 times (8 of each symbol). Thus, for all the data rates (9.6, 4.8, 2.4, and 1.2 kbps), symbol repetition will result in a constant modulation symbol rate of 19.2 ksps. For a sync channel, each encoded symbol is repeated once (two of each symbol) and the modulation symbol rate is 4800 sps.

All symbols after repetition on the sync channel, paging channel, and forward traffic channel are block interleaved. The purpose of using block interleaving is to protect the data from burst errors while sending them through a multipath fading environment. After interleaving, each code channel in the forward CDMA channel is orthogonally spread by one of 64 Walsh functions and is then spread by a quadrature pair of pilot PN sequences at a fixed chip rate of 1.2288 Mcps. The binary data ('0's and '1's) spread in quadrature are applied to the baseband filters. Following the baseband filtering, the forward CDMA channel combines the respective Binary Phase Shift Keying (BPSK) data modulated with the carriers to produce Quadrature Phase Shift Keying (QPSK) just before transmission.

Data scrambling applies to the paging channel and the forward traffic channel as well. Data scrambling is performed on the block interleaver output at the modulation symbol rate 19.2 ksps. Referring to Fig. 1.2, data scrambling is accomplished by performing the modulo-2 addition of the interleaver output with the decimated binary value of the long code. The long code is a PN sequence with period $2^{42}-1$ that is used for *scrambling* on the forward CDMA channel (i.e., paging and forward traffic channels) and *spreading* on the reverse CDMA channel (i.e., access and reverse traffic channels).

The long code mask is a 42-bit binary number that creates the unique identity of the long code. Each PN chip of the long code is generated by the modulo-2 inner product of a 42-bit mask and the 42-bit LFSR stage in the long code generator. The long code operating at 1.2288 MHz clock rate is equivalent to the PN chip sequence which is the output of the long code generator. Note that a PN chip is defined as one bit in the PN sequence. When the long code is divided into every 64 bits (or chips), the first bit of every 64 bits is used for data scrambling at a 19.2 ksps rate. The function of the decimator in Fig. 1.2 is to reduce the size of the long code by taking one out of every 64 (= $1.2288 \times 10^6/192 \times 10^2$) bits.

The base station does not insert a power control subchannel on the paging channel. But a power control subchannel on the forward traffic channel continuously transmits the power control bits at a rate of 800 bps, i.e., one bit ('0' or '1') every 1.25 ms (=1/800). A '0' power control bit indicates to the mobile station that it should increase the mean output power control level, and a '1' power control bit indicates to the mobile station that it should decrease the mean output power level. Thus, the mobile station will adjust its mean output power level in response to each valid power control bit received on the forward traffic channel.

The reverse CDMA channel is composed of access channels and reverse traffic channels. Figure 1.3 shows the overall structure of the reverse CDMA channel. Data transmitted on the reverse CDMA channel is grouped into 20 ms frames. All data transmitted on the reverse CDMA channel is convolutionally encoded for random-error correction, block interleaved for protection from burst errors, modulated by the 64-ary Walsh codes, consisting of each 64 chips long, and direct-sequence spread by the long code of period $2^{42}-1$ chips prior to transmission.

The data burst randomizer is not used when the mobile station transmits on the access channel. But in the reverse traffic channel, the data burst randomizer generates a masking pattern of 0s and 1s that randomly masks out the redundant data generated by the code repetition. The reverse traffic channel and the access channel are direct-sequence spread by the long code. This spreading operation involves modulo-2 addition of the output stream from the data burst randomizer and the long code. Following the direct sequence spreading, the reverse traffic channel and access channel are spread in quadrature as shown in Fig. 1.3. Notice that the Q-channel data spread by the quadrature-phase pilot PN sequence is delayed by half of a chip time (406.9 ns) with respect to the I-channel data spread by the in-phase pilot PN sequence.

A frame is defined as a basic timing interval in the system. For the access channel, paging channel, and forward/reverse traffic channels, a frame is 20 ms long. For the sync channel, a frame is 26.666 ms in length.

The frame quality indicator is the CRC check applied to 9600 bps and 4800 bps traffic channel frames. The cyclic redundancy code (CRC) is a class of linear error detecting codes which generate parity check bits by finding the remainder of a polynomial division. The frame quality indicator (CRC) supports two functions at the receiver. The first function is to determine the transmission rate of the frame, the second is to determine whether the frame is in error.

Encoder tail bits represent a fixed sequence of bits added to the end of a data frame to reset the convolutional encoder to a known state.

Data transmitted on either the reverse CDMA channel or the forward CDMA channel are grouped in 20 ms frames.

The frame structures on both the forward traffic channel and the reverse traffic channel are described as follows:

(I) Each reverse traffic channel frame sent at:

1. The 9600 bps data rate consists of 192 bits which are composed of 172 information bits followed by 12 frame quality indicator (CRC) bits and 8 encoder tail bits.
2. The 4800 bps data rate consists of 96 bits which are composed of 80 information bits followed by 8 CRC bits and 8 encoder tail bits.
3. The 2400 bps rate consists of 48 bits which are composed of 40 information bits followed by 8 encoder tail bits. The 8 CRC bits are not used for the reverse traffic channel frame at the 2400 bps data rate.
4. The 1200 bps data rate consists of 24 bits which are composed of 16 information bits followed by 8 encoder tail bits. The 8 CRC bits are not used for the reverse traffic channel frame at the 1200 bps data rate.
5. Each access channel frame contains 96 bits (20 ms frame at 4800 bps rate) which consist of 88 information bits and 8 encoder tail bits. The 8-bit CRC are not used for the access channel frame.

(II) Each forward traffic channel frame sent at:

1. The 9600 bps data rate consists of 192 bits which are composed of 172 information bits followed by 12 frame quality indicator (CRC) bits and 8 encoder tail bits.
2. The 4800 bps data rate consists of 96 bits which are composed of 80 information bits followed by 8 frame quality indicator (CRC) bits and 8 encoder tail bits.
3. The 2400 bps data rate consists of 48 bits which are composed of 40 information bits followed by 8 encoder tail bits. The 8-bit CRC is not used for the forward traffic channel frame at the 2400 bps data rate.
4. The 1200 bps data rate consists of 24 bits which are composed of 16 information bits followed by 8 encoder tail bits. The 8-bit CRC is not used for the forward traffic channel frame at the 1200 bps data rate.

The forward traffic channel is 20 ms in length. The data rate shall be selected on a frame-by-frame (i.e., 20 ms) basis. Although the data rate may vary on a frame-by-frame basis, the modulation rate is kept constant by code repetition at 19.2 ksps. The bit rate for the sync channel is 1200 bps. A sync channel frame is 26.66 ms in length. The paging channel frame is 20 ms in length.

1.3 CHARACTERISTICS AND FUNCTIONS OF CDMA CHANNEL LINK

Since CDMA uses forward and reverse channels, functional relationships exist between the base station and the mobile station in a cell coverage area. The following summary (Table 1.1) explains the functions of the base/mobile stations and the characteristics of all the code channels in the CDMA channel. This brief survey includes system acquisition, timing, synchronization, interleaving, orthogonal channelization, spreading techniques, power control, call processing, handoff procedures, authentication, and message privacy.

Table 1.1   CDMA Cellular System

Four overhead messages are conveyed between the base station and mobile station in order to meet the required functions in the CDMA cellular system.

Overhead Messages	
System parameter message:	Contains paging channel, registration parameters, parameter to aid pilot acquisition, etc.
Access parameter message:	Contains access channel and control parameters. Some of these control parameters provide dynamic feedback to the mobile station to control its transmission rate, and thus serves to stabilize the access channel.
Neighbor list message:	Contains information to speed handoff to a neighboring base station, including the time offset of the pilot PN and the basic neighbor configuration.
CDMA channel list message:	Lists CDMA frequency assignments that contain paging channels. This allows the mobile station to correctly determine where to find its paging channel.

Mobile Station	Channels	Base Station
Pilot Channel		
•The mobile station monitors the pilot channel at all times except when not receiving in the slotted mode. The slotted mode is an operation mode for which the mobile station monitors only selected slots on the paging channel when in the mobile station idle state.	•The pilot channel is a reference channel which the mobile station uses for acquisition, timing, and as a phase reference for coherent demodulation.	•The base station continually transmits a pilot channel for every CDMA channel supported by the base station.
•A mobile station uses the pilot channel for synchronization.	•A pilot is transmitted at all times by the base station on each active forward CDMA channel. The pilot channel is an unmodulated spread spectrum signal that is used for synchronization by a mobile station operating within the coverage area of the base station.	•During pilot and sync channel processing, the base station transmits the pilot channel and sync channel which the mobile station uses to acquire and synchronize to the CDMA system.

Table 1.1   CDMA Cellular System *(continued)*

Pilot Channel *(continued)*		
•The mobile station acquires the pilot channel of a CDMA system and provides a phase reference for coherent demodulation.		
Sync Channel		
•Receives sync channel message.	•Sync channel is used for the system acquisition.	•Sends sync channel message.
•Adjusts its timing to normal system timing.	•Once the mobile station acquires the system, it will not normally reuse the sync channel.	•Only one message is sent on the sync channel, i.e., sync channel message.
•Determines and begins monitoring its paging channel.	•Since the pilot PN sequences are offset differently for each base station, the flaming of the sync channel is different for every base station.	•Sync channel frame is the length of pilot PN sequence.
•Sync channel is an encoded, interleaved, spread, and modulated spread spectrum signal that is used by mobile stations to acquire initial time synchronization.	•Sync channel transports the synchronization message to the mobile station.	•Transmits the sync channel frame time-aligned with the pilot PN sequence.
Paging Channel		
•Monitors only a single paging channel.	•A paging channel is determined by hashing over all the available paging channels.	•The base station may also assign a mobile station to a particular paging channel.
•Orders (a broad class of messages) are used to control a particular mobile station.	•Data rates: 2.4, 4.8, 9.6 kbps (paging channels on different CDMA frequency assignments).	•Page message contains pages to one or more mobile stations.
•Orders are used for everything from acknowledging registration to locking or preventing an errant mobile station from transmitting.	•CDMA frequency assignments listed in CDMA channel list message allow the mobile station to correctly determine where to find its paging channel.	•Sends pages when the base station receives a call for the mobile station and pages are usually sent by several different base stations.

Table 1.1   CDMA Cellular System *(continued)*

Paging Channel *(continued)*		
•The channel assignment message allows the base station to assign a mobile station to the traffic channel, change it paging channel assignment, or direct the mobile station to use the analog FM system.	•Paging channel messages convey information from the base station to the mobile station.	•The paging channel is an encoded, interleaved, spread and modulated spread spectrum signal used for transmission of control information and pages from a base station to a mobile station.
•Data scrambling applies to the paging channel by performing the modulo-2 addition of interleaver output with the long code.	•For major types of messages: overhead, paging, order and channel assignment.	
Access Channel		
•Responds to a paging channel message by transmitting on one of the associated access channels.	•The access channel is a random access CDMA channel.	•Responds to transmissions on a particular access channel by a message on an associated paging channel.
•Chooses randomly both an access channel from the set of available access channels and a PN time alignment from the set of PN time alignments.	•Provides communications from the mobile station to the base station when the mobile station is not using a traffic channel.	•Unless two or more mobile stations choose the same access channel and the same PN time alignment, the base station is able to receive simultaneous transmissions.
•Provides communications to the base station when a traffic channel is not used at the mobile station.	•Uses only the special 4800 bps mode in our case.	•Controls the rate of access channel transmissions to prevent too many simultaneous transmissions by multiple mobile stations.
•Distinguished by a different long PN code.	•Provides for: Call originations Responses to pages Orders Registrations	•Control of access channel transmissions is accomplished through the parameters contained in the access parameters message which is sent on the paging channel.
	•One or more access channels is paired with every paging channel.	

Table 1.1   CDMA Cellular System *(continued)*

Access Channel *(continued)*		
	•The access channel transmission rate can be varied for different types of transmissions and for different classes of mobile stations. But one particular data rate should be chosen.	
Traffic Channel		
•When the mobile station has been assigned to a traffic channel, signaling (either blank-and-burst or dim-and-burst) occurs directly on the traffic channel.	•Both the forward and reverse traffic channels use a similar control structure consisting of 20 ms frames. Frames can be sent at either 9.6, 4.8, 2.4, or 1.2 kbps.	•During traffic channel processing, the base station uses the forward and reverse traffic channels to communicate with the mobile station while the mobile station is in the mobile station control on the traffic channel state.
•Blank-and-burst signaling is sent at 9.6 kbps and replaces one or more frames of primary traffic data, typically vocoded voice, with signaling data, as in the analog FM system.	•There are five types of control messages on the traffic channel: Messages controlling the call itself Messages controlling handoff Messages controlling the forward line power Messages for security and authentication Messages eliciting or supplying special information from or to the mobile station.	•Traffic channel processing consists of the following substates: Traffic channel initialization substate—In this substate, the base station begins transmitting on the forward traffic channel and receiving on the reverse traffic channel. Waiting for order substate—In this substate, the base station sends the alert with information message to the mobile station. Conversation substate—In this substate, the base station exchanges primary traffic bits with the mobile station's primary service option application. Release substate—In this substate, the base station disconnects the call.

Table 1.1   CDMA Cellular System *(continued)*

Traffic Channel *(continued)*		
•Dim-and-burst signaling sends both signaling and primary traffic data in a frame using the 9.6 kbps transmission rate. Dim-and-burst signaling has an immense advantage over blank-and-burst degradation in voice quality being essentially undetectable.	•The forward traffic channel is used for transmission of user and signaling traffic from the base station to a specific mobile station during a call.	•During traffic channel operation, the base station and mobile station may support primary traffic services. Each such service option has a set of requirements that govern the way in which the primary traffic bits from forward and reverse traffic channel frames are processed by the base station and mobile station. Either the base station or mobile station can request a service option. The base station can request a particular service option when paging the mobile station or during traffic channel operation.
•The mobile station supports the following handoff procedures: In case of commencing communications with a new base station without interrupting communications with the old base station (Soft handoff). In case of transitions between disjoint sets of base stations, different frequency assignments, or different frame offset (Hard handoff).	•A reverse traffic channel is used to transport user and single traffic from a signaling mobile station to one or more base stations.	
•Encryption and Authentication: Authentication is achieved only when the base station possesses identical sets of shared secret data (SSD) with the mobile staiton. A 128-bit SSD is stored in the mobile station and it is portitioned into distinct subsets: SSD-A and SSD-B. The 64-bit SSD-A is used to support the authentication. The 64-bit SSD-B is used for CDMA voice privacy and date confidentiality.	•Data scrambling applies to the forward traffic channel. Data spreading applies to the reverse traffic channel.	

1.4 CALL PROCESSING

Call processing can be separated into two parts, mobile station call processing and base station call processing. Call processing refers to the technique of message flow protocols between the mobile station and the base station.

1.4.1 Mobile Station Call Processing

As illustrated in Fig. 1.4, mobile station call processing consists of the following four states.

1. Mobile station initialization state

   In this state, the mobile station must:
   - Select which system to use (Analog or CDMA operation).
   - Acquire the pilot channel of the selected CDMA system within 20 ms.
   - Receive and process the sync channel message to obtain system configuration and timing information.
   - Synchronize its long code timing and system timing to those of the CDMA system.

2. Mobile station idle state

   In this state, the mobile station monitors messages on the paging channel. The mobile station can receive messages, receive an incoming call, initiate an originating call, initiate a registration, or initiate a message transmission.

3. System access state

   In this state, the mobile station sends messages to the base station on the access channel and receives messages from the base station on the paging channel. The system access state consists of the following substates and the mobile station must:
   - Monitor the paging channel until it has received a current set of configuration messages.
   - Send an origination message to the base station.
   - Send a paging response message to the base station.
   - Send a response to a message received from the base station.
   - Send a registration message to the base station.
   - Send a data burst message to the base station.

   The mobile station transmits on the access channel using a random access procedure. The entire process of sending one message and receiving an acknowledgment for that message is called an access attempt. Each transmission in the access attempt is called an access probe. The mobile station transmits the same message in each access probe in an access attempt. Each access probe consists of an access channel preamble and an access channel message capsule.

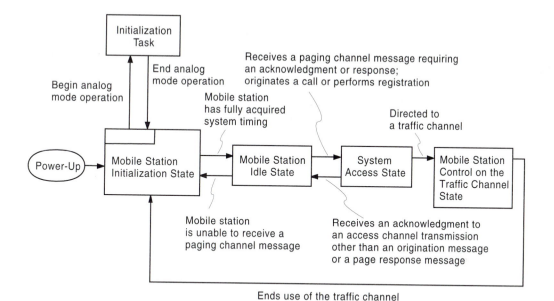

Figure 1.4 Mobile station call processing states

4. Mobile station control on the traffic channel state

In this state, the mobile station communicates with the base station using the forward and reverse traffic channels.

This state consists of the following substates and the mobile station must:

- Verify that it can receive the forward traffic channel and begin transmitting on the reverse traffic channel.
- Wait for an order on an alert with information message.
- Wait for the user to answer the call.
- Exchange primary traffic packets with the base station under its primary service option application.
- Disconnect the call in this release substate.

1.4.2 Base Station Call Processing

Base station call processing refers to the method relating to the message flow between the base station and the mobile station.

Base station call processing consists of the following types of processing.

1. Pilot and sync channel processing

   During pilot and sync channel processing, the base station transmits the pilot channel and sync channel which the mobile station uses to acquire and synchronize to the CDMA system while the mobile station is in the mobile station initialization state.

2. Paging channel processing

   During paging channel processing, the base station transmits the paging channel which the mobile station monitors to receive messages while the mobile station is in the mobile station idle state and system access state.

3. Access channel processing

   During access channel processing, the base station monitors the access channel to receive messages which the mobile station sends while the mobile station is in the system access state.

4. Traffic channel processing

   During traffic channel processing, the base station uses the forward and reverse traffic channels to communicate with the mobile station while the mobile station is in the mobile station control on the traffic channel state.

1.5 Authentication and Message Confidentiality

Authentication refers to the process by which the base station confirms the identity of the mobile station. The successful authentication can be achieved only when the base station possesses identical sets of shared secret data (SSD) with the mobile station. SSD is a 128-bit pattern stored in the mobile station and it is partitioned into two distinct subsets, that is, SSD-A and SSD-B. The 64-bit SSD-A is used to support the authentication and the 64-bit SSD-B is used for CDMA voice privacy and data confidentiality.

The base station is equipped with a database that includes unique mobile station authentication keys (A-keys) and shared secret data (SSD) for each registered mobile station in the cell system. This database is used for authentication of mobile stations. If the base station supports mobile station authentication, the base station may send and receive authentication messages and perform the authentication calculations. SSD is updated using the SSD-generation procedure initialized with the mobile station specific information (ESN), random data (RANDSSD), and the mobile station's A-key. This 64-bit A-key is stored in the mobile station's permanent security and identification memory and is known only to the mobile station and to its associated Home Location Register/Authentication Center (HLR/AC).[1]

The SSD update procedure is described as shown in Fig. 1.5 and performed as follows:

The base station sends an SSD update message order on either the paging channel or the forward traffic channel. Upon receipt of the SSD update message, the mobile station sets the

---

1.   HLR is the home location register to which a mobile station identification number is assigned for record purposes, such as subscriber information. AC is the authentication center that manages the authentication information related to the mobile station.

input parameters (RANDSSD, ESN, A-key) to the SSD-generation algorithm. The mobile station then executes the SSD-generation procedure. SSD-A NEW and SSD-B NEW are generated as the outputs of the SSD-generation procedure. The mobile station then selects a 32-bit random number (RANDBS), and sends it to the base station in a base station challenge order on the access channel or reverse traffic channel.

　　Both the mobile station and the base station then set the input parameters (RANDBS, ESN, MIN1, SSD-A-NEW) of the Auth-Signature procedure as illustrated in Fig. 1.5 and execute the Auth-Signature procedure.

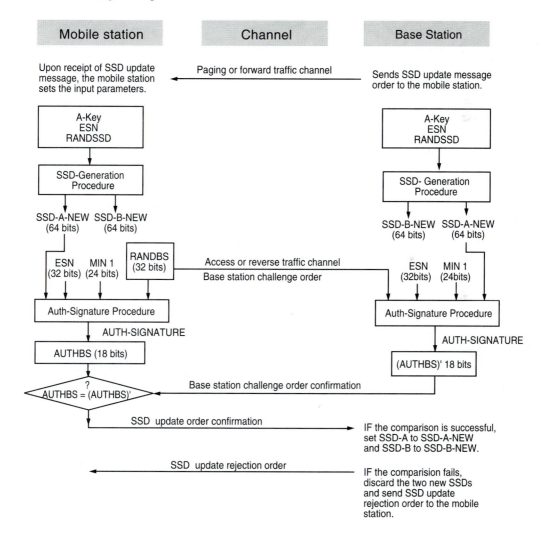

Figure 1.5 SSD update message flow

AUTHBS is set to the 18-bit result AUTH-SIGNATURE. The base station sends its computed value of AUTHBS to the mobile station in a base station challenge confirmation order on the paging channel or the forward traffic channel. Upon receipt of the base station challenge confirmation order, the mobile station compares the received value of AUTHBS to its internally generated value.

If the comparison is successful, the mobile station executes the SSD-Update procedure to set SSD-A and SSD-B to SSD-A-NEW and SSD-B-NEW, respectively. The mobile station then sends an SSD update confirmation order to the base station, indicating successful completion of the SSD update.

If the comparison has failed, the mobile station discards SSD-A-NEW and SSD-B-NEW. The mobile station then sends an SSD update rejection order to the base station, indicating unsuccessful completion of the SSD update.

Upon receipt of the SSD update confirmation order, the base station sets SSD-A and SSD-B to the values computed by the HLR/AC.

SSD updates are carried out only in the mobile station and its associated HLR/AC, not in the serving system. The serving system obtains a copy of the SSD computed by the HLR/AC via intersystem communication with the mobile station's HLR/AL.

In an effort to protect sensitive subscriber information, a method has been devised to encrypt certain fields of selected traffic channel signaling messages. If the base station supports mobile station authentication, it may also support message encryption by sending encryption control messages and performing the operations of encryption and decryption. However, the encryption algorithm is not available in TIA/EIA/IS-95 because the algorithm is governed and regulated under the U.S. International Traffic and Arms Regulation (ITAR) for a description of how the algorithm is initialized and applied.

Signaling should not be encrypted if authentication is not performed. Signaling message encryption is individually controlled for each call.

Protocols for details of the initialization and use of the encryption procedure will be discussed in Chapter 8.

Voice privacy is provided in the CDMA system by means of the private long code mask used for PN spreading. Voice privacy can be applied on the traffic channels only. All calls are initiated using the public long code mask for PN spreading. The mobile station user may request voice privacy during call setup using the origination message or page response message, and during traffic channel operation using the long code transition request order.

The transition to private long code mask will not be performed if authentication is not performed. To initiate a transition to the private or public long code mask, either the base station or the mobile station sends a long code transition request order on the traffic channel.

1.6 WIDEBAND CDMA CHANNELS

Sections 1.2 and 1.3 covered CDMA links based on TIA/EIA/IS-95. Modulation parameters of both the reverse and forward channels are shown as listed in Table 1.2.

Table 1.2   Reverse and Forward Traffic Channel Modulation Parameters

Parameters \ Channel	Reverse Traffic Channel	Forward Traffic Channel
Data rates (bps)	9600, 4800, 2400, 1200	9600, 4800, 2400, 1200
PN chip rate (Mcps)	1.2288	1.2288
Code symbols rate (sps)	28800	19200

On the other hand, JTC (AIR)/95-01-30 titled "Proposed Wideband CDMA PCS Standard" was distributed by TAG-7 in 1995. The Joint Technical Committee on Wireless Access (JTC) was formed to evaluate eight different PCS proposals which are based on the IS-95 system with the option of an extended system that offers additional services. Modulation parameters of both the reverse and forward information channels of respective traffic channels are shown in Table 1.3

Table 1.3   Modulation Parameters for Reverse and Forward Information Channels of Respective Traffic Channels

Parameters \ Channel	Reverse Information Channel	Forward Information Channel
Data rates (bps)	64000, 32000, 16000	64000, 32000, 16000
PN chip rate (Mcps)	4.096	4.096
Code symbols rate (sps)	64000	64000

1.6.1   Reverse W-CDMA Channel

The reverse W-CDMA channel is the communication link from the personal station to the base station. The reverse link is composed of access and reverse traffic channels. Each channel includes a pilot channel. The reverse traffic channel also includes a reverse signaling channel. A personal station transmits a reverse pilot channel, which is synchronized to the pilot channel from the base station.

The access channel consists of two channels, i.e., the reverse pilot channel and the reverse access channel. The reverse traffic channel consists of three channels whose types are the reverse pilot channel, reverse information channel, and reverse signaling channel, as shown in the following diagram.

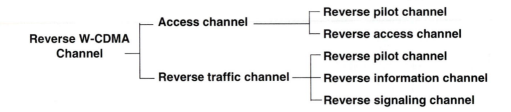

These channels share the same CDMA frequency assignment using direct-sequence CDMA techniques. Each channel is identified by a distinct user long code sequence.

All data transmitted on the reverse traffic channel is convolutionally encoded, interleaved, and modulated by direct-sequence spreading prior to transmission. The reverse CDMA channel structure for the reverse traffic channel is shown in Fig. 1.6.

1.6.2   Forward W-CDMA Channel

The forward W-CDMA channel consists of the following code channels: (1) one pilot channel, (2) one sync channel, (3) up to eight paging channels, and (4) a number of forward traffic chan-

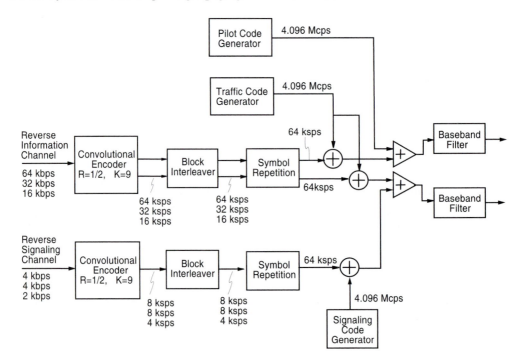

Figure 1.6 Reverse W-CDMA channel structure for reverse traffic channel

nels. The forward traffic channel consists of a forward information channel and a forward signaling channel as shown below.

Each of these code channels is orthogonally spread by the appropriate Walsh code and is then spread by a PN sequence at a fixed chip rate of 4.096 Mcps. The forward traffic channel structure for the forward information channel and forward signaling channel is shown in Fig. 1.7.

- A pilot channel is transmitted at all times by the base station when it is active. The pilot channel is an unmodulated spread spectrum signal that is used for synchronization by a personal station operating within the coverage area of the base station. The pilot channel shall be orthogonally spread with Walsh code index zero prior to transmission.
- The sync channel is convolutionally encoded, interleaved, spread, and modulated to become a spread spectrum signal that is used by personal stations operating within the coverage area of the base station to acquire initial frame synchronization. The bit rate for

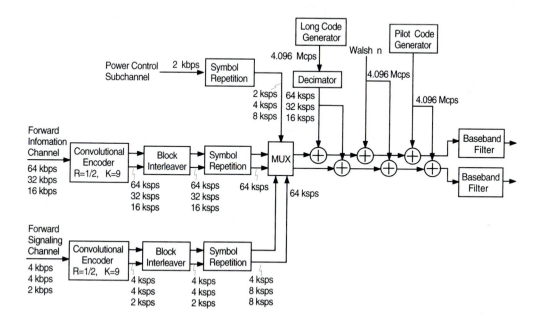

Figure 1.7 Forward W-CDMA channel structure for forward traffic channel

the sync channel is 16 kbps. The in-phase and quadrature-phase pilot PN sequences for the sync channel use the same pilot PN sequence offset as the pilot channel for a given station.

- The paging channel is encoded, interleaved, spread, and modulated to become a spread spectrum signal that is monitored by personal stations operating within the coverage area of the base station. The base station uses multiple paging channels to transmit system information and personal station specific messages.

- The forward traffic channel is used for the transmission of user and signaling information to a specific personal station during a call. The maximum number of forward traffic channels that can be simultaneously supported by a given forward channel is equal to 64 for 64 kbps, 128 for 32 kbps, and 256 for 16 kbps. The number of pilot channel, sync channel, and paging channels operating on the same forward channel must be subtracted from these numbers.

- The forward information channel of the forward traffic channel frames sent at the 64, 32, and 16 kbps data rate will consist of 320, 160, and 80 bits, respectively. The forward traffic channel consists of a forward information channel and a forward signaling channel. The forward signaling channel is convolutionally encoded, block interleaved, direct-sequence spread by the long code, orthogonally spread with a Walsh function, quadrature modulated by a pilot PN sequence at a fixed chip rate of 4.096 Mcps, baseband filtered, and finally transmitted by QPSK waveform.

Detailed discussion on W-CDMA channels will be presented in Chapters 9 and 10.

CHAPTER 2

Elements Required for CDMA Channel Operation

T he forward CDMA channel consists of the pilot channel, one sync channel, paging channels, and forward traffic channels. Each of these code channels is orthogonally spread by the appropriate Walsh function and is then spread by a quadrature pair of pilot PN sequences at a fixed chip rate of 1.2288 Mcps.

The reverse CDMA channel is composed of access channels and reverse traffic channels. All data transmitted on the reverse CDMA channel is convolutionally encoded, block interleaved, modulated by the 64-ary orthogonal modulation, and direct-sequence spread by the long code prior to transmission.

This chapter defines several prerequisites that are essential elements for CDMA channel operation.

2.1 CONVOLUTIONAL ENCODING

Modern digital communication systems are often designed to transmit at very high data rates. Convolutional codes have been applied in many diverse systems. For example, convolutional encoding/decoding has found application in not only CDMA mobile communications, but many space and satellite communications. To protect such systems from errors, convolutional codes are often used. The information data sequence divides it into much smaller blocks of length k and is encoded into codeword symbols of length n. An (n, k, m) convolutional code is implemented with a k-input, n-output linear sequential circuit with the memory order m. In general, n and k are small integers with $k < n$, but m is relatively large. In particular, when $k=1$, the information sequence is not divided into blocks so that the data sequence can be processed continuously. Therefore, the advance of convolutional coding spawned a number of practical applications to digital transmission over wire and radio (wireless) communication channels.

An (n, k, m) convolutional code designates the code rate $R=k/n$ with encoder stages of $m=K-1$, where K is the constraint length of the code. The encoder memory order m is equal to the length of the data sequence delay. A set of n generator sequences for an m-stage encoder is generally described by

$$g_i^{(j)} = (g_{i,0}^{(j)}, g_{i,1}^{(j)}, \cdots, g_{i,m}^{(j)}) \tag{2.1}$$

where $i = 1, 2, \cdots, k$ stands for the number of input terminals and $j = 1, 2, \cdots, n$ for the number of modulo-2 adders (output terminals).

Equation 2.1 also can be expressed in the polynomial form as

$$g_i^{(j)}(D) = \sum_{\lambda=0}^{m} g_{i,\lambda}^{(j)} D^\lambda \tag{2.2}$$

where D is the delay operator, and the power of D of each term corresponds to the number of units of delay for that term.

Each generator sequence is directly determined by the sequence of connections from the encoder stages to the respective modulo-2 adder, 1 representing a connection and 0 representing no connection. The components of each generator sequence consist of $m+1$ binary digits. If the information sequence $d^{(i)} = (d_0^{(i)}, d_1^{(i)}, d_2^{(i)}, \cdots)$ enters the encoder one bit at a time, then the encoder output sequence $c^{(j)} = (c_0^{(j)}, c_1^{(j)}, c_2^{(j)}, \cdots)$ can be obtained by combining the discrete convolution of $d^{(i)}$ with $g_i^{(j)}$ such that

$$c^{(j)} = \sum_{i=1}^{k} d^{(i)} * g_i^{(j)}, \; j = 1, 2, \cdots, n \tag{2.3}$$

where

$$c_\lambda^{(j)} = \sum_{l=1}^{m} \sum_{i=1}^{k} d_{\lambda-l}^{(i)} g_{i,l}^{(j)} 3 \; l = 0,1, \cdots \lambda \tag{2.4}$$

The base station convolutionally encodes the data transmitted on the forward CDMA channel, i.e., sync, paging, and forward traffic channels. The forward CDMA channel uses the (2,1,8) convolutional code representing a code rate $R=1/2$ and a constraint length of 9.

The generator sequences for this code are $g_1^{(1)} = 753$ (octal) = (111101011) (binary) and $g_1^{(2)} = 561$ (octal) = (101110001) (binary). Since a code rate is 1/2, two code symbols (output) are generated from each input data bit to the encoder.

The first output symbol $c_\lambda^{(1)}$ after initialization is a code symbol encoded with the generator sequence $g_1^{(1)}$ and the second output symbol $c_\lambda^{(2)}$ is a code symbol encoded with the generator sequence $g_1^{(2)}$. As shown in Fig. 2.1, convolutional encoding involves the modulo-2 addition of selected taps of a serially time-delayed data sequence.

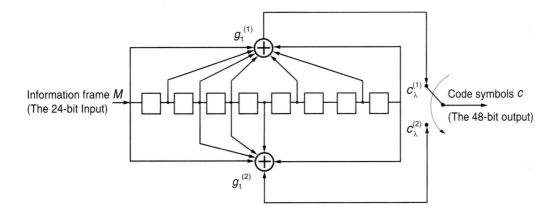

$g_1^{(1)}$

Information frame M
(The 24-bit Input)

$c_\lambda^{(1)}$

Code symbols c

(The 48-bit output)

$c_\lambda^{(2)}$

$g_1^{(2)}$

Figure 2.1 The (2,1,8) convolutional encoder with the 24-bit input

Example 2.1   The forward traffic channel structure at the 1.2 kbps transmission rate consists of 24 bits (20ms). These 24 bits are composed of 16 information bits followed by 8 encoder tail bits, as shown in Fig. 2.2. The last eight bits of the encoder tail are all set to 0. If the information sequence is expressed as d=(1010100100000101), the corresponding data polynomial is $d(x)$ = $1+x^2 + x^4 + x^7 + x^{13} + x^{15}$. Since the 16-bit information is followed by the 8-bit encoder tail, the forward traffic channel frame is expressed as M = (101010010000010100000000) or in the polynomial form, we have

$$M(x) = 1 + x^2 + x^4 + x^7 + x^{13} + x^{15}.$$

24-bit frame (20 *ms*)	
16	8

Information bits   Tail bits

Figure 2.2 Forward traffic channel frame sent
at the 1200 bps transmission rate

Figure 2.1 illustrates the (2,1,8) convolutional encoder that is used for this channel. Since $m = 8$ and $n = 2$, this encoder consists of an 8-stage shift register coupled with two modulo-2 adders and a commutator switch for serializing the encoder outputs. For the convolutional encoder shown in Fig. 2.1, two generator sequences are $g_1^{(1)} = (g_{1,0}^{(1)}, g_{1,1}^{(1)}, g_{1,2}^{(1)}, g_{1,3}^{(1)}, g_{1,4}^{(1)}, g_{1,5}^{(1)}, g_{1,6}^{(1)}, g_{1,7}^{(1)}, g_{1,8}^{(1)}) = (111101011)$ and $g_1^{(2)} = (g_{1,0}^{(2)}, g_{1,1}^{(2)}, g_{1,2}^{(2)}, g_{1,3}^{(2)}, g_{1,4}^{(2)}, g_{1,5}^{(2)}, g_{1,6}^{(2)}, g_{1,7}^{(2)}, g_{1,8}^{(2)}) = (101110001)$, respectively.

Using Eq. 2.4, for $i = 1$ and $j = 1, 2$ we have

$$c_\lambda^{(1)} = \sum_{i=0}^{8} d_{\lambda-1}^{(1)} g_{i,\,l}^{(1)}$$

$$\text{and} \qquad c_\lambda^{(2)} = \sum_{l=0}^{8} d_{\lambda-1}^{(1)} g_{i,\,l}^{(2)} \qquad\qquad (2.5)$$

Using the components $g_{1,\iota}^{(1)}$ and $g_{1,\iota}^{(2)}$, $0 \le \iota \le 8$, of generator sequences $g_1^{(1)}$ and $g_1^{(2)}$, the outputs from each modulo-2 adder are, respectively,

$$c_\lambda^{(1)} = d_\lambda^{(1)} + d_{\lambda_1}^{(1)} + d_{\lambda_2}^{(1)} + d_{\lambda_3}^{(1)} + d_{\lambda_5}^{(1)} + d_{\lambda_7}^{(1)} + d_{\lambda_8}^{(1)}$$

$$\text{and} \ \ c_\lambda^{(2)} = d_\lambda^{(2)} + d_{\lambda_2}^{(2)} + d_{\lambda_3}^{(2)} + d_{\lambda_4}^{(2)} + d_{\lambda_8}^{(2)}$$

Thus, the code symbols corresponding to the frame input M are computed as follows:

λ	M	Shift register contents	$(c_\lambda^{(1)}, c_\lambda^{(2)})$, $0 \le \lambda \le 23$
0	1	1 0 0 0 0 0 0 0 0	(1 1)
1	0	0 1 0 0 0 0 0 0 0	(1 0)
2	1	1 0 1 0 0 0 0 0 0	(0 0)
3	0	0 1 0 1 0 0 0 0 0	(0 1)
4	1	0 0 1 0 1 0 0 0 0	(1 0)
5	0	0 0 0 1 0 1 0 0 0	(0 1)
6	0	0 0 0 0 1 0 1 0 0	(0 1)
7	1	0 0 0 0 0 1 0 1 0	(0 0)
8	0	1 0 0 0 0 0 1 0 1	(0 0)
9	0	0 1 0 0 0 0 0 1 0	(0 0)
10	0	0 0 1 0 0 0 0 0 1	(0 0)
11	0	1 0 0 1 0 0 0 0 0	(0 0)
12	0	0 1 0 0 1 0 0 0 0	(1 1)
13	1	1 0 1 0 0 1 0 0 0	(1 0)
14	0	0 1 0 1 0 0 1 0 0	(0 1)
15	1	1 0 1 0 1 0 0 1 0	(1 1)
16	0	0 1 0 1 0 1 0 0 1	(0 0)
17	0	0 0 1 0 1 0 1 0 0	(1 0)
18	0	0 0 0 1 0 1 0 1 0	(1 1)
19	0	0 0 0 0 1 0 1 0 1	(1 0)
20	0	0 0 0 0 0 1 0 1 0	(0 0)
21	0	0 0 0 0 0 0 1 0 1	(1 1)
22	0	0 0 0 0 0 0 0 1 0	(1 0)
23	0	0 0 0 0 0 0 0 0 1	(1 1)

The output symbols $(c_\lambda^{(1)}, c_\lambda^{(2)})$ are concatenated into a single sequence that is the code symbol sequence to be sent to the block interleaver, which is then given by

$$c = (c_0^{(1)} c_0^{(2)}, c_1^{(1)} c_1^{(2)}, \cdots, c_{23}^{(1)} c_{23}^{(2)}) \tag{2.6}$$

$$= (111000011001010000000000011100111001011110001110 11)$$

For the forward CDMA channel, each convolutionally encoded symbol is repeated k times prior to block interleaving whenever the information rate is lower than 9600 bps, as shown below.

Channel	Information rate (bps)	Repeating times (k)	Code repetition
Sync	1200	1	2
Paging	4800	1	2
	9600	0	1
Forward traffic	1200	7	8
	2400	3	4
	4800	1	2
	9600	0	1

2.2 BLOCK INTERLEAVING

Direct sequence spread spectrum CDMA supports simultaneous services for digital communication among a considerably higher community of users than any single user. This will be reflected in how this excess of dimensionality or redundancy is exploited to improve performance. Two processing techniques are considered to achieve improvements: interleaving for the excess redundancy and forward error-correcting coding.

Interleaving is the process of permuting a sequence of symbols. This reordering procedure to achieve time diversity is called interleaving and can be considered in two ways: block interleaving and convolutional interleaving. Interleaving also can be an effective technique to design codes for correcting multiple bursts or long bursts.

All symbols after repetition on the sync, paging, and forward traffic channels are block interleaved. The sync channel uses a block interleaver where the input symbol sequence is given in Table 2.1. Since Table 2.1 consists of 16 rows by 8 columns, it contains 128 modulation symbols at the modulation symbol rate of 4800 sps. Hence the interleaver time span is 128/4800 = 26.66 ms. Table 2.1 is read down by columns from the left to the right. The first input symbol '1' is at the top left, the second input symbol '1' is just below the first input symbol, the 17th input symbol '9' is just to the right of the first input symbol, and the last input symbol '64' is located at the rightmost bottom corner. On the other hand, the output symbol sequence is given in Table

2.2. The table is read the same way as Table 2.1. The first output symbol '1' is at the top left, the second output symbol '33' is just below the first output symbol, the 17th output '3' is just to the right of the first output symbol, and the last output symbol '64' is the last symbol located at the rightmost bottom corner. The repeated code symbols are written into the interleaver by columns filling the complete 16 x 8 matrix. This sync channel interleaver is an array with 16 rows and 8 columns, i.e., 128 cells. Thus, we can say that the sync channel uses a block interleaver spanning 26.66 ms which is equivalent to 128 modulation symbols at the symbol rate of 4800 sps.

Table 2.1   Sync Channel Interleaver Input (Array Write Operation)

1	9	17	25	33	41	49	57
1	9	17	25	33	41	49	57
2	10	18	26	34	42	50	58
2	10	18	26	34	42	50	58
3	11	19	27	35	43	51	59
3	11	19	27	35	43	51	59
4	12	20	28	36	44	52	60
4	12	20	28	36	44	52	60
5	13	21	29	37	45	53	61
5	13	21	29	37	45	53	61
6	14	22	30	38	46	54	62
6	14	22	30	38	46	54	62
7	15	23	31	39	47	55	63
7	15	23	31	39	47	55	63
8	16	24	32	40	48	56	64
8	16	24	32	40	48	56	64

Similarly, the paging and forward traffic channels use a block interleaver matrix of 24 rows by 16 columns equivalent to 384 modulation symbols at the modulation symbol rate of 19200 sps. Now we see that the interleaver time span is 384/19200 = 20 ms. This will be explained further in Chapter 4 (Tables 4.9 through 4.12).

Example 2.2   Consider the forward traffic channel at the 1200 bps data rate consisting of the 24-bit frame structure. This channel frame of M = (101010010000010100000000) is assumed as the input to the (2,1,8) convolutional encoder (the same as Example 2.1). The output code sequence of this encoder (from before) is c = (1110000110010100000000000111001 110010111000111011) which is the input sequence to the block interleaver after repetition.

The forward traffic channel interleaver output operation at 1200 bps is shown in Table 4.18. Using Table 4.18, the interleaver output w is resulted from the leftmost two columns by reading down from the left to the right as follows:

Table 2.2 Sync Channel Interleaver Output (Array Read Operation)

1	3	2	4	1	3	2	4
33	35	34	36	33	35	34	36
17	19	18	20	17	19	18	20
49	51	50	52	49	51	50	52
9	11	10	12	9	11	10	12
41	43	42	44	41	43	42	44
25	27	26	28	25	27	26	28
57	59	58	60	57	59	58	60
5	7	6	8	5	7	6	8
37	39	38	40	37	39	38	40
21	23	22	24	21	23	22	24
53	55	54	56	53	55	54	56
13	15	14	16	13	15	14	16
45	47	46	48	45	47	46	48
29	31	30	32	29	31	30	32
61	63	62	64	61	63	62	64

1 9 17 25 33 41 5 13 21 29 37 45 3 11 19 27 35 43 7 15 23 31 39 47
2 10 18 26 34 42 6 14 22 30 38 46 4 12 20 28 36 44 8 16 24 32 40 48
$w = (1101000000111001110001111001000101100100011001011)$

Since the symbols with the same number denote repeated code symbols, the interleaver output shall be the repeated seven more identical sequences of w, as shown in Table 2.3.

Table 2.3 Forward Traffic Channel Interleaver Output at 1200 bps

1	1	1	1	1	1	1	1	1	1	1	1	1	1	1	1
1	0	1	0	1	0	1	0	1	0	1	0	1	0	1	0
0	0	0	0	0	0	0	0	0	0	0	0	0	0	0	0
1	1	1	1	1	1	1	1	1	1	1	1	1	1	1	1
0	0	0	0	0	0	0	0	0	0	0	0	0	0	0	0
0	0	0	0	0	0	0	0	0	0	0	0	0	0	0	0
0	0	0	0	0	0	0	0	0	0	0	0	0	0	0	0
0	1	0	1	0	1	0	1	0	1	0	1	0	1	0	1
0	0	0	0	0	0	0	0	0	0	0	0	0	0	0	0

Table 2.3   Forward Traffic Channel Interleaver Output at 1200 bps *(continued)*

0	1	0	1	0	1	0	1	0	1	0	1	0	1	0	1
1	1	1	1	1	1	1	1	1	1	1	1	1	1	1	1
1	0	1	0	1	0	1	0	1	0	1	0	1	0	1	0
1	0	1	0	1	0	1	0	1	0	1	0	1	0	1	0
0	1	0	1	0	1	0	1	0	1	0	1	0	1	0	1
0	0	0	0	0	0	0	0	0	0	0	0	0	0	0	0
1	0	1	0	1	0	1	0	1	0	1	0	1	0	1	0
1	0	1	0	1	0	1	0	1	0	1	0	1	0	1	0
1	1	1	1	1	1	1	1	1	1	1	1	1	1	1	1
0	1	0	1	0	1	0	1	0	1	0	1	0	1	0	1
0	0	0	0	0	0	0	0	0	0	0	0	0	0	0	0
0	0	0	0	0	0	0	0	0	0	0	0	0	0	0	0
1	1	1	1	1	1	1	1	1	1	1	1	1	1	1	1
1	0	1	0	1	0	1	0	1	0	1	0	1	0	1	0
1	1	1	1	1	1	1	1	1	1	1	1	1	1	1	1

Block interleaving is the data reordering procedure which can be considered as follows: Interleaved output symbols as long with i times as many coded symbols from the convolutional encoder can be archived by interleaving (or interlacing), as shown in Tables 2.1 and 2.2. This is simply done by arranging j coded symbols into i rows of a rectangular array and then transmitting them by column after column (or vice versa) as shown in Fig. 2.3. If the convolutional code corrects random errors, the interleaved code corrects single bursts of length j or less.

2.3 ORTHOGONAL SPREADING USING WALSH FUNCTIONS

Each code channel transmitted on the forward CDMA channel is spread with a Walsh function at a fixed chip rate of 1.2288 Mcps to provide orthogonal channelization among all code channels. Modulation for the reverse CDMA channel is 64-ary orthogonal modulation. The modulation symbol is one of 64 mutually orthogonal waveforms generated using Walsh functions. One of 64 possible modulation symbols is transmitted for each six symbols, i.e., c_i, $0 \leq i \leq 5$.

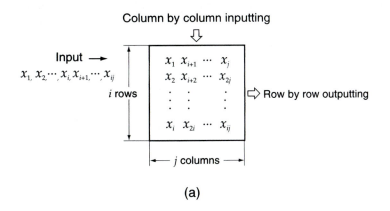

(a)

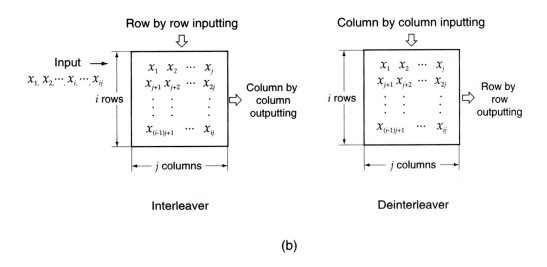

(b)

Figure 2.3 Data reordering procedure by block interleaving

Using a Hadamard matrix H_n, where n is a power of 2, i.e., $n=2^m$, Walsh functions are constructed to be shown as follows: A Hadamard matrix is an orthogonal $n \times n$ matrix of the entries +1 and −1 with the property that any row differs from any other row in exactly $n/2$ positions. One row of the matrix contains all +1s. The other rows contain evenly the +1s and −1s of $n/2$ each. Futhermore, all the entries in the first row and the first column of H_n have all +1s. The $n \times n$ Hadamard matrices can exist only if n is a power of 2.

By changing the +1 to 0s and the −1 to 1s, the Hadamard matrix for $n=2$ can be expressed by

$$H_2 = \begin{bmatrix} +1+1 \\ +1-1 \end{bmatrix} = \begin{bmatrix} 00 \\ 01 \end{bmatrix}$$

The Hadamard matrix for $n = 2^6$ is the 64 x 64 orthogonal Walsh function shown in Table 2.4. This 64 by 64 matrix can be generated by means of the following recursive procedure: For $n=2^m$, $0 \leq m$ (a positive integer),

$$H_1 = 0 \quad for \quad m = 0$$

$$H_2 = \begin{bmatrix} H_1 & H_1 \\ H_1 & \overline{H_1} \end{bmatrix} = \begin{bmatrix} 00 \\ 01 \end{bmatrix} \quad for \quad m = 1$$

$$H_4 = \begin{bmatrix} H_2 & H_2 \\ H_2 & \overline{H_2} \end{bmatrix} = \begin{bmatrix} 0000 \\ 0101 \\ 0011 \\ 0110 \end{bmatrix} \quad for \quad m = 2$$

$$H_8 = \begin{bmatrix} H_4 & H_4 \\ H_4 & \overline{H_4} \end{bmatrix} = \begin{bmatrix} 00000000 \\ 01010101 \\ 00110011 \\ 01100110 \\ 00001111 \\ 01101010 \\ 00111100 \\ 01101001 \end{bmatrix} \quad for \quad m = 3$$

.
.
.

$$H_{2n} = \begin{bmatrix} H_n & H_n \\ H_n & \overline{H_n} \end{bmatrix}$$

where $2n = 2^{m+1}$ and $\overline{H_n}$ denotes the binary complement of H_n. The rows of Walsh matrix H_{64} form a liner binary code of block length $n = 64$ and the minimum distance $d_{min} = n/2 = 32$.

Each code channel transmitted on the forward CDMA channel is spread with a Walsh function at a fixed chip rate of 1.2288 Mcps to provide orthogonal channelization among all code channels. The Walsh function spreading sequence is repeated with a period of 52.083 (=64/1.2288 Mcps) μs which is equal to the duration of one forward traffic channel modulation symbol.

Therefore, a code channel spread by using the Walsh function n (the row n of Hadamard-Walsh matrix) is assigned to code channel number n $(0 \leq n \leq 63)$. Code channel number zero is always assigned to the pilot channel. If the sync channel is present, it is assigned code channel number 32. If paging channels are present, they are assigned to code channel numbers one through seven in sequence. The remaining code channels are available for assignment to the forward traffic channels. Thus, each of these code channels is orthogonally spread by the appropriate Walsh function and is then spread by the quadrature pair of pilot PN sequences at a fixed chip rate 1.2288 Mcps.

Table 2.4   64-ary Orthogonal Symbol Set

Walsh Chip within Symbol

Modulation Symbol Index	0123	4567	11 8901	1111 2345	1111 6789	2222 0123	2222 4567	2233 8901	3333 2345	3333 6789	4444 0123	4444 4567	4455 8901	5555 2345	5555 6789	6666 0123
0	0000	0000	0000	0000	0000	0000	0000	0000	0000	0000	0000	0000	0000	0000	0000	0000
1	0101	0101	0101	0101	0101	0101	0101	0101	0101	0101	0101	0101	0101	0101	0101	0101
2	0011	0011	0011	0011	0011	0011	0011	0011	0011	0011	0011	0011	0011	0011	0011	0011
3	0110	0110	0110	0110	0110	0110	0110	0110	0110	0110	0110	0110	0110	0110	0110	0110
4	0000	1111	0000	1111	0000	1111	0000	1111	0000	1111	0000	1111	0000	1111	0000	1111
5	0101	1010	0101	1010	0101	1010	0101	1010	0101	1010	0101	1010	0101	1010	0101	1010
6	0011	1100	0011	1100	0011	1100	0011	1100	0011	1100	0011	1100	0011	1100	0011	1100
7	0110	1001	0110	1001	0110	1001	0110	1001	0110	1001	0110	1001	0110	1001	0110	1001
8	0000	0000	1111	1111	0000	0000	1111	1111	0000	0000	1111	1111	0000	0000	1111	1111
9	0101	0101	1010	1010	0101	0101	1010	1010	0101	0101	1010	1010	0101	0101	1010	1010
10	0011	0011	1100	1100	0011	0011	1100	1100	0011	0011	1100	1100	0011	0011	1100	1100
11	0110	0110	1001	1001	0110	0110	1001	1001	0110	0110	1001	1001	0110	0110	1001	1001
12	0000	1111	1111	0000	0000	1111	1111	0000	0000	1111	1111	0000	0000	1111	1111	0000
13	0101	1010	1010	0101	0101	1010	1010	0101	0101	1010	1010	0101	0101	1010	1010	0101
14	0011	1100	1100	0011	0011	1100	1100	0011	0011	1100	1100	0011	0011	1100	1100	0011
15	0110	1001	1001	0110	0110	1001	1001	0110	0110	1001	1001	0110	0110	1001	1001	0110
16	0000	0000	0000	0000	1111	1111	1111	1111	0000	0000	0000	0000	1111	1111	1111	1111
17	0101	0101	0101	0101	1010	1010	1010	1010	0101	0101	0101	0101	1010	1010	1010	1010
18	0011	0011	0011	0011	1100	1100	1100	1100	0011	0011	0011	0011	1100	1100	1100	1100
19	0110	0110	0110	0110	1001	1001	1001	1001	0110	0110	0110	0110	1001	1001	1001	1001

Table 2.4   64-ary Orthogonal Symbol Set *(continued)*

Walsh Chip within Symbol

	0123	4567	11 8901	1111 2345	1111 6789	2222 0123	2222 4567	2233 8901	3333 2345	3333 6789	4444 0123	4444 4567	4455 8901	5555 2345	5555 6789	6666 0123
20	0000	1111	0000	1111	1111	0000	1111	0000	0000	1111	0000	1111	1111	0000	1111	0000
21	0101	1010	0101	1010	1010	0101	1010	0101	0101	1010	0101	1010	1010	0101	1010	0101
22	0011	1100	0011	1100	1100	0011	1100	0011	0011	1100	0011	1100	1100	0011	1100	0011
23	0110	1001	0110	1001	1001	0110	1001	0110	0110	1001	0110	1001	1001	0110	1001	0110
24	0000	0000	1111	1111	1111	1111	0000	0000	0000	0000	1111	1111	1111	1111	0000	0000
25	0101	0101	1010	1010	1010	1010	0101	0101	0101	0101	1010	1010	1010	1010	0101	0101
26	0011	0011	1100	1100	1100	1100	0011	0011	0011	0011	1100	1100	1100	1100	0011	0011
27	0110	0110	1001	1001	1001	1001	0110	0110	0110	0110	1001	1001	1001	1001	0110	0110
28	0000	1111	1111	0000	1111	0000	0000	1111	0000	1111	1111	0000	1111	0000	0000	1111
29	0101	1010	1010	0101	1010	0101	0101	1010	0101	1010	1010	0101	1010	0101	0101	1010
30	0011	1100	1100	0011	1100	0011	0011	1100	0011	1100	1100	0011	1100	0011	0011	1100
31	0110	1001	1001	0110	1001	0110	0110	1001	0110	1001	1001	0110	1001	0110	0110	1001
32	0000	0000	0000	0000	0000	0000	0000	0000	1111	1111	1111	1111	1111	1111	1111	1111
33	0101	0101	0101	0101	0101	0101	0101	0101	1010	1010	1010	1010	1010	1010	1010	1010
34	0011	0011	0011	0011	0011	0011	0011	0011	1100	1100	1100	1100	1100	1100	1100	1100
35	0110	0110	0110	0110	0110	0110	0110	0110	1001	1001	1001	1001	1001	1001	1001	1001
36	0000	1111	0000	1111	0000	1111	0000	1111	1111	0000	1111	0000	1111	0000	1111	0000
37	0101	1010	0101	1010	0101	1010	0101	1010	1010	0101	1010	0101	1010	0101	1010	0101
38	0011	1100	0011	1100	0011	1100	0011	1100	1100	0011	1100	0011	1100	0011	1100	0011
39	0110	1001	0110	1001	0110	1001	0110	1001	1001	0110	1001	0110	1001	0110	1001	0110
40	0000	0000	1111	1111	0000	0000	1111	1111	1111	1111	0000	0000	1111	1111	0000	0000
41	0101	0101	1010	1010	0101	0101	1010	1010	1010	1010	0101	0101	1010	1010	0101	0101
42	0011	0011	1100	1100	0011	0011	1100	1100	1100	1100	001i	0011	1100	1100	0011	0011
43	0110	0110	1001	1001	0110	0110	1001	1001	1001	1001	0110	0110	1001	1001	0110	0110
44	0000	1111	1111	0000	0000	1111	1111	0000	1111	0000	0000	1111	1111	0000	0000	1111
45	0101	1010	1010	0101	0101	1010	1010	0101	1010	0101	0101	1010	1010	0101	0101	1010
46	0011	1100	1100	0011	0011	1100	1100	0011	1100	0011	0011	1100	1100	0011	0011	1100
47	0110	1001	1001	0110	0110	1001	1001	0110	1001	0110	0110	1001	1001	0110	0110	1001
48	0000	0000	0000	0000	1111	1111	1111	1111	1111	1111	1111	1111	0000	0000	0000	0000
49	0101	0101	0101	0101	1010	1010	1010	1010	1010	1010	1010	1010	0101	0101	0101	0101
50	0011	0011	0011	0011	1100	1100	1100	1100	1100	1100	1100	1100	0011	0011	0011	0011
51	0110	0110	0110	0110	1001	1001	1001	1001	1001	1001	1001	1001	0110	0110	0110	0110
52	0000	1111	0000	1111	1111	0000	1111	0000	1111	0000	1111	0000	0000	1111	0000	1111
53	0101	1010	0101	1010	1010	0101	1010	0101	1010	0101	1010	0101	0101	1010	0101	1010
54	0011	1100	0011	1100	1100	0011	1100	0011	1100	0011	1100	0011	0011	1100	0011	1100
55	0110	1001	0110	1001	1001	0110	1001	0110	1001	0110	1001	0110	0110	1001	0110	1001

Modulation Symbol Index

Table 2.4 64-ary Orthogonal Symbol Set *(continued)*

Walsh Chip within Symbol

		11	1111	1111	2222	2222	2233	3333	3333	4444	4444	4455	5555	5555	6666
0123	4567	8901	2345	6789	0123	4567	8901	2345	6789	0123	4567	8901	2345	6789	0123
56	0000	0000	1111	1111	1111	1111	0000	0000	1111	1111	0000	0000	0000	0000	1111	1111
57	0101	0101	1010	1010	1010	1010	0101	0101	1010	1010	0101	0101	0101	0101	1010	1010
58	0011	0011	1100	1100	1100	1100	0011	0011	1100	1100	0011	0011	0011	0011	1100	1100
59	0110	0110	1001	1001	1001	1001	0110	0110	1001	1001	0110	0110	0110	0110	1001	1001
60	0000	1111	1111	0000	1111	0000	0000	1111	1111	0000	0000	1111	0000	1111	1111	0000
61	0101	1010	1010	0101	1010	0101	0101	1010	1010	0101	0101	1010	0101	1010	1010	0101
62	0011	1100	1100	0011	1100	0011	0011	1100	1100	0011	0011	1100	0011	1100	1100	0011
63	0110	1001	1001	0110	1001	0110	0110	1001	1001	0110	0110	1001	0110	1001	1001	0110

For the reverse CDMA channel, the modulation symbol will be one of 64 mutually orthogonal waveforms generated using Walsh functions. Since the code symbol rate at the interleaver output is 28.8 ksps, the symbol rate at the orthogonal modulator output becomes 4.8 ksps or 307.2 kcps. These modulation symbols numbered 0 through 63 are selected according to the following formula:

$$MSI = c_0 + 2c_1 + 4c_2 + 8c_3 + 16c_4 + 32c_5$$

where MSI denotes the modulation symbol index; and c_i, $0 \le i \le 5$, represent each group of six code symbols output from the block interleaver. Six code symbols are associated with one modulation symbol. The period of time required to transmit a single modulation symbol is equal to $1/4800 = 208.333\mu s$. The period of time associated with 1/64 of the modulation symbol is referred to as a Walsh chip and will be equal to $1/(4800 \times 64) = 1/307200 = 3.255\mu s$. Within a Walsh code, Walsh chips are transmitted in the order of 0, 1, 2, $\cdots$, 63.

Example 2.3 The reverse traffic channel frame sent at the 1.2 kbps transmission rate is composed of the 16-bit information followed by the 8-bit encoder tail. The mobile station is convolutionally encoded by the 24-bit frame data transmitted on either an access channel or a reverse traffic channel.

The generator sequences for the (3, 1, 8) convolutional encoder are $g_1^{(1)} = (111001001)$, $g_1(2) = (110110011)$, and $g_1(3) = (111001001)$. Suppose the 24-bit frame sequence input into the convolutional encoder is

1 1 1 0 0 0 1 0 1 1 0 1 0 0 1 1   0 0 0 0 0 0 0 0

← Information bits ——→ ←Encoder tail bits (all zero)——→

Since the code rate is R=1/3, the encoder output terminals generate the three code symbols (c_0, c_1, c_2) for each input frame bit. The generated code symbols at each modulo-2 adder are, respectively

$c_0 = (1\ 1\ 0\ 0\ 0\ 0\ 1\ 1\ 1\ 0\ 0\ 0\ 0\ 1\ 1\ 1\ 0\ 0\ 0\ 0\ 0\ 0\ 0\ 1)$

$c_1 = (1\ 0\ 0\ 0\ 0\ 0\ 0\ 0\ 1\ 1\ 1\ 0\ 1\ 0\ 1\ 0\ 1\ 0\ 1\ 0\ 0\ 1\ 0\ 1)$

$c_2 = (1\ 0\ 1\ 0\ 1\ 1\ 0\ 0\ 1\ 1\ 1\ 1\ 1\ 0\ 1\ 0\ 0\ 0\ 0\ 0\ 1\ 0\ 1\ 1)$

Thus, the serialized code symbol output can be produced as shown below:

$c = (1\ 1\ 1\ 1\ 0\ 0\ 0\ 0\ 1\ 0\quad 0\ 0\ 0\ 0\ 1\ 0\ 0\ 1\ 1\ 0\quad 0\ 1\ 0\ 0\ 1\ 1\ 1\ 0\ 1\ 1\quad 0\ 1\ 1\ 0\ 0\ 1\ 0\ 1\ 1\ 1$

$\qquad 0\ 0\ 1\ 1\ 1\ 1\ 0\ 0\ 0\ 1\quad 0\ 0\ 0\ 0\ 0\ 1\ 0\ 0\ 0\ 0\quad 0\ 0\ 1\ 0\ 1\ 0\ 0\ 0\ 1\ 1\quad 1\ 1)$

This is the input symbols to the interleaver. A block interleaver spanning 20 ms will interleave all code symbols in c prior to orthogonal modulation.

The interleaver forms a 576-cell array with 32 rows and 18 columns, as shown in Table 2.5.

Table 2.5 Reverse Traffic Channel Interleaver Memory (1200 bps) (TIA/EIA/IS-95)

1	5	9	13	17	21	25	29	33	37	41	45	49	53	57	61	65	69
1	5	9	13	17	21	25	29	33	37	41	45	49	53	57	61	65	69
1	5	9	13	17	21	25	29	33	37	41	45	49	53	57	61	65	69
1	5	9	13	17	21	25	29	33	37	41	45	49	53	57	61	65	69
1	5	9	13	17	21	25	29	33	37	41	45	49	53	57	61	65	69
1	5	9	13	17	21	25	29	33	37	41	45	49	53	57	61	65	69
1	5	9	13	17	21	25	29	33	37	41	45	49	53	57	61	65	69
1	5	9	13	17	21	25	29	33	37	41	45	49	53	57	61	65	69
2	6	10	14	18	22	26	30	34	38	42	46	50	54	58	62	66	70
2	6	10	14	18	22	26	30	34	38	42	46	50	54	58	62	66	70
2	6	10	14	18	22	26	30	34	38	42	46	50	54	58	62	66	70
2	6	10	14	18	22	26	30	34	38	42	46	50	54	58	62	66	70
2	6	10	14	18	22	26	30	34	38	42	46	50	54	58	62	66	70
2	6	10	14	18	22	26	30	34	38	42	46	50	54	58	62	66	70
2	6	10	14	18	22	26	30	34	38	42	46	50	54	58	62	66	70
2	6	10	14	18	22	26	30	34	38	42	46	50	54	58	62	66	70
3	7	11	15	19	23	27	31	35	39	43	47	51	55	59	63	67	71

Table 2.5   Reverse Traffic Channel Interleaver Memory (1200 bps)
(TIA/EIA/IS-95) *(continued)*

3	7	11	15	19	23	27	31	35	39	43	47	51	55	59	63	67	71
3	7	11	15	19	23	27	31	35	39	43	47	51	55	59	63	67	71
3	7	11	15	19	23	27	31	35	39	43	47	51	55	59	63	67	71
3	7	11	15	19	23	27	31	35	39	43	47	51	55	59	63	67	71
3	7	11	15	19	23	27	31	35	39	43	47	51	55	59	63	67	71
3	7	11	15	19	23	27	31	35	39	43	47	51	55	59	63	67	71
3	7	11	15	19	23	27	31	35	39	43	47	51	55	59	63	67	71
4	8	12	16	20	24	28	32	36	40	44	48	52	56	60	64	68	72
4	8	12	16	20	24	28	32	36	40	44	48	52	56	60	64	68	72
4	8	12	16	20	24	28	32	36	40	44	48	52	56	60	64	68	72
4	8	12	16	20	24	28	32	36	40	44	48	52	56	60	64	68	72
4	8	12	16	20	24	28	32	36	40	44	48	52	56	60	64	68	72
4	8	12	16	20	24	28	32	36	40	44	48	52	56	60	64	68	72
4	8	12	16	20	24	28	32	36	40	44	48	52	56	60	64	68	72
4	8	12	16	20	24	28	32	36	40	44	48	52	56	60	64	68	72

The interleaver output w at the 1200 bps transmission rate can be computed according to the following order of interleaver rows:

Row Numbers

1	9	2	10	3	11	4	12	5	13	6	14	7	15	8	16
17	25	18	26	19	27	20	28	21	29	22	30	23	31	24	32

For example, using the interleaver input c, the interleaver output w corresponding to the row number 1 will be computed as 101000111001000011. By making use of the interleaver input c and the row numbers in Table 2.5, the interleaver output w can be obtained as shown in Table 2.6.

Considering the first two rows 1 and 9 in Table 2.6, Row 1 is computed as 101000111001000011 and Row 9 represents 100011110101100001.

Table 2.6   Interleaver Output Corresponding to Row Numbers

Row Number	Interleaver Output	Row Number	Interleaver Output
[1]	101000111001000011	[17]	100110100110000101
[9]	100011110101100001	[25]	100000011110010001
[2]	101000111001000011	[18]	100110100110000101
[10]	100011110101100001	[26]	100000011110010001
[3]	101000111001000011	[19]	100110100110000101
[11]	100011110101100001	[27]	100000011110010001
[4]	101000111001000011	[20]	100110100110000101
[12]	100011110101100001	[28]	100000011110010001
[5]	101000111001000011	[21]	100110100110000101
[13]	100011110101100001	[29]	100000011110010001
[6]	101000111001000011	[22]	100110100110000101
[14]	100011110101100001	[30]	100000011110010001
[7]	101000111001000011	[23]	100110100110000101
[15]	100011110101100001	[31]	100000011110010001
[8]	101000111001000011	[24]	100110100110000101
[16]	100011110101100001	[32]	100000011110010001

Let us show how to compute modulation symbols corresponding to Row 1 and 9 in the following:

1. Computation for Row 1
   - For the 6-symbol input $1\ 0\ 1\ 0\ 0\ 0$ ($c_0=1$, $c_1=0$, $c_2=1$, $c_3=0$, $c_4=0$, and $c_5=0$), we get MSI = 5. Using Table 2.4, the modulation symbol is obtained as
     010110100101101001011010010110100101101001011010
   - For the next 6 symbols 111001, we have MSI = 39. The modulation symbol corresponding to MSI = 39 is
     011010010110100101101001011010011001011010010110100101101001011010010110

- For the last 6 symbols 0 0 0 0 1 1 , we get MSI = 48. The corresponding modulation symbol is

  000000000000000011111111111111111
  11111111111111111100000000000000000

2. Computation for Row 9
   - Input symbol = 1000011, MSI = 49
     Modulation symbol = 0101010101010101101010101010101010
     1010101010101010100101010101010101

   - Input symbol = 110101, MSI = 43
     Modulation symbol = 0110011010011001011001101001100110
     1001100101100110100110010110011010

   - Input symbol = 100001, MSI = 33
     Modulation symbol = 0101010101010101010101010101010101
     1010101010101010101010101010101010

Thus, the interleaver output stream relating to Rows 1 and 9 is

w = 1 0 1 0 0 0 1 1 1 0 0 1 0 0 0 0 1 1 1 0 0 0 1 1 1 1 0 1 0 1 0 1 1 0 0 0 0 1

Using Table 2.4, the modulation symbols corresponding Rows 1 and 9 are computed as
0101101001011010010110100101101001011010010110100101101001011010010110100101101001011010
0110100101101001011010010110100110010110100101101001011010010110
0000000000000000111111111111111111111111111111111100000000000000000
0101010101010101101010101010101010101010101010101001010101010101101
0110011010011001011001101001100110011001011001101001100101100110
0101010101010101010101010101010110101010101010101010101010101010

A total of 3072 x 8 = 24576 symbols will be produced as the Walsh function output sequence. The period of time required to transmit 24576 symbols is equal to 24576/1.2288 x 10^6 = 20 ms, as expected.

2.4 DIRECT SEQUENCE SPREADING

The reverse traffic channel and the access channel on the reverse CDMA channel will be direct sequence spread by the long code to provide limited privacy. For the reverse traffic channel, the direct sequence (DS) spreading operation involves modulo-2 addition of the data burst randomizer output and the long code. The data burst randomizer generates a masking pattern of 0s and 1s that randomly masks out the redundant data generated by the code repetition. The masking pattern is determined by the data rate of the frame and by a block of the last 14 bits taken from the long code. For the access channel, the DS spreading operation involves modulo-2 addition of the 64-ary orthogonal modulator output and the long code.

Let $d_w(t)$ be the data sequence modulated by Walsh chips and T_b be the data bit time interval. The Walsh modulated data sequence is modulo-2 added by the spreading PN chips of the long code $c(t)$. Each pulse of $c(t)$ is called a chip and T_c denotes the chip time interval such that $T_b = 4T_c$. The rate of the spreading PN sequence is fixed at 1.2288 Mcps. Since six code symbols are modulated by one of 64 time-orthogonal Walsh functions, the modulated symbol transmission rate is fixed at 28.8/6 = 4.8 ksps. Therefore, each Walsh chip is spread by four PN chips, that is, $1.2288 \times 10^6 / 307.2 \times 10^3 = 4$. The direct sequence spreading of $d_w(t)$ by the long code PN chips of 1.2288 Mcps is shown in Fig. 2.4.

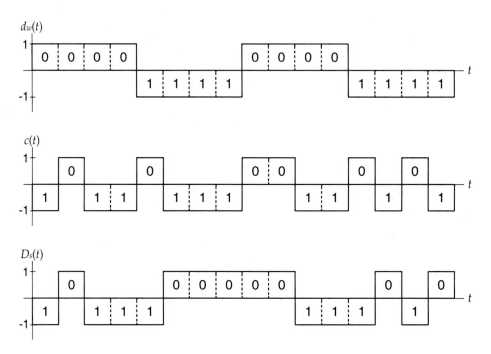

Figure 2.4 Direct sequence spreading of $d_w(t)$ by the long code PN chips $c(t)$

2.5 QPSK AND OFFSET QPSK MODULATION

A spectrally efficient modulation technique for CDMA channels to maximize bandwidth efficiency requires simultaneous transmission on two carriers which are in phase quadrature. Quadrature modulation is of primary importance in a spread-spectrum system and is less sensitive to some types of jamming.

Let $d(t) = d_0, d_1, d_2, \cdots$ be the original data stream. $d(t)$ is bipolar pulses representing binary one (-1) or zero ($+1$) as shown in Fig.2.5 (a). This pulse data stream is divided into an in-phase stream $d_I(t) = d_0, d_2, d_4, \cdots$ (even bits) and a quadrature-phase stream $d_Q(t) = d_1, d_3, d_5, \cdots$ (odd bits) as illustrated in Figs. 2.5 (b) and 2.5 (c). We notice here that $d_I(t)$ and $d_Q(t)$ have half the bit rate of $d(t)$, respectively.

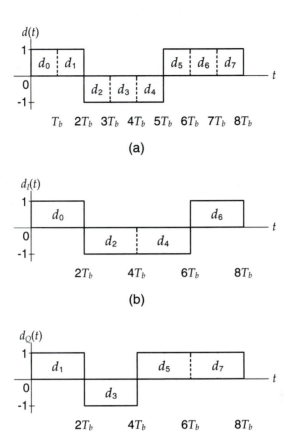

Figure 2.5 (a) Original data stream $d(t)$,
(b) in-phase stream $d_I(t)$, and
(c) quadrature-phase stream $d_Q(t)$

An orthogonal QPSK waveform $s(t)$ is obtained by amplitude modulating $d_I(t)$ and $d_Q(t)$ each onto the cosine and sine functions of a carrier wave as shown in Fig. 2.6.

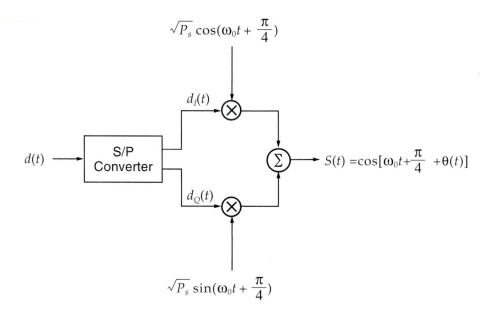

Figure 2.6 QPSK spread-spectrum modulator

$$s(t) = \sqrt{P_s}\, d_I(t)\, \cos(\omega_0 t + \frac{\pi}{4}) + \sqrt{P_s}\, d_Q(t)\, \sin(\omega_0 t + \frac{\pi}{4})$$

$$= \sqrt{2P_s}\, (\cos(\omega_0 t + \frac{\pi}{4})\, \cos\,\theta(t) - \sin(\omega_0 t + \frac{\pi}{4})\, \sin\,\theta(t))$$

$$= \sqrt{2P_s}\, \cos\,[\omega_0 t + \frac{\pi}{4} + \theta(t)\,]$$

(2.7)

where $\cos\,\theta(t) = d_I(t)\,/\sqrt{2}$ and $\sin\,\theta(t) = -\,d_Q(t)\,/\sqrt{2}$,
whence $\theta(t) = -\,\tan^{-1}(d_Q(t)/d_I(t))$.

The in-phase stream $d_I(t)$ amplitude-modulates the cosine function with an amplitude of +1 (binary zero) or −1 (binary one), which produces a BPSK waveform. Similarly, the quadrature-phase stream $d_Q(t)$ modulates the sine function, resulting in a BPSK waveform orthogonal to that due to the cosine function. Thus, the summation of these two orthogonal components of the carrier yields the QPSK waveform.

Since both spreading waveforms, $d_I(t)$ and $d_Q(t)$, are only taken on values of ± 1, these spreading waveforms are assumed to be chip synchronous but otherwise totally independent of one another. The value of $\theta(t)$ will correspond to one of the four possible combinations of $d_I(t)$ and $d_Q(t)$. The output $s(t)$ of the QPSK modulator corresponding to the specific values of $d_I(t)$ and $d_Q(t)$ can be determined with respect to the phase-offset of $\theta(t)$ as follows:

Using $d_I(t) = \sqrt{2} \cos\theta(t)$ and $d_Q(t) = -\sqrt{2} \sin\theta(t)$, $s(t)$ can be determined as follows:

1. For $\theta(t) = -\dfrac{\pi}{4}$, it gives $d_I(t) = \sqrt{2} \cos(-\dfrac{\pi}{4}) = 1$ and

   $d_Q(t) = -\sqrt{2} \sin(-\dfrac{\pi}{4}) = 1$, whence $s(t) = \sqrt{2P_S} \cos\omega_0 t$.

2. For $\theta(t) = \dfrac{\pi}{4}$, it follows $d_I(t) = \sqrt{2} \cos(\dfrac{\pi}{4}) = 1$ and

   $d_Q(t) = -\sqrt{2} \sin(\dfrac{\pi}{4}) = -1$, whence $s(t) = -\sqrt{2P_S} \cos\omega_0 t$. (2.8)

3. For $\theta(t) = \dfrac{5\pi}{4}$, we have $d_I(t) = \sqrt{2} \cos(\dfrac{5\pi}{4}) = -1$ and

   $d_Q(t) = -\sqrt{2} \sin(\dfrac{5\pi}{4}) = 1$, whence $s(t) = \sqrt{2P_S} \sin\omega_0 t$.

4. For $\theta(t) = \dfrac{3\pi}{4}$, we have $d_I(t) = \sqrt{2} \cos(\dfrac{3\pi}{4}) = -1$ and

   $d_Q(t) = -\sqrt{2} \sin(\dfrac{3\pi}{4}) = -1$, whence $s(t) = -\sqrt{2P_S} \cos\omega_0 t$.

Thus, the signal space for QPSK can be sketched as shown in Fig 2.7.

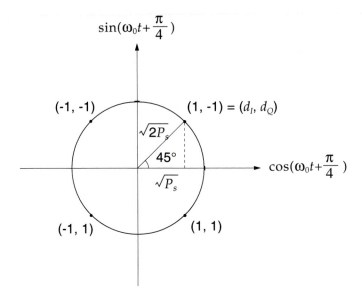

Figure 2.7 Signal space for QPSK

The original data stream $d(t)$ with the duration T_b can be divided into the in-phase stream $d_I(t)$ and the quadrature-phase stream $d_Q(t)$ with each duration $2T_b$ by S/P converter (see Fig. 2.6). In QPSK, the in-phase and quadrature-phase data streams are both transmitted at the rate of $1/2T_b$ bps and are synchronously aligned such that their transitions coincide in time. In non-offset QPSK, the two pulse streams of $d_I(t)$ and $d_Q(t)$ coincide in time, as illustrated in Fig.2.8. Due to this coincident alignment of $d_I(t)$ and $d_Q(t)$, the carrier phase changes only once every $2T_b$, resulting in any one of the four possible phases as shown in Fig. 2.7.

Equation 2.7 also can be used for Offset QPSK (OQPSK) signaling. OQPSK is different from standard non-offset QPSK in the alignment of the two baseband waveforms. Two pulse streams $d_I(t)$ and $d_Q(t)$ are staggered and thus do not change stages simultaneously. The difference between the two modulation schemes is that the timing of pulse streams $d_I(t)$ and $d_Q(t)$ is shifted such that the alignment of the two streams is offset by T_b, as shown in Fig. 2.9 (a). The possibility of the carrier changing phase is limited to $0°$ and $±90°$ every T_b seconds. Whereas in non-offset QPSK, the two pulse streams of $d_I(t)$ and $d_Q(t)$ coincide in time and the carrier phase can change only once every $2T_b$ as shown in Fig. 2.8. A typical OQPSK waveform corresponding to OQPSK data streams is sketched in Fig. 2.9 (b).

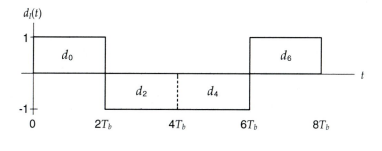

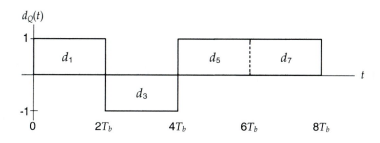

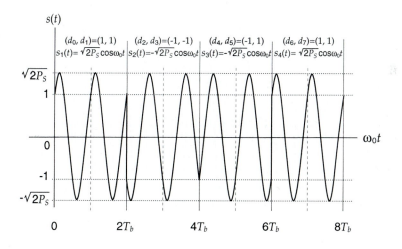

Figure 2.8 The QPSK waveform $s(t)$ resulted from the coincident
alignment of $d_I(t)$ and $d_Q(t)$

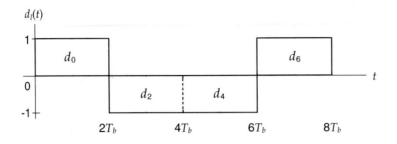

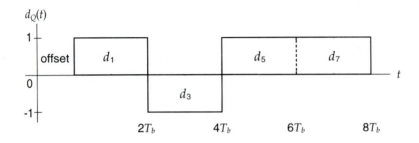

(a) Partitioned offset QPSK data streams

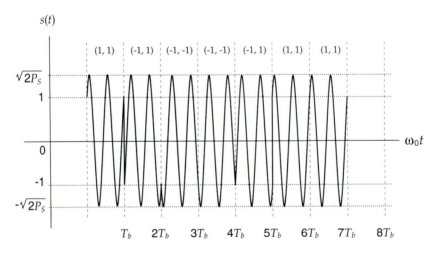

(b) Offset QPSK waveform

Figure 2.9 The OQPSK waveform $s(t)$ resulted from the
staggered pulse streams of $d_I(t)$ and $d_Q(t)$

2.6 LONG CODE GENERATION

The long code provides limited privacy. The long code is a PN sequence with $2^{42}-1$ that is used for scrambling on the forward CDMA channel and spreading the reverse CDMA channel. The long code uniquely identifies a mobile station on both the reverse traffic channel and the forward traffic channel. The long code is characterized by the long code mask that is used to form either the public long code or the private long code. The long code also separates multiple access channels on the same CDMA channel.

When transmitting on the access channel, direct sequence spread by the long code is applied before transmission. This spreading operation involves modulo-2 addition of the 64-ary orthogonal modulator output sequence and the long code, as shown in Fig. 2.10.

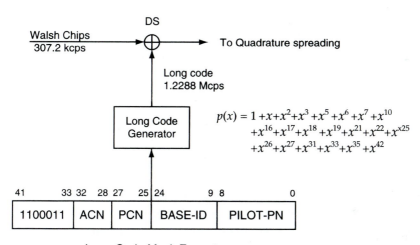

Long Code Mask Format

ACN: Access Channel Number
PCN: Paging Channel Number
BASE-ID: Base Station Identification
PILOT-PN: PN offset for Forward CDMA Channel
1100011: Long Code Mask Header

Figure 2.10 Access channel long code

The long code is periodic with period $2^{42}-1$ chips specifying by the LFSR tap polynomial $p(x)$ of the code generator:

$$p(x) = 1 + x + x^2 + x^3 + x^5 + x^6 + x^7 + x^{10} + x^{16} + x^{17} + x^{18} + x^{19} + x^{21} + x^{22}$$
$$+ x^{25} + x^{26} + x^{27} + x^{31} + x^{33} + x^{35} + x^{42} \tag{2.9}$$

Each PN chip of the long code is generated by EX-ORing all AND gates resulting from a 42-bit mask and each output from LFSR 42 stages, as illustrated in Fig. 2.11.

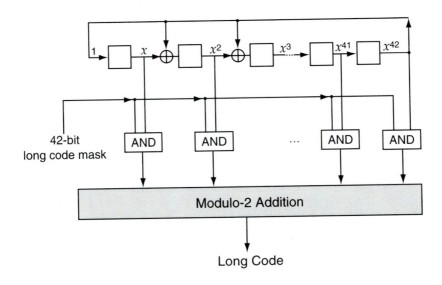

Figure 2.11 Long code generator

The long code mask consists of a 42-bit binary sequence that creates the unique identity of the long code. The long code mask varies depending on the channel type on which the mobile station is transmitting. The long code generator, activated with the long code mask, which produces PN chips (1.2288 kcps) is shown in Fig. 2.11.

2.7 DATA SCRAMBLING

Data scrambling is applied to the forward traffic channel as well as the paging channel. Data scrambling is performed on the block interleaver output at the modulation symbol rate of 19.2 ksps. Data scrambling is accomplished by performing the modulo-2 addition of the block interleaver output with the decimated sampling values (0 or 1) of the long code PN chips. This PN sequence is equivalent to the long code operation at 1.2288 MHz clock rate where only the first output bit of every 64 bits is used for the data scrambling at 19.2 ksps.

The forward traffic channel data is scrambled utilizing the long code generator. The data scrambling mechanism is illustrated in Fig. 2.12. The long code operation at 1.2288 MHz clock rate will generate the PN chip sequence which is the output of the long code generator.

When the long code is divided into the 64-bit length each, every first bit per 64 bits is the decimated sampling value and these decimated values are used for data scrambling at a 19.2 ksps rate. The function of the decimator is to reduce the size of long code by taking one bit from every 64-bit output.

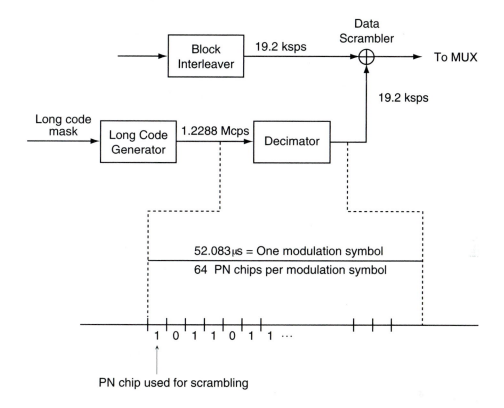

Figure 2.12 Data scrambling mechanism

Voice privacy is provided in the CDMA system by means of the private long code mask being used for PN spreading. Voice privacy control is provided on the traffic channels only. All calls are initialized using the public long code mask for PN spreading. The transition to private long code mask must not be performed if authentication is not performed. To initiate a transition to the private or public code mask, either the base station or the mobile station sends a long code transition request order on the traffic channel. The mobile station or the base station takes action in response to receipt of this order, respectively. The base station also can cause a transition to the public long code mask by sending the handoff direction message with the PRIVATE-LCN bit appropriately set.

2.8 CDMA CODE CHANNEL QUADRATURE SPREADING

Following direct sequence spreading, the reverse traffic channel and the access channel are spread in quadrature.

On the other hand, following orthogonal spreading on the forward CDMA channel, each code channel (pilot, sync, paging, or forward traffic channel) is spread in quadrature.

The spreading sequence is a quadrature sequence of length 2^{15} or 32768 PN chips in length. This quadrature sequence is called the pilot PN sequence and is based on the following tap polynomials, respectively:

$$P_I(x) = 1 + x^5 + x^7 + x^8 + x^9 + x^{13} + x^{15} \tag{2.10}$$
for the in-phase I sequence

and

$$P_Q(x) = 1 + x^3 + x^4 + x^5 + x^6 + x^{10} + x^{11} + x^{12} + x^{15} \tag{2.11}$$
for the quadrature-phase Q sequence.

The I-phase and Q-phase pilot PN sequences also are obtained by means of the following reciprocal polynomials:

$$\begin{aligned} i(x) &= x^{15} P_I(x^{-1}) \\ &= 1 + x^2 + x^6 + x^7 + x^8 + x^{10} + x^{15} \end{aligned} \tag{2.12}$$

$$\begin{aligned} \text{and } q(x) &= x^{15} P_Q(x^{-1}) \\ &= 1 + x^3 + x^4 + x^5 + x^9 + x^{10} + x^{11} + x^{12} + x^{15} \end{aligned} \tag{2.13}$$

The maximum length LFSR sequences based on Eqs. 2.12 and 2.13 are of length $2^{15}-1$ and can be generated by the following linear recursive operation:

$$i(n) = i(n\text{-}15) \oplus i(n\text{-}10) \oplus i(n\text{-}8) \oplus i(n\text{-}7) \oplus i(n\text{-}6) \oplus i(n\text{-}2) \tag{2.14}$$

$$\text{and } q(n) = q(n\text{-}15) \oplus q(n\text{-}12) \oplus q(n\text{-}11) \oplus q(n\text{-}10) \oplus q(n\text{-}9) \oplus q(n\text{-}5) \oplus q(n\text{-}4)$$

$$\oplus\, q(n\text{-}3) \tag{2.15}$$

where $i(n)$ and $q(n)$ for $1 \leq n \leq 32767$ are binary-valued 0 or 1 and $i(15)=1$ and $q(15)=1$, and $\oplus$ denotes the modulo-2 addition.

The I and Q pilot PN sequences, $\overline{P}_I$ and $\overline{P}_Q$, are then generated from $i(n)$ and $q(n)$ under the initial content of LFSR, IC = (1000000000000000). $\overline{P}_I$ and $\overline{P}_Q$ can be computed using the polynomials $i(n)$ and $q(n)$ of length $2^{15}-1$. That is,

1. Using $i(n) = i(n-15) = 0$ for $1 \leq n \leq 14$, $i(15) = \underline{1}$, and $i(n) = \displaystyle\sum_{k=15,10,8,7,6,2} i(n-k)$ $16 \leq n \leq 32767$, the in-phase pilot PN sequence $\overline{P}_I$ is found by computing $i(n)$ first and then inserting '0' in $i(n)$ after 14 consecutive 0's.

2. Similarly, using $q(n) = q(n-15) = 0$ for $1 \leq n \leq 14$, $q(15) = 1$, and $q(n)$ $= \displaystyle\sum_{k=15,12,11,10,9,5,4,3} q(n-k)$ for $16 \leq n \leq 32767$, from which the quadrature-

phase pilot PN sequence $\overline{P}_Q$ is found by inserting '0' at the 15th place in $q(n)$.

2.9 ORTHOGONAL CHANNELIZATION AMONG ALL CODE CHANNELS

For the W-CDMA system, the reverse information channel and signaling channel of the reverse traffic channel are direct sequence spread by the pilot code sequence, traffic code I sequence, traffic code Q sequence, and the signaling code sequence prior to transmission. These code sequences are generated by a code sequence with a period of 81920 chips, generated from the modulo-2 addition of the long code generator and the appropriate Walsh function, as shown in Fig. 2.13. As illustrated in Fig. 2.13 (b), the Walsh function with index zero is assigned to the pilot code. The Walsh function with index one is assigned to the traffic I and Q and access codes. The Walsh function with index two is assigned to the signaling code. Each code sequence results in a period of 81920 chips with 20 ms. PN code generation for the long code is devised as shown in Fig. 2.13 (a).

The long code seed is sent in the message of the paging channel from the base station. This long code is a shortened linear code with period of $2^{23}-1$ chips and will satisfy the linear recursion specified by the following generator polynomial:

$$p(x) = 1 + x + x^2 + x^{22} + x^{32} \tag{2.16}$$

The Walsh function at a fixed chip rate of 4.096 Mcps provides orthogonal channelization among all code channels (reverse pilot channel, reverse information channel and reverse signaling channel), on the reverse traffic channel or access channel (see Fig. 2.13 (b)). Three of the 64 Walsh functions, as defined in Table 2.7, will be used.

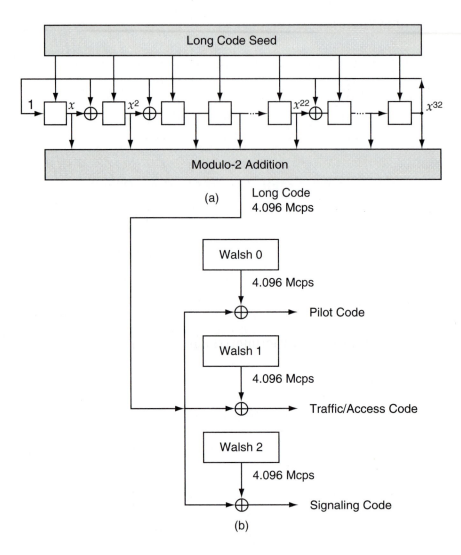

Figure 2.13 Orthogonal channelization among all code channels
             (a) PN code generator for long code
             (b) PN sequence generator for long code

Table 2.7   Walsh Functions for Reverse Link

Walsh chip within a Walsh function

Walsh index	0123	4567	11 8901	1111 2345	1111 6789	2222 0123	2222 4567	2233 8901	3333 2345	3333 6789	4444 0123	4444 4567	4455 8901	5555 2345	5555 6789	6666 0123
0	0000	0000	0000	0000	0000	0000	0000	0000	0000	0000	0000	0000	0000	0000	0000	0000
1	0101	0101	0101	0101	0101	0101	0101	0101	0101	0101	0101	0101	0101	0101	0101	0101
2	0011	0011	0011	0011	0011	0011	0011	0011	0011	0011	0011	0011	0011	0011	0011	0011

Walsh function time alignment is such that the first Walsh chip, indicated by 0 in the column heading of Table 2.7, begins at the first chip of each frame with length 5 ms.

Example 2.4   Referring to Fig. 2.13 (a), the tap polynomial of PN code generator, $g(x) = 1+x+x^2+x^{22}+x^{32}$, whose vector form is 1110000000 0000000000 0010000000 01; while the initial seed vector is assumed to be 1001001100 1010011011 0101111000 11. Using these two vectors, the long code sequence can be generated as shown in Table 2.8.

Table 2.8   Long Code Computation Using PN Code Generator

Shift No.	Register Contents	Σ '1's	Long Code Bit
0	10010011001010011011010111100011	17	1
1	10101001100101001101100011110001	16	0
2	10110100110010100110111001111000	17	1
3	01011010011001010011011100111100	17	1
4	00101101001100101001101110011110	17	1
5	00010110100110010100110111001111	17	1
6	11101011010011001010010011100111	18	0
7	10010101101001100101000001110011	15	1
8	10101010110100110010101000111001	16	0
9	10110101011010011001011100011100	17	1
10	01011010101101001100101110001110	17	1

Thus, the generated long code sequence can be shown as below:

```
1011110101110100 0010111010110010 1001111011101011 0111001111000001
0010100011011010 0110110111011100 1010110100101100 0100101000000101
1100001010110101 0101101101001111 1001111011101101 0011101110100000
0011000010001111 1000000110000000 1111001110010010 0110000000011010
1101110001001000 0011001001001000 1000011111011001 0110100110100001
1111111101110001 1100000000010101 1111100100101011 0000101010101011
0100010100000000 0111010101001101 1000100001100011 0101001000110100
0011010000100101 1011011111100001 0111010010100101 0010111011101101
1111010000000110 1100100100110011 1000000011110000 0100001110001010
0110000001001001 11011000
```

Example 2.5   As illustrated in Fig. 2.13 (b), the pilot code, traffic/access code, and signaling code are generated from the modulo-2 addition of the long code sequence and the appropriate Walsh function. Using three of the 64 Walsh functions (see Table 2.7), these code sequences are generated by a code sequence with a period of 81920 chips.

1. Pilot code generation
   The Walsh function with index zero, W0, is assigned to the pilot code.
   Pilot code = Long code ⊕ Walsh 0

```
Long code:   1011110101110100 0010111010110010 1001111011101011 011100111100001
      ⊕
Walsh 0 :    0000000000000000 0000000000000000 0000000000000000 000000000000000
Pilot code : 1011110101110100 0010111010110010 1001111011101011 011100111100001
```

This sample computation uses only the first row of the long code sequence.

Thus, the pilot code is generated from the modulo-2 addition of the long code sequence and the Walsh function with index zero as follows:

```
1011110101110100 0010111010110010 1001111011101011 0111001111000001
0010100011011010 0110110111011100 1010110100101100 0100101000000101
1100001010110101 0101101101001111 1001111011101101 0011101110100000
0011000010001111 1000000110000000 1111001110010010 0110000000011010
1101110001001000 0011001001001000 1000011111011001 0110100110100001
1111111101110001 1100000000010101 1111100100101011 0000101010101011
0100010100000000 0111010101001101 1000100001100011 0101001000110100
0011010000100101 1011011111100001 0111010010100101 0010111011101101
```

The pilot code is a reproduction of the long code because it is obvious that Walsh 0 is all zeros.

2. Traffic/Access code generation

The Walsh function with index one, W1, is assigned to traffic/access codes.

Traffic/Access codes = Long code $\oplus$ Walsh 1

Long code: 1011110101110100 0010111010110010 1001111011101011 0111001111000001
$\oplus$

Walsh 1:   0101010101010101 0101010101010101 0101010101010101 0101010101010101

Traffic/Access

codes: 1110100000100001 0111101111100111 1100101110111110 0010011010010100

The Traffic/Access codes are generated from the modulo-2 addition of the long code sequence and the Walsh function with index one, as shown below:

1110100000100001	0111101111100111	1100101110111110	0010011010010100
0111110110001111	0011100010001001	1111100001111001	0001111101010000
1001011111100000	0000111000011010	1100101110111000	0110111011110101
0110010111011010	1101010011010101	1010011011000111	0011010101001111
1000100100011101	0110011100011101	1101001010001100	0011110011110100
1010101000100100	1001010101000000	1010110001111110	0101111111111110
0001000001010101	0010000000011000	1101110100110110	0000011101100001
0110000101110000	1110001010110100	0010000111110000	0111101110111000

3. Signaling code generation

The Walsh function with index two, W2, is assigned to the signaling code.

Signaling code = Long code $\oplus$ Walsh 2

Long code: 1011110101110100 0010111010110010 1001111011101011 0111001111000001
$\oplus$

Walsh 2:   0011001100110011 0011001100110011 0011001100110011 0011001100110011

Signaling

code: 1000111001000111 0001110110000001 1010110111011000 0100000011110010

The signaling code is generated from the modulo-2 addition of the long code sequence and the Walsh function with index two, as listed below:

1000111001000111	0001110110000001	1010110111011000	0100000011110010
0001101111101001	0101111011101111	1001111000011111	0111100100110110
1111000110000110	0110100001111100	1010110111011110	0000100010010011
0000001110111100	1011001010110011	1100000010100001	0101001100101001
1110111101111011	0000000101111011	1011010011101010	0101101010010010
1100110001000010	1111001100100110	1100101000011000	0011100110011000
0111011000110011	0100011001111110	1011101101010000	0110000100000111
0000011100010110	1000010011010010	0100011110010110	0001110111011110

Prior to transmission, the reverse information channel and signaling channel are direct-sequence spread by the traffic code I sequence, traffic code Q sequence, and the signaling code sequence.

Data transmitted on the reverse W-CDMA channel is grouped into 5 ms frames. The reverse information channel frames sent at the 64, 32, and 16 kbps transmission rates consist of 320, 160, and 80 bits, respectively.

2.10 AUTHENTICATION AND PRIVACY

The mobile station should operate in conjunction with the base station to authenticate the identity of the mobile station. Authentication is the process by which information is exchanged between a mobile station and base station for the purpose of confirming the identity of the mobile station. A successful outcome of the authentication process will occur only when it can be demonstrated that the mobile station and base station process identical sets of shared secret data (SSD). For example, in a CDMA authentication protocol, a mobile station and base station each have matching sealed authenticators (i.e., identical SSD), actually a short message digest of symbols produced and distributed by the authentication algorithm.

2.10.1 Authentication

The shared secret data (SSD) is a 128-bit pattern stored in the mobile station (in semi-permanent memory) and readily available to the base station. SSD is partitioned into two distinct subsets. Each subset is used to support a different process: SSD-A is used to support the authentication procedure, and SSD-B is used to support voice privacy and message confidentiality.

In computer-communication networks, it is often necessary for communication parties to verify one another's identity. One practical way is the use of cryptographic authentication protocols employing a one-way hash function. Authentication in the CDMA system is the process of confirming the identity of the mobile station by exchanging information between a mobile station and base station.

One possible authentication scheme may be considered for the case of any block cipher. It can be possible to use a symmetric block cipher algorithm (such as DES) in order to compute the 18-bit hash value. If the block algorithm is secure, then it is assumed that the one-way hash function is also secure.

The 152-bit message block M is input into the authentication algorithm device. Using DES, M is first broken down into the 64-bit blocks, so that $M = M_1, M_2, M_3, \cdots$. The first message block M_1 becomes the DES key. Division into 64-bit blocks can be accomplished by mapping the 152-bit message value onto the 192-bit value by padding with 40-bit zeros. Appropriate padding is needed to devise the message to conveniently divide into certain fixed lengths. Our authentication scheme is to generate an 18-bit authentication data from a message length of 192 bits.

The 18-bit message digest for authentication data (AUTHR) also can be derived by hashing the 176-bit padded message by appending the 24-bit padding.

Example 2.6   Figure 2.15 illustrates a typical example for authentication data computation. First, enlarge the 152-bit message into 192 bits by appending the 40-bit padding. Second, divide this padded message of 192 bits into three blocks : $M_1 = 64$ bits, $M_2 = 64$ bits, and $M_3 = 64$ bits. Use M_1 as the 64-bit key which is shown in Fig. 2.14. The first message block M_1 is applicable to the gray block in Fig. 2.15 and can be hashed as the 48-bit encryption keys. The detailed analysis for this problem will be discussed in Section 8.11.2.

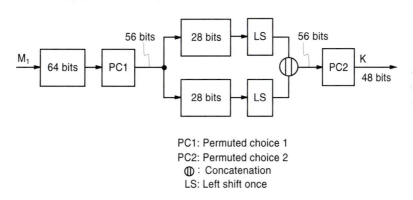

PC1: Permuted choice 1
PC2: Permuted choice 2
ⅅ : Concatenation
LS: Left shift once

Figure 2.14 Key generation based on DES key schedule

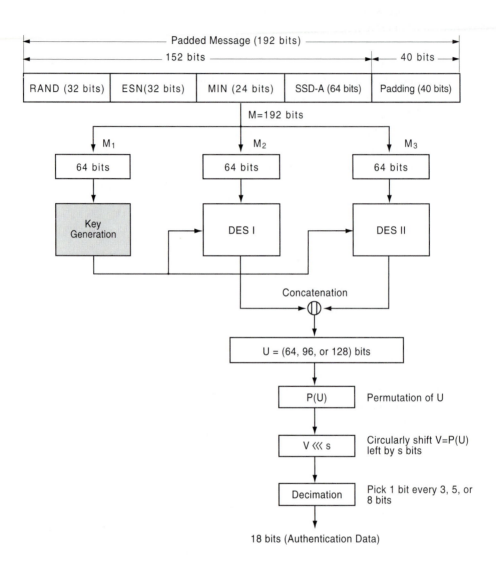

Figure 2.15 Computation of authentications of mobile station at
registrations, unique challenge-response procedure,
originations, or terminations

There are many ways to consider the computation of authentication data (AUTHR) as will be described in Chapter 8.

For computation of the 18-bit authentication data as described above, there are two cases to consider: (1) the 176-bit padded message (the 152-bit message plus the 24-bit padding); and (2) the 192-bit padded message appended by the 40-bit padding.

In addition to this, there are a few more techniques to compute the 18-bit message digest, including the DM scheme. The message to be hashed is first divided into fixed length blocks : M_1, M_2,$\cdots$, M_t. The message block M_i, $1 \leq i \leq t$, will be used as the encryption key. The previous message block is encrypted using that key and then EX-ORed with itself. The resulted data becomes the input to the next round. Thus, at the last round, the length of message becomes the last M_i. The DM scheme is devised using DES block cipher to construct a hash function such that $H_i = E_{M_i}(H_{i-1}) \oplus H_{i-1}$, $1 \leq i \leq t$, where H_0 is an initializing vector.

Another scheme to consider is the modified CBC mode. The encryption steps for the modified CBC scheme are described as

$$H_1 = E_{H_0 \oplus M_1}(H_0) \oplus H_0$$

$$H_2 = E_{H_1 \oplus M_2}(H_1) \oplus H_1$$

$$H_3 = E_{H_2 \oplus M_3}(H_2) \oplus H_2$$

where H_0 denotes an initializing vector and $H_i \oplus M_j$ denotes the enciphering key.

The MD5 algorithm is used for computation of SSD-A and SSD-B. MD5 takes a message of arbitrary length and produces a 128-bit message digest as output. MD5 processes the input message in 512-bit blocks, which can be divided into sixteen 32-bit subblocks. The message digest is a set of four 32-bit blocks, which concatenate to form a single 128-bit hash code. Since SSD update is accomplished by a 128-bit pattern stored in the mobile station, it can be partitioned into two 64-bit distinct substates, i.e., SSD-A-NEW and SSD-B-NEW.

2.10.2 Message Encryption and Information Security

In an effort to enhance the authentication process and to protect sensitive subscriber information (such as PIN), a method is provided to encrypt certain fields of selected traffic channel signaling messages. However, TIA/EIA/IS-95 neither discusses nor lists messages and fields to be encrypted due to the fact that the availability of encryption algorithm information is entirely governed by the U.S.A. International Traffic and Arms Regulation (ITAR) and the Export Administration Regulations.

Messages are not encrypted if authentication is not performed (i.e., AUTH field equal to '00' in the access parameters message). Signaling message encryption is controlled for each call individually. The initial encryption mode for the call is established by the value of the signaling encryption field in the encryption message at the channel assignment. If the signaling encryption field is set to '00', message encryption is off. To turn encryption on after channel assignment, the base station sends the encryption message with the signaling field set to '01'.

To turn signaling message encryption off, the base station sends an encryption message with the signaling encryption field set to '00'.

Every reverse traffic channel message contains an encryption field which identifies the message encryption mode active at the time the message was created.

Data transmitted on the reverse traffic channel is grouped into 20 ms frames. Data frames may be transmitted on the reverse traffic channel at variable data rates of 9600, 4800, 2400, and 1200 bps. The reverse traffic channel is used for the transmission of user and signaling information to the base station during a call.

Each data frame is 20 ms in duration and consists of either information-CRC-tail bits or information and tail bits depending on transmission rates. For example, for 4800 bps, the frame bits (96 bits) consist of the 80-bit information, 8-bit CRC, and 8-bit encoder tail.

Message encryption can be considered in two ways: external encryption and internal encryption as shown in Fig. 2.16.

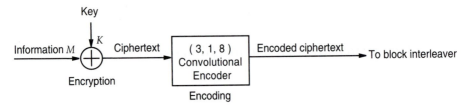

(a) External encryption

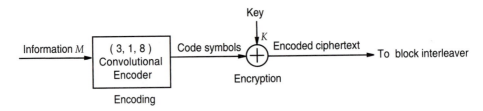

(b) Internal encryption

Figure 2.16 Message encryption scheme

The message information M is first enciphered with the key stream k and then encoded with the (3, 1, 8) convolutional encoder. We shall call this kind of encryption scheme "external encryption." The second scheme to be considered is illustrated as shown in Fig. 2.16(b). Encoding precedes encryption and decryption precedes decoding. We call this kind of encryption scheme "internal encryption."

Example 2.7   Consider the external encryption case. Assume the 80-bit information data for the 4800-bps frame (20 ms) as shown below:

> 1010110011 1001101111 0010010100 0100110011
> 0011001000 1110010110 1001100110 0110011110

CRC computation is accomplished using the generator polynomial

$g(x) = 1 + x + x^3 + x^4 + x^7 + x^8$ and the all-one initial contents of the register. The CRC bits are computed as

> CRC = 10011101

Thus, the 96-bit frame data for the 4800 bps rate is simply computed by concatenating the encoder tail bits (extra all-zero bits) as follows:

> 1010110011 1001101111 0010010100 0100110011 0011001000
> 1110010110 1001100110 0110011110 1001110100 000000

This is the data sequence to be enciphered with the enciphering key which is assumed as follows:

> 0100010111 1010101100 1010110011 0111000100 1010100110
> 1010010011 1100110101 0011100100 1100000010 100110

EX-ORing the information frame data with the key sequence generates the following ciphertext data sequence:

> 1110100100 0011000011 1000100111 0011110111 100110110
> 0100000101 0101010011 0101111010 0101110110 100110

This is the encrypted frame data to be inputted to the (3, 1, 8) convolutional encoder.

Reverse CDMA Channel

$\mathbf{C}$ode Division Multiple Access (CDMA) designates a technique for spread-spectrum multiple-access digital communications that uses unique code sequences. A CDMA channel consists of the forward CDMA channel and the reverse CDMA channel between the base station and the mobile station. The former is the one from a base station to a mobile station. The latter is the one from the mobile station to the base station.

The reverse CDMA channel is used by mobile stations for communicating to the base station while sharing the same CDMA frequency assignment using direct-sequence spreading prior to transmission. The reverse CDMA channel is the reverse link (RL) from the mobile station to the base station. Data transmitted on the reverse CDMA channel is grouped into 20 ms frames. The reverse CDMA channel is composed of access channels and reverse traffic channels. An access channel is used for short signaling message exchanges by providing for call originations, response to pages, orders, and registrations. A reverse traffic channel is used to transport user and signaling traffic from a single mobile station to one or more base stations. Figure 3.1 shows an example of the sum of all mobile station transmissions received by a base station on the logical reverse CDMA channel.

3.1 ACCESS CHANNEL

This section describes the message sent by the mobile station on the access channel. The overall structure of the access channel is shown in Fig. 3.2. The access channel is a reverse CDMA code channel used by the mobile station to initiate communication with the base station and to respond to paging channel messages. One or more access channels are paired with every paging channel. The data rate of the access channel is fixed at 4800 bps. An access channel frame is 20 ms in duration and begins only when system time is an integral multiple of 20 ms. An access channel transmission consists of the access channel preamble and the access channel message capsule. As shown in Fig. 3.2, an access channel transmission is a coded, interleaved, and modulated spread-spectrum signal. Each access channel is distinguished by a different long PN code.

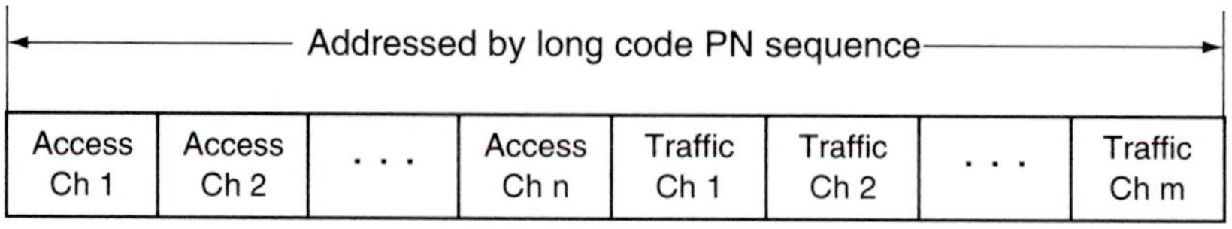

Figure 3.1 An example of reverse CDMA code channels received at a base station (1.23 MHz channel received by the base station)

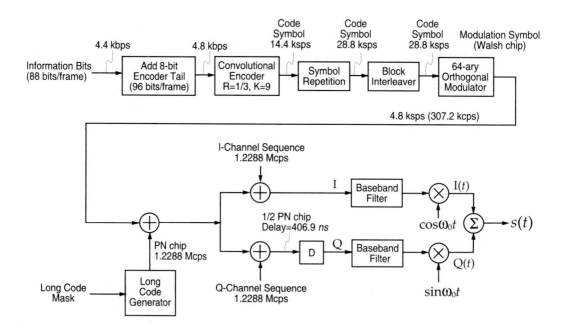

Figure 3.2 Access channel structure

3.1.1 Access Channel Frame Structure

Each access channel frame contains 96 bits (20 ms frame at 4.8 kbps).

Each access channel frame format consists of 88 information bits and 8 encoder tail bits, as shown in Fig. 3.3.

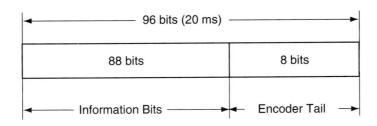

Figure 3.3 Access channel frame structure

Example 3.1 Consider the increment of data rate by adding the encoder tail of 8 bits to the information bits. Since the 8-bit encoder tail is added to the 88-bit information, an access channel becomes 96 bits per frame in 20 ms duration. Thus the data rate 4.4 kbps is to be increased by $96/20 \times 10^{-3} = 4800$ bps. Thus, we can say that the data rate of any access channel is fixed at 4.8 kbps. On the other hand, for the 88-bit information per frame, we have $4400/88 = 50$ bps which is equivalent to 1 bit per frame. Thus, increment by 8-bit encoder tail becomes $50 \times 8 = 0.4$ kbps. It is then proved that the data rate is increased by $4.4 + 0.4 = 4.8$ kbps.

3.1.2 Convolutional Encoding

An (n, k, m) convolutional code designates the code rate $R=k/n$ with encoder stages of $m = K-1$, where K is the constraint length of the code.

The mobile station convolutionally encodes the data transmitted on the access channel prior to block interleaving. The reverse CDMA channel uses the $(3, 1, 8)$ convolutional code representing a code rate $R = 1/3$ and a constraint length of 9. Let $g_k^{(n)}$ denote the generator sequence, where $k = 1$ and $n = 1, 2, 3$ in this case. Thus, the generator sequences for this code are $g_1^{(1)} = 557$ (octal) = 101101111 (binary), $g_1^{(2)} = 663$ (octal) = 110110011 (binary), and $g_1^{(3)} = 711$ (octal)=111001001 (binary). Since $k=1$, $n=3$ and $m=8$, this convolutional encoder consists of a single input terminal, an 8-stage shift register coupled with three modulo-2 adders and a commutator switch for serializing the encoder outputs, as illustrated in Fig. 3.4.

Three code symbols are generated for each input information bit to the encoder. These code symbols will be output such that the code symbol c_0 encoded with the generator sequence $g_1^{(1)}$ is the first output, the code symbol c_1 encoded with the generator sequence $g_1^{(2)}$ is the second output, and the code symbol c_2 encoded with the generator sequence $g_1^{(3)}$ will be the last output.

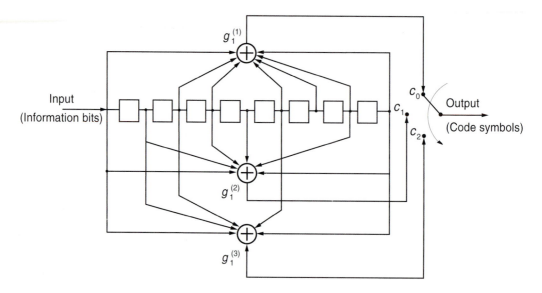

Figure 3.4 The (3,1,8) convolutional encoder

The initial state of the convolutional encoder is assumed to be all-zero. The first code symbol output after initialization will then be a code symbol encoded with the generator sequence $g_1^{(1)}$.

Example 3.2 For the convolutional encoder illustrated in Fig. 3.4, three generator sequences are $g_1^{(1)} = (101101111)$, $g_1^{(2)} = (110110011)$, and $g_1^{(3)} = (111001001)$. Each access channel frame contains 96 bits (20 ms at 4.8 kbps) consisting of 88 information bits and 8 encoder tail bits as shown Fig. 3.3.

Suppose the information sequence (88 bits) with the encoder tail (all-zero 8 bits) is expressed as shown below:

Frame data sequence (96 bits)

MSB

| 1 0 0 0 1 0 1 1 1 1 0 0 1 0 1 0 0 0 1 1 0 0 0 0 1 0 0 0 0 0 1 1 1 1 1 1 0 1 0 1   0 1 1 0 0 1 1 0 |
| 1 1 1 0 1 1 0 1 0 0 1 0 0 1 1 1 0 0 0 1 0 1 1 1 1 0 0 1 0 1 0 0 0 1 1 0 0 0 0 1   0 0 0 0 0 0 0 0 |

LSB

◄──────── Information Sequence (88 bits) ────────► ◄─ Encoder Tail ─►
                                                         (8 bits)

This access channel frame is subject to the input to the convolutional encoder. The order of input bits is started from the most significant bit (MSB) which designates the leftmost bit and ends the least significant bit (LSB) in the rightmost position in the frame data sequence.

Encoder output symbols c_0, c_1, and c_2 are computed from

$$c_0 = d_8 + d_6 + d_5 + d_3 + d_2 + d_1 + d_0 \qquad using\ g_1^{(1)} = (\ 101101111)$$
$$c_1 = d_8 + d_7 + d_5 + d_4 + d_1 + d_0 \qquad using\ g_1^{(2)} = (\ 110110011) \qquad (3.1)$$
$$c_2 = d_8 + d_7 + d_6 + d_3 + d_0 \qquad using\ g_1^{(3)} = (\ 111001001)$$

The code symbol c_0 encoded with $g_1^{(1)} = 557$ (octal) is the first output symbol, the code symbol c_1 encoded with $g_1^{(2)} = 663$ (octal) is the second output symbol, and the code symbol c_2 encoded with $g_1^{(3)} = 771$ (octal) is the third output symbol.

Using Eq. 3.1, the code symbols (c_0, c_1, c_2) corresponding to the 96-bit frame input M are computed as tabulated in Table 3.1.

Table 3.1 The Code Symbol Output C Corresponding to the 96-bit Frame Input M

Frame input bit M	Convolutional encoder contents	Output code Symbol (c_0, c_1, c_2)	Frame input bit M	Convolutional encoder contents	Output code Symbol (c_0, c_1, c_2)
1	100000000	111	1	101100110	110
0	010000000	011	1	110110011	001
0	001000000	101	1	111011001	001
0	000100000	110	0	111101100	001
1	100010000	101	1	101110110	100
0	010001000	110	1	110111011	100
1	101000100	110	0	011011101	010
1	110100010	100	1	101101110	011
1	111010001	100	0	010110111	010
1	111101000	010	0	001011011	011
0	011110100	110	1	100101101	111
0	001111010	010	0	010010110	011
1	100111101	101	0	001001011	001
0	010011110	110	1	100100101	010
1	101001111	010	1	110010010	000
0	010100111	000	1	111001001	011
0	001010011	110	0	011100100	100
0	000101001	100	0	001110010	111
1	100010100	001	0	000111001	110
1	110001010	111	1	100011100	100
0	011000101	101	0	010001110	100
0	001100010	101	1	101000111	111
0	000110001	011	1	110100011	011
0	000011000	111	1	111010001	100
1	100001100	110	1	111101000	010
0	010000110	001	0	011110100	110
0	001000011	100	0	101111010	010

Table 3.1   The Code Symbol Output C Corresponding to the 96-bit Frame Input M

Frame input bit M	Convolutional encoder contents	Output code Symbol (c_0, c_1, c_2)	Frame input bit M	Convolutional encoder contents	Output code Symbol (c_0, c_1, c_2)
0	000100001	001	1	100111101	101
0	000010000	010	0	010011110	110
0	000001000	101	1	101001111	010
1	100000100	011	0	010100111	000
1	110000010	010	0	001010011	110
1	111000001	110	0	000101001	100
1	111100000	111	1	100010100	001
1	111110000	101	1	110001010	111
1	111111000	000	0	011000101	101
0	011111100	011	0	001100010	101
1	101111110	001	0	000110001	011
0	010111111	111	0	000011000	111
1	101011111	000	1	100001100	110
0	010101111	101	0	010000110	001
1	101010111	101	0	001000011	100
1	110101011	110	0	000100001	001
0	011010101	111	0	000010000	010
0	001101010	000	0	000001000	101
1	100110101	000	0	000000100	100
1	110011010	101	0	000000010	110
0	011001101	000	0	000000001	111

Concatenation of all the 96 code symbols output (c_0, c_1, c_2) s from the modulo-2 adders results in a single output sequence which is input to the symbol repeater. Thus, the convolutional code symbol output (288 bits) can be written as

```
1 1 1 0 1 1 1 0 1 1 1 0 1 0 1 1 1 0 1 1 0 1 0 0 1 0 0 0 1 0 1 1 0 0 1 0 1 0 1 1 1 0 0 1 0 0 0 0
1 1 0 1 0 0 0 0 1 1 1 1 1 0 1 1 0 1 0 1 1 1 1 1 1 0 0 0 1 1 0 0 0 0 1 0 1 0 1 0 1 0 1 1 0 1 0
1 1 0 1 1 1 1 0 1 0 0 0 0 1 1 0 0 1 1 1 1 0 0 0 1 0 1 1 0 1 1 1 0 1 1 1 0 0 0 0 0 0 1 0 1 0 0 0
1 1 0 0 0 1 0 0 1 0 0 1 1 0 0 1 0 0 0 1 0 0 1 1 0 1 0 0 1 1 1 1 0 1 1 0 0 1 0 1 0 0 0 0 0 1 1
1 0 0 1 1 1 1 1 0 1 0 0 1 0 0 1 1 1 0 1 1 1 0 0 0 1 0 1 1 0 0 1 0 1 0 1 1 1 0 0 1 0 0 0 0 1 1 0
1 0 0 0 0 1 1 1 1 1 0 1 1 0 1 0 1 1 1 1 1 1 1 0 0 0 1 1 0 0 0 0 1 0 1 0 1 0 1 1 0 0 1 1 0 1 1 1
```

3.1.3　Code Symbol Repetition

The data rate supporting the access channel is fixed at 4.8 ksps. The code symbol output sequence from the convolutional encoder will be repeated before being interleaved. Each code symbol at the fixed data rate of 4800 sps on the access channel is repeated once; that is, each symbol occurs 2 consecutive times. These repeated code symbols will be input to the block interleaver discussed in Section 3.1.4.

Repeated code symbols which occur 2 consecutive times, as shown below, are block-interleaved.

```
111111001111110011111100110011111100111100110000
110000001100111100001100110011111100001100000000
111100110000000011111111110011110011001111111111
111100000011100000000110011001100110011110011001100
111100111111110011000000001111000011111111000000
110011110011111100111110000000000011001100000
111100000011000011000011110000110000001100001111
001100001111111110011110000110011000000000001111
110000111111111100110000110000111111001111110000
001100111100001100110011111100001100000000111100
110000000011111111110011110011001111111111111100
000011110000000011001100110011110000111100111111
```

3.1.4　Block Interleaving

The purpose of using block interleaving is not only to correct burst errors while sending the data through a multipath fading environment, but also to achieve excess redundancy for improving performance. The mobile station interleaves all code symbols on the access channel prior to modulation and transmission. A block interleaver framing 20 ms in duration is used. The block interleaver is a 576-cell array having 32 rows and 18 columns (see Table 3.2).

The access channel code symbols will be the output from the interleaver by rows in the following order:

　1　17　9　25　5　21　13　29　3　19　11　27　7　23　15　31
　2　18　10　26　6　22　14　30　4　20　12　28　8　24　16　32

Table 3.2   4800 bps Interleaver Memory Applicable to Access Channel or Reverse Traffic Channel (After TIA/EIA/IS-95)

1	17	33	49	65	81	97	113	129	145	161	177	193	209	225	241	257	273
1	17	33	49	65	81	97	113	129	145	161	177	193	209	225	241	257	273
2	18	34	50	66	82	98	114	130	146	162	178	194	210	226	242	258	274
2	18	34	50	66	82	98	114	130	146	162	178	194	210	226	242	258	274
3	19	35	51	67	83	99	115	131	147	163	179	195	211	227	243	259	275
3	19	35	51	67	83	99	115	131	147	163	179	195	211	227	243	259	275
4	20	36	52	68	84	100	116	132	148	164	180	196	212	228	244	260	276
4	20	36	52	68	84	100	116	132	148	164	180	196	212	228	244	260	276
5	21	37	53	69	85	101	117	133	149	165	181	197	213	229	245	261	277
5	21	37	53	69	85	101	117	133	149	165	181	197	213	229	245	261	277
6	22	38	54	70	86	102	118	134	150	166	182	198	214	230	246	262	278
6	22	38	54	70	86	102	118	134	150	166	182	198	214	230	246	262	278
7	23	39	55	71	87	103	119	135	151	167	183	199	215	231	247	263	279
7	23	39	55	71	87	103	119	135	151	167	183	199	215	231	247	263	279
8	24	40	56	72	88	104	120	136	152	168	184	200	216	232	248	264	280
8	24	40	56	72	88	104	120	136	152	168	184	200	216	232	248	264	280
9	25	41	57	73	89	105	121	137	153	169	185	201	217	233	249	265	281
9	25	41	57	73	89	105	121	137	153	169	185	201	217	233	249	265	281
10	26	42	58	74	90	106	122	138	154	170	186	202	218	234	250	266	282
10	26	42	58	74	90	106	122	138	154	170	186	202	218	234	250	266	282
11	27	43	59	75	91	107	123	139	155	171	187	203	219	235	251	267	283
11	27	43	59	75	91	107	123	139	155	171	187	203	219	235	251	267	283
12	28	44	60	76	92	108	124	140	156	172	188	204	220	236	252	268	284
12	28	44	60	76	92	108	124	140	156	172	188	204	220	236	252	268	284
13	29	45	61	77	93	109	125	141	157	173	189	205	221	237	253	269	285
13	29	45	61	77	93	109	125	141	157	173	189	205	221	237	253	269	285
14	30	46	62	78	94	110	126	142	158	174	190	206	222	238	254	270	286
14	30	46	62	78	94	110	126	142	158	174	190	206	222	238	254	270	286
15	31	47	63	79	95	111	127	143	159	175	191	207	223	239	255	271	287
15	31	47	63	79	95	111	127	143	159	175	191	207	223	239	255	271	287
16	32	48	64	80	96	112	128	144	160	176	192	208	224	240	256	272	288
16	32	48	64	80	96	112	128	144	160	176	192	208	224	240	256	272	288

Code symbols should be written into the block interleaver by columns filling the complete 32 x 18 matrix shown in Table 3.2.

For convenience, we arrange the interleaver input as shown below:

Symbol number	Block interleaver input			
01–32	11101110	11101011	10110100	10001011
33–64	00101011	10010000	11010000	11111011
65–96	01011111	11000110	00010101	01011010
97–128	11011110	10000110	01111000	10110111
129–160	01110000	00101000	11000100	10011001
161–192	00010011	01001111	10110010	10000011
193–224	10011111	01001001	11011100	01011001
225–256	01011100	10000110	10000111	11011010

The block interleaver output created by rows can be computed using the interleaver input symbols, as shown in Example 3.3.

Example 3.3. Using the order number $\lceil \lambda \rceil$ of interleaver rows, we can compute the output symbols as follows:

$\lambda = 1,$	1	17	33	49	65	81	97	113	129	145	161	177	193	209	225	241	257	273
	1	1	0	1	0	0	1	0	0	1	0	1	1	1	0	1	1	1
$\lambda = 17,$	9	25	41	57	73	89	105	121	137	153	169	185	201	217	233	249	265	281
	1	1	1	1	1	0	1	1	0	1	0	1	0	0	1	1	0	0
$\lambda = 9,$	5	21	37	53	69	85	101	117	133	149	165	181	197	213	229	245	261	277
	1	0	1	0	1	0	1	1	0	0	0	0	1	1	1	0	1	1

.

.

.

Thus, the block interleaver output corresponding to the row-order number λ can be computed as shown below:

λ	Interleaver output symbols	λ	Interleaver output symbols
[1]	1 1 0 1 0 0 1 0 0 1 0 1 1 1 0 1 1 1	[2]	1 1 0 1 0 0 1 0 0 1 0 1 1 1 0 1 1 1
[17]	1 1 1 1 1 0 1 1 0 1 0 1 0 0 1 1 0 0	[18]	1 1 1 1 1 0 1 1 0 1 0 1 0 0 1 1 0 0
[9]	1 0 1 0 1 0 1 1 0 0 0 0 1 1 1 0 1 1	[10]	1 0 1 0 1 0 1 1 0 0 0 0 1 1 1 0 1 1
[25]	1 1 0 1 0 1 0 0 1 1 1 0 1 1 0 1 0 0	[26]	1 1 0 1 0 1 0 0 1 1 1 0 1 1 0 1 0 0
[5]	1 1 1 0 0 0 0 1 1 0 0 1 0 0 0 0 1 1	[6]	1 1 1 0 0 0 0 1 1 0 0 1 0 0 0 0 1 1
[21]	1 0 0 1 0 0 0 1 1 0 0 0 0 0 0 0 1 1	[22]	1 0 0 1 0 0 0 1 1 0 0 0 0 0 0 0 1 1
[13]	1 0 1 0 1 0 1 0 0 0 1 1 1 0 0 1 1 1	[14]	1 0 1 0 1 0 1 0 0 0 1 1 1 0 0 1 1 1
[29]	1 1 0 1 1 1 1 1 0 0 1 1 0 0 1 1 0 1	[30]	1 1 0 1 1 1 1 1 0 0 1 1 0 0 1 1 0 1
[3]	1 0 0 1 1 0 1 1 1 1 0 0 0 1 1 0 1 0	[4]	1 0 0 1 1 0 1 1 1 1 0 0 0 1 1 0 1 0
[19]	1 0 0 1 1 1 0 0 0 0 1 0 1 1 0 1 0 0	[20]	1 0 0 1 1 1 0 0 0 0 1 0 1 1 0 1 0 0
[11]	1 1 0 0 1 1 1 0 0 1 0 0 1 1 1 1 1 0	[12]	1 1 0 0 1 1 1 0 0 1 0 0 1 1 1 1 1 0
[27]	0 0 0 0 1 0 1 1 0 0 1 0 0 0 1 0 0 1	[28]	0 0 0 0 1 0 1 1 0 0 1 0 0 0 1 0 0 1
[7]	0 1 0 1 1 1 1 1 1 0 1 1 1 1 1 0 1 0	[8]	0 1 0 1 1 1 1 1 1 0 1 1 1 1 1 0 1 0
[23]	0 0 1 1 0 1 0 1 0 1 0 0 0 1 0 1 1 1	[24]	0 0 1 1 0 1 0 1 0 1 0 0 0 1 0 1 1 1
[15]	0 0 1 0 1 1 0 0 0 0 1 0 1 0 0 1 0 1	[16]	0 0 1 0 1 1 0 0 0 0 1 0 1 0 0 1 0 1
[31]	1 1 0 1 0 0 0 1 0 1 1 1 1 1 0 0 0 1	[32]	1 1 0 1 0 0 0 1 0 1 1 1 1 1 0 0 0 1

Finally the interleaver output sequence can be obtained by concatenating all the rows corresponding to the order number λ. They are,

```
1 1 0 1 0 0 1 0 0 1 0 1 1 1 0 1 1 1 1 1 1 1 1 0 1 1 0 1 0 1 0 0 1 1 0 0 1 0 1 0 1 0 1 1 0 0 0 0
1 1 1 0 1 1 1 1 0 1 0 1 0 0 1 1 1 0 1 1 0 1 0 0 1 1 1 0 0 0 0 1 1 0 0 1 0 0 0 0 1 1 1 0 0 1 0 0
0 1 1 0 0 0 0 0 0 0 1 1 1 0 1 0 1 0 1 0 0 0 1 1 1 0 0 1 1 1 1 1 0 1 1 1 1 1 0 0 1 1 0 0 1 1 0 1
1 0 0 1 1 0 1 1 1 1 0 0 0 1 1 0 1 0 1 0 0 1 1 1 0 0 0 0 1 0 1 1 0 1 0 0 1 1 0 0 1 1 1 1 0 0 1 0 0
1 1 1 1 1 0 0 0 0 0 1 0 1 1 0 0 1 0 0 0 1 0 0 1 0 1 0 1 1 1 1 1 1 0 1 1 1 1 1 0 1 0 0 0 1 1 0 1
0 1 0 1 0 0 0 1 0 1 1 1 0 0 1 0 1 1 0 0 0 0 1 0 1 0 0 1 0 1 1 1 0 1 0 0 0 1 0 1 1 1 1 1 0 0 0 1
1 1 0 1 0 0 1 0 0 1 0 1 1 1 0 1 1 1 1 1 1 1 1 0 1 1 0 1 0 1 0 0 1 1 0 0 1 0 1 0 1 0 1 1 0 0 0 0
1 1 1 0 1 1 1 1 0 1 0 1 0 0 1 1 1 0 1 1 0 1 0 0 1 1 1 0 0 0 0 1 1 0 0 1 0 0 0 0 1 1 1 0 0 1 0 0
0 1 1 0 0 0 0 0 0 0 1 1 1 0 1 0 1 0 1 0 0 0 1 1 1 0 0 1 1 1 1 1 0 1 1 1 1 1 0 0 1 1 0 0 1 1 0 1
1 0 0 1 1 0 1 1 1 1 0 0 0 1 1 0 1 0 1 0 0 1 1 1 0 0 0 0 1 0 1 1 0 1 0 0 1 1 0 0 1 1 1 1 0 0 1 0 0
1 1 1 1 1 0 0 0 0 0 1 0 1 1 0 0 1 0 0 0 1 0 0 1 0 1 0 1 1 1 1 1 1 0 1 1 1 1 1 0 1 0 0 0 1 1 0 1
0 1 0 1 0 0 0 1 0 1 1 1 0 0 1 0 1 1 0 0 0 0 1 0 1 0 0 1 0 1 1 1 0 1 0 0 0 1 0 1 1 1 1 1 0 0 0 1
```

This binary sequence represents the interleaver output which is the orthogonal modulation input. As mentioned previously, the code symbols are repeated once prior to transmission on the access channel, as indicated in Table 3.2.

These repeated code symbols are transmitted following the transmission structure as illustrated in Fig. 3.5.

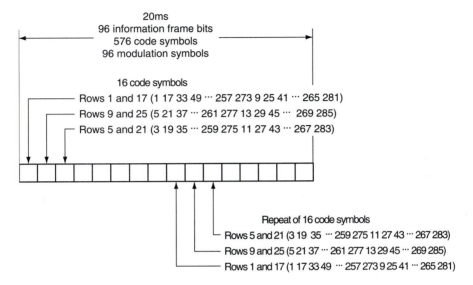

Figure 3.5 Access channel transmission structure
(use Table 3.2 and interleaver's row order)

3.1.5   64-ary Orthogonal Modulation

Modulation for the access channel is 64-ary orthogonal modulation. The modulation symbol is one of 64 mutually orthogonal waveforms generated using Walsh functions. These modulation symbols are given in Table 2.4.

One of 64 possible modulation symbols is transmitted for each six code symbols. Observing from Fig. 3.6, since the code symbol rate at the block interleaver output is 28.8 ksps, the symbol rate at the orthogonal modulator output will become 28.8/6=4.8 ksps or 4.8 x 64=307.2 kcps.

Modulation symbols are selected according to the following formula for a modulation symbol index (MSI).

$$\text{MSI} = c_0 + 2c_1 + 4c_2 + 8c_3 + 16c_4 + 32c_5$$

where c_i, $0 \le i \le 5$, represent the binary valued code symbols that form a modulation symbol index.

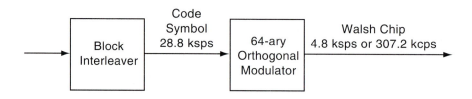

Figure 3.6 Orthogonal modulation

The time period required to transmit a single modulation symbol is equal to 1/4800 =208.333μs. A Walsh chip is defined as the time period associated with 1/64 of a modulation symbol and a Walsh chip period is then equal to 208.333/64=3.2552μs.

Example 3.4 Since any one of 64 possible modulation symbols is transmitted for each six code symbols, it is better to divide the input to the 64-ary orthogonal modulator into the 6-symbol equal length as shown below.

Orthogonal Modulator Input (576 symbols)

```
110100 100101 110111 111110 110101 001100 101010 110000
111011 110101 001110 110100 111000 011001 000011 100100
011000 000011 101010 100011 100111 110111 110011 001101
100110 111100 011010 100111 000010 110100 110011 100100
111110 000010 110010 001001 010111 111011 111010 001101
010100 010111 001011 000010 100101 110100 010111 110001
110100 100101 110111 111110 110101 001100 101010 110000
111011 110101 001110 110100 111000 011001 000011 100100
011000 000011 101010 100011 100111 110111 110011 001101
100110 111100 011010 100111 000010 110100 110011 100100
111110 000010 110010 001001 010111 111011 111010 001101
010100 010111 001011 000010 100101 110100 010111 110001
```

Let us next show how to compute the orthogonal modulator output (only 512 symbols out of 6144 symbols).

Using $MSI = \sum_{i=0}^{5} 2^i c_i$, the orthogonal modulator output (512 symbols) corresponding to the top row (48 symbols) of the orthogonal input (576 symbols) is used to present a simple demonstration.

6-symbol input	MSI	64-symbol output
110100	11	0110011010011001011001101001100101100110100110010110011010011001
100101	41	0101010110101010010101011010101010101010010101011010101001010101
110111	59	0110011010011001100110010110011010011001011001100110011010011001
111110	31	0110100110010110100101100110100101101001100101101001011001101001
110101	43	0110011010011001011001101001100110011001011001101001100101100110
001100	12	0000111111110000000011111111000000001111111100000000111111110000
101010	21	0101101001011010101001011010010101011010010110101010010110100101
110000	3	0110011001100110011001100110011001100110011001100110011001100110

Therefore, the 512-symbol partial output out of the 6114-symbol orthogonal modulator output is tabulated as shown below:

```
0110011010011001011001101001100101100110100110010110011010011001
0101010110101010010101011010101010101010010101011010101001010101
0110011010011001100110010110011010011001011001100110011010011001
0110100110010110100101100110100101101001100101101001011001101001
0110011010011001011001101001100110011001011001101001100101100110
0000111111110000000011111111000000001111111100000000111111110000
0101101001011010101001011010010101011010010110101010010110100101
0110011001100110011001100110011001100110011001100110011001100110
```

3.1.6   Direct Sequence Spreading by the Long Code

When transmitting on the access channel, direct sequence spread by the long code is applied before transmission. This spreading operation involves modulo-2 addition of the 64-ary orthogonal modulator output sequence and the long PN code (see Fig. 3.7). The long code is periodic with period $2^{42}-1$ chips specifying by the LFSR tap polynomial $p(x)$ of the code generator:

$$p(x) = 1 + x + x^2 + x^3 + x^5 + x^6 + x^7 + x^{10} + x^{16} + x^{17} + x^{18} + x^{19} + x^{21} + x^{22} + x^{25}$$
$$+ x^{26} + x^{27} + x^{31} + x^{33} + x^{35} + x^{42} \tag{3.2}$$

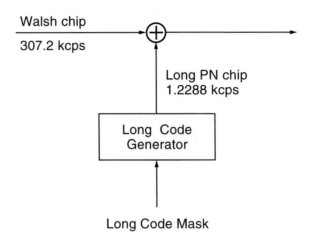

Figure 3.7 Direct spreading operation through the modulo-2 addition

of the 64-ary orthogonal modulator output and the long code

Each PN chip of the long code is generated by the modulo-2 sum of the AND gates resulting from a 42-bit mask and the 42-stage LFSR of the sequence generator as shown in Fig. 3.8.

The long code mask varies depending on the channel type on which the mobile station is transmitting. The access channel long code mask is shown in Fig. 3.9.

When transmitting on the access channel, the mask format sets as follows:

M_{41} - M_{33}: Set to 110001111

M_{32} - M_{28}: Set to the access channel number chosen

M_{27} - M_{25}: Set to the code channel number for the associated paging channel.

M_{24} - M_9: Set to the BASE-ID value for the current base station.

M_8 - M_0: Set to the PILOT-PN value for the current CDMA channel.

Next, it is appropriate to discuss EX-ORing the Walsh chip from the orthogonal modulator and the long PN chip generated from the long code generator, as illustrated in Fig. 3.7.

Each PN chip of the long code is generated by the modulo-2 sum of a 42-bit mask and the 42-bit stage generator. The long code generator, activated with the long code mask, which produces PN chips (1.2288 kcps) is shown in Fig. 3.8.

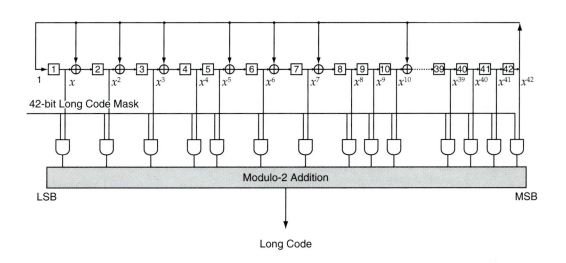

Figure 3.8 Long code generator (After TIA/EIA/IS-95)

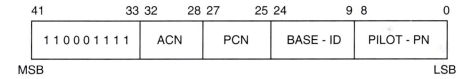

ACN: Access Channel Number
PCN: Paging Channel Number
BASE-ID: Base Station Identification
PILOT-PN: PN offset for Forward CDMA Channel

Figure 3.9 Access channel long code mask

Example 3.5   The long code mask header is set to (110001111). Assume also that ACN is set to (00010), PCN (011), BASE-ID (0100011010011101), and PILOT-PN (100111001). Then the 42-bit long code mask can be arranged as (100111001101110010110001011001000111100011) from LSB to MSB.

Referring to Eq. 3.2, the 42-bit generator tap sequence can be represented as (111101110010000011110110011100010101000000).

Suppose the initial contents of the long code generator are set to (100000000 0000000000000000000000000000000). Each bit of the long code mask and each bit coming out from each stage of the generator should be passed through the AND gates. The outputs reading out from the AND gates should be EX-ORing by the modulo-2 adder. The long PN sequence can then be produced by the steps taken as illustrated in Table 3.3.

Table 3.3 Generating Steps for Long PN Code

Shift No.	Generator Contents	No. of 1s after AND operation	Output
0	1000000000000000000000000000000000000000000	1	1
1	0100000000000000000000000000000000000000000	0	0
2	0010000000000000000000000000000000000000000	0	0
3	0001000000000000000000000000000000000000000	1	1
4	0000100000000000000000000000000000000000000	1	1
5	0000010000000000000000000000000000000000000	1	1
6	0000001000000000000000000000000000000000000	0	0
7	0000000100000000000000000000000000000000000	0	0
8	0000000010000000000000000000000000000000000	1	1
9	0000000001000000000000000000000000000000000	1	1
10	0000000000100000000000000000000000000000000	0	0
11	0000000000010000000000000000000000000000000	1	1
12	0000000000001000000000000000000000000000000	1	1
13	0000000000000100000000000000000000000000000	1	1
14	0000000000000010000000000000000000000000000	0	0
15	0000000000000001000000000000000000000000000	0	0
16	0000000000000000100000000000000000000000000	1	1
17	0000000000000000010000000000000000000000000	0	0
18	0000000000000000001000000000000000000000000	1	1
19	0000000000000000000100000000000000000000000	1	1
20	0000000000000000000010000000000000000000000	0	0
21	0000000000000000000001000000000000000000000	0	0
22	0000000000000000000000100000000000000000000	0	0
23	0000000000000000000000010000000000000000000	1	1
24	0000000000000000000000001000000000000000000	0	0
25	0000000000000000000000000100000000000000000	1	1
26	0000000000000000000000000010000000000000000	1	1
27	0000000000000000000000000001000000000000000	0	0
28	0000000000000000000000000000100000000000000	0	0
29	0000000000000000000000000000010000000000000	1	1

Table 3.3 Generating Steps for Long PN Code *(continued)*

Shift No.	Generator Contents	No. of 1s after AND operation	Output
30	00000000000000000000000000000100000000000	0	0
31	00000000000000000000000000000010000000000	0	0
32	00000000000000000000000000000001000000000	0	0
33	00000000000000000000000000000000100000000	1	1
34	00000000000000000000000000000000010000000	1	1
35	00000000000000000000000000000000001000000	1	1
36	00000000000000000000000000000000000100000	1	1
37	00000000000000000000000000000000000010000	0	0
38	00000000000000000000000000000000000001000	0	0
39	00000000000000000000000000000000000000100	0	0
40	00000000000000000000000000000000000000010	1	1
41	00000000000000000000000000000000000000001	1	1
42	11110111001000001111011001110001010 1000000	10	0
43	01111011100100000111101100111000010 10100000	10	0
44	00111101110010000011110110011100010 1010000	12	0
45	00011110111001000001111011001110001 0101000	11	1
46	00001111011100100000111101100111000 1010100	9	1
47	00000111101110010000011110110011100 0101010	8	0
48	00000011110111001000001111011001110 0010101	10	0
49	11110110110011010110111100111011011 001010	15	1
50	01111011011001110101101111001110110 1100101	12	0
51	11001010100100110101101110010110001 1110010	11	1
52	01100101010010011010110111001011000 1111001	10	0
53	11000101100001000010000010010100110 1111100	9	1
54	01100010110000100001000001001010011 0111110	8	0
55	00110001011000010000100000100101001 1011111	8	0
56	11101111100100000111001001100011110 0101111	13	1
57	10000000111010001100111101000000101 1010111	11	1
58	10110111010101001001000111010001000 0101011	13	1
59	10101100100010101011110100110011101 010101	11	1
60	10100001011001011010100100111011011 101010	12	0
61	01010000101100101101010010011110110 1110101	10	0
62	11011111011110011001110000111110001 1111010	15	1
63	01101111101111001100111000011111000 1111101	11	1
64	11000000111111101001000101111110110 1111110	16	0
65	01100000011111110100100010111111101 10111111	11	1

Table 3.3   Generating Steps for Long PN Code *(continued)*

Shift No.	Generator Contents	No. of 1s after AND operation	Output
66	11000111000111110101001000101110111100111111	12	0
67	10010100101011110101111101100110001000111111	14	0
68	10111101011101110101100111000010010000001111	13	1
69	10101001100110110101101010010000011100011	11	1
70	10100011111011010101011011001110010110100001	12	0
71	10100110110101100101101111101101111100010000	13	1
72	01010011011010110010110111110110111001000	11	1
73	00101001101101011001011011111011011111100100	12	0
74	00010100110110101100101101111101101111110010	15	1
75	00001010011011010110010110111111011011111001	12	0
76	11110010000101100100010010101110001111111100	9	1
77	01111001000010110010001001010111100011111110	9	1
78	00111100100001011001000100101011110001111111	12	0
79	11101001011000100011111011100100100101011111	11	1
80	10000011100100011110100100000011000110011111	10	0
81	10110110111010000000001011110000110111001111	13	1
82	10101100010101001111011100001001001111011	15	1
83	10100001000010101000110111110101110011101	10	0
84	10100111101001011011000010001011101101111010	10	0
85	01010011110100101101100001000101110110111	13	1
86	11011110110010011001101001010011101111110111	15	1
87	10011000010001000011101101011000100011011	12	0
88	10111011000000101110101111011101000101011101	10	0
89	10101010101000011000001110011111110110110	10	0
90	01010101010100001100000111001111111101101111	13	1
91	11011101100010001001011010010110101001011011	12	0
92	10011001111001001011101001110100000001101	12	0
93	10111011110100101010100011101100010100110	14	0
94	01011101111010010101010001110110001010100011	14	0
95	11011001110101000101110001001010010010001	11	1
96	10011011110010101101100001010100011001000	13	1
97	01001101111001010110110000101010001110010	10	0
98	00100110111100101011011000010101000111001	11	1
99	00010011011110010101101100001010100011100	8	0
.	.	.	.
.	.	.	.
.	.	.	.

Thus, the long code sequence can be computed as

10011100110111001011000101100100011110001100011001010100111100110100110110101101011001100010000110010 ... ...

Since the PN chip rate is 1.2288 Mcps and Walsh chip rate is 307.20 kcps, we can see that the ratio is PN chips/Walsh chip = 1.2288 x 10^6/307.20 x 10^3=4.

Therefore, for EX-ORing the orthogonal modulator output (O.M) and the long code output (L.C), it is convenient to divide the long code sequence into the 4-bit equal length as shown below.

O.M	0	1	1	0	0	1	1	0	1	0	0	1	$\cdots$
L.C	1001	1100	1101	1100	1011	0001	0110	0100	0111	1000	1100	0110	$\cdots$
EX-OR($\oplus$)	1001	0011	0010	1100	1011	1110	1001	0100	1000	1000	1100	1001	$\cdots$

Thus, the output sequence of modulo-2 adder for direct sequence spreading is

10010011 00101100 10111110 10010100 10001000 11001001 . . . . . .

3.1.7   Quadrature Spreading

Following the direct sequence spreading, the access channel is spread in quadrature, as shown in Fig. 3.10. The sequences used for this spreading are the zero-offset I and Q pilot PN sequences. These PN sequences are periodic with period 2^{15} chips and will be based on the following LFSR tap polynomials, respectively.

For the in-phase sequence, it is given as

$$P_I(x) = 1 + x^5 + x^7 + x^8 + x^9 + x^{13} + x^{15} \tag{3.3}$$

and for the quadrature-phase sequence, it gives

$$P_Q(x) = 1 + x^3 + x^4 + x^5 + x^6 + x^{10} + x^{11} + x^{12} + x^{15} \tag{3.4}$$

The 15-stage LFSRs based on Eqs. 3.3 and 3.4 can be sketched as shown in Figs 3.11 and 3.12, respectively.

Example 3. 6   Compute the data spread by the I and Q pilot PN sequences.

The data sequences I and Q are the output sequences which are the resulting sequences by EX-ORing the I and Q pilot PN sequences and the input data resulted from the direct sequence spreading by the long code.

Consider first the pilot PN code generation. From Eqs. 3.3 and 3.4, the tap sequences are $P_I = (1\,0\,0\,0\,0\,1\,0\,1\,1\,1\,0\,0\,0\,1\,0)$ and $P_Q = (1\,0\,0\,1\,1\,1\,1\,0\,0\,0\,1\,1\,1\,0\,0)$, respectively.

Assuming the initial contents of the generator as (100000000000000), the shifted P_I and P_Q sequences are generated from the LFSRs as shown in Table 3.4.

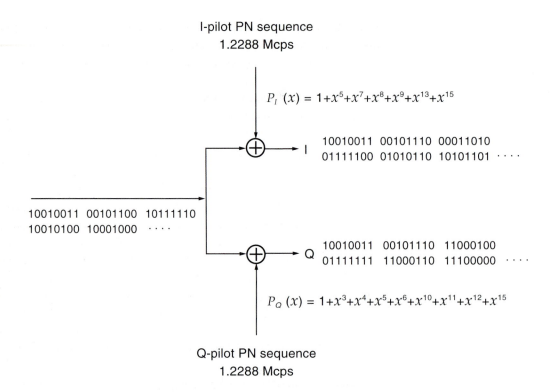

I-pilot PN sequence
1.2288 Mcps

$$P_I(x) = 1 + x^5 + x^7 + x^8 + x^9 + x^{13} + x^{15}$$

I   10010011 00101110 00011010
01111100 01010110 10101101 · · · ·

10010011 00101100 10111110
10010100 10001000 · · · ·

Q   10010011 00101110 11000100
01111111 11000110 11100000 · · · ·

$$P_Q(x) = 1 + x^3 + x^4 + x^5 + x^6 + x^{10} + x^{11} + x^{12} + x^{15}$$

Q-pilot PN sequence
1.2288 Mcps

Figure 3.10 Spreading in quadrature

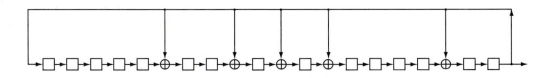

Figure 3.11 A 15-stage LFSR based on the tap polynomial
$$P_I(x) = 1 + x^5 + x^7 + x^8 + x^9 + x^{13} + x^{15}$$

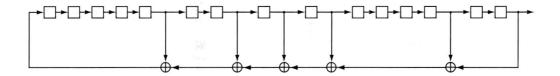

Figure 3.12 A 15-stage LFSR based on the reciprocal tap polynomial
$$i(x) = 1 + x^2 + x^6 + x^7 + x^8 + x^{10} + x^{15}$$

Table 3.4   Generation of the Quadrature Pilot PN Sequences P_I and P_Q

Shift No.	LFSR contents	P_I	Shift No.	LFSR contents	P_Q
0	100000000000000		0	100000000000000	
1	010000000000000	0	1	010000000000000	0
2	001000000000000	0	2	001000000000000	0
3	000100000000000	0	3	000100000000000	0
4	000010000000000	0	4	000010000000000	0
5	000001000000000	0	5	000001000000000	0
6	000000100000000	0	6	000000100000000	0
7	000000010000000	0	7	000000010000000	0
8	000000001000000	0	8	000000001000000	0
9	000000000100000	0	9	000000000100000	0
10	000000000010000	0	10	000000000010000	0
11	000000000001000	0	11	000000000001000	0
12	000000000000100	0	12	000000000000100	0
13	000000000000010	0	13	000000000000010	0
14	000000000000001	0	14	000000000000001	0
15	100001011100010	1	15	100111100011100	1
16	010000101110001	0	16	010011110001110	0
17	101001001011010	1	17	001001111000111	0
18	010100100101101	0	18	100011011111111	1
19	101011001110100	1	19	110110001100011	1
20	010101100111010	0	20	111100100101101	1
21	001010110011101	0	21	111001110001010	1
22	100100000101100	1	22	011100111000101	0
23	010010000010110	0	23	101001111111110	1
24	001001000001011	0	24	010100111111111	0
25	100101111100111	1	25	101101111100011	1

Table 3.4   Generation of the Quadrature Pilot PN Sequences P_I and P_Q *(continued)*

Shift No.	LFSR contents	P_I	Shift No.	LFSR contents	P_Q
26	110011100010001	1	26	110001011101101	1
27	111000101101010	1	27	111111001101011	1
28	011100010110101	0	28	011111100110101	0
29	101111010111000	1	29	101000010000110	1
30	010111101011100	0	30	010100001000011	0
31	001011110101110	0	31	101101100111101	1
32	000101111010111	0	32	110001010000010	1
33	100011100001001	1	33	011000101000001	0
34	110000101100110	1	34	101011110111100	1
35	011000010110011	0	35	010101111011110	0
36	101101010111011	1	36	001010111101111	0
37	110111110111111	1	37	100010111101011	1
38	111010100111101	1	38	110110111101001	1
39	111100001111100	1	39	111100111101000	1
40	011110000111110	0	40	011110011110100	0
41	001111000011111	0	41	001111001111010	0
42	100110111101101	1	42	000111100111101	0
43	110010000010100	1	43	100100010000010	1
44	011001000001010	0	44	010010001000001	0
45	001100100000101	0	45	101110100111100	1
46	100111001100000	1	46	010111010011110	0
47	010011100110000	0	47	001011101001111	0
48	001001110011000	0	48	100010010111011	1
49	000100111001100	0	49	110110101000001	1
50	000010011100110	0	50	111100110111100	1
51	000001001110011	0	51	011110011011110	0
52	100001111011011	1	52	001111001101111	0
53	110001100001111	1	53	100000000101011	1
54	111001101100101	1	54	110111100001001	1
55	111101101010000	1	55	111100010011000	1
56	011110110101000	0	56	011110001001100	0
57	001111011010100	0	57	001111000100110	0
58	000111101101010	0	58	000111100010011	0
59	000011110110101	0	59	100100010010101	1
60	100000100111000	1	60	110101101010110	1
61	010000010011100	0	61	011010110101011	0
62	001000001001110	0	62	101010111001001	1
63	000100000100111	0	63	110010111111000	1

Table 3.4   Generation of the Quadrature Pilot PN Sequences P_I and P_Q *(continued)*

Shift No.	LFSR contents	P_I	Shift No.	LFSR contents	P_Q
64	100011011110001	1	64	011001011111100	0
65	110000110011010	1	65	001100101111110	0
66	011000011001101	0	66	000110010111111	0
67	101101010000100	1	67	100100101000011	1
68	010110101000010	0	68	110101110111101	1
69	001011010100001	0	69	111101011000010	1
70	100100110110010	1	70	011110101100001	0
71	010010011011001	0	71	101000110101100	1
72	101000010001110	1	72	010100011010110	0
73	010100001000111	0	73	001010001101011	0
74	101011011000001	1	74	100010100101001	1
75	110100110000010	1	75	110110110001000	1
76	011010011000001	0	76	011011011000100	0
77	101100010000010	1	77	001101101100010	0
78	010110001000001	0	78	000110110110001	0
79	101010011000010	1	79	100100111000100	1
80	010101001100001	0	80	010010011100010	0
81	101011111010010	1	81	001001001110001	0
82	010101111101001	0	82	100011000100100	1
83	101011100010110	1	83	010001100010010	0
84	010101110001011	0	84	001000110001001	0
85	101011100100111	1	85	100011111011000	1
86	110100101110001	1	86	010001111101100	0
87	111011001011010	1	87	001000111110110	0
88	011101100101101	0	88	000100011111011	0
89	101111101110100	1	89	100101101100001	1
90	010111110111010	0	90	110101010101100	1
91	001011111011101	0	91	011010101010110	0
92	100100100001100	1	92	001101010101011	0
93	010010010000110	0	93	100001001001001	1
94	001001001000011	0	94	110111000111000	1
95	100101111000011	1	95	011011100011100	0
96	110011100000011	1	96	001101110001110	0
97	111000101100011	1	97	000110111000111	0
98	111101001010011	1	98	100100111111111	1
99	111111111001011	1	99	110101111100011	1
.			.		
.			.		
.			.		

In order to obtain the I and Q pilot PN sequences, a 0 (zero) is inserted in P_I and P_Q after 14 consecutive 0 outputs (see the arrow mark ($\downarrow$)). For the purpose of alignment, P_I and P_Q must insert 1 after the 15 consecutive 0s.

Thus, the in-phase pilot PN sequence P_I is

$$\downarrow$$

P_I : 000000000000000101010010011101000110111110011001000
        0011110000100011010010101101010101011101001001111

and the quadrature-phase pilot PN sequence P_Q is

$$\downarrow$$

P_Q : 000000000000000100111101011101011010011100010110011
        10011100011011000111010011000100100100011001100011

Finally, EX-ORing the DS spreading sequence and the pilot PN sequences, P_I and P_Q, the quadrature sequences I and Q can be computed as follows:

1. For the in-phase sequence I, it follows

DSS   10010011 00101100 10111110 10010100 10001000 11001001 $\cdots$
$\oplus$

P_I 00000000 00000001 01010010 01110100 01101111 00110010

I	10010011 00101101 11101100 11100000 11100111 11111011 $\cdots$

2. For the quadrature sequence Q, it follows

DSS   10010011 00101100 10111110 10010100 10001000 11001001$\cdots$
$\oplus$

P_Q 00000000 00000001 00111101 01110101 10100111 00010100

Q	10010011 00101101 10000011 11100001 00101111 11011101 $\cdots$

The data impulse Q spread by the Q-pilot PN sequence is delayed by half a PN chip time, that is $1/2 (1.2288 \times 10^6) = 406.9$ ns, with respect to the data impulse I spread by the I-pilot PN sequence.

These I and Q impulses are applied to the inputs of I and Q baseband filters as shown in Fig. 3.2. After baseband filters, the binary data of I and Q are mapped into phase transition as discussed in Section 3.1.9.

The zero-offset I and Q pilot PN sequences are periodic with period 2^{15} chips are based on LFSR tap polynomials $P_I(x)$ and $P_Q(x)$ of Eqs. 3.3 and 3.4.

The same I and Q pilot PN sequences are also generated by the reciprocal polynomial $i(x)$ $= x^n P_I(x^{-1})$ and $q(x) = x^n P_Q(x^{-1})$, respectively.

Example 3.7 Consider a $(2^n-1, n)$ maximum-length code with minimum distance $d_{min} = 2^{n-1}$. This code is the dual of Hamming code of length 2^n-1, which is obtained by taking a primitive polynomial $P(x)$ of degree n as its generator polynomial. The dual code of a $(2^n-1, n)$ maximum length code is generated by the reciprocal polynomial $i(x) = x^n P(x^{-1})$ which is also a primitive polynomial of degree n.

Example 3.8 The LFSRs based on the tap polynomials $P_I(x)$ for the in-phase sequence and $P_Q(x)$ for the quadrature-phase sequence are shown in Figs. 3.11 and 3.12, respectively.

The reciprocal polynomials $i(x) = x^n P_I(x^{-1})$ and $q(x) = x^n P_Q(x^{-1})$ can be computed as

$$
\begin{aligned}
i(x) &= x^{15}(1 + x^{-5} + x^{-7} + x^{-8} + x^{-9} + x^{-13} + x^{-15}) \\
&= 1 + x^2 + x^6 + x^7 + x^8 + x^{10} + x^{15}
\end{aligned}
\tag{3.5}
$$

$$
\begin{aligned}
q(x) &= x^{15}(1 + x^{-3} + x^{-4} + x^{-5} + x^{-6} + x^{-10} + x^{-11} + x^{-12} + x^{-15}) \\
&= 1 + x^3 + x^4 + x^5 + x^9 + x^{10} + x^{11} + x^{12} + x^{15}
\end{aligned}
\tag{3.6}
$$

Using the reciprocal tap polynomials $i(x)$ and $q(x)$, the LFSRs based on Eqs. 3.5 and 3.6 can be sketched as shown in Figs. 3.13 and 3.14.

Example 3.9 Let $i(n)$ and $q(n)$ be the maximum length LFSR sequences based on the tap polynomials $P_I(x)$ or $i(x)$ and $P_Q(x)$ or $q(x)$, respectively.

These sequences $i(n)$ and $q(n)$ are then called the LFSR sequences that are of period $2^{15}-1$ and can be generated by using the following linear recursive operations:

$$
i(n) = i(n-15) \oplus i(n-10) \oplus i(n-8) \oplus i(n-7) \oplus i(n-6) \oplus i(n-2)
\tag{3.7}
$$

and $q(n) = q(n-15) \oplus q(n-12) \oplus q(n-11) \oplus q(n-10) \oplus q(n-9) - q(n-5)$ (3.8)
$\oplus q(n-4) \oplus q(n-3)$

where $i(n)$ and $q(n)$ for $1 \le n \le 32767$ represent 0 or 1, $i(15)=1$ and $q(15)=1$, and $\oplus$ denotes the modulo-2 addition.

Since $i(n)=0$ for $1 \le n \le 14$ and $i(15)=1$, $i(n)$ for $16 \le n \le 32767$ can be computed as follows:

For $n=16$, $0 \oplus 0 \oplus 0 \oplus 0 \oplus 0 \oplus 0 = 0$
                $i(1) \oplus i(6) \oplus i(8) \oplus i(9) \oplus i(10) \oplus i(14)=0,\ i(16)=0$

For $n=17$, $0 \oplus 0 \oplus 0 \oplus 0 \oplus 0 \oplus 1 = 1$
                $i(2) \oplus i(7) \oplus i(9) \oplus i(10) \oplus i(11) \oplus i(15)=1,\ i(17)=1$

For $n=18$, $0 \oplus 0 \oplus 0 \oplus 0 \oplus 0 \oplus 0 = 0$
                $i(3) \oplus i(8) \oplus i(10) \oplus i(11) \oplus i(12) \oplus i(16)=0,\ i(18)=0$

.
.
.

    etc

Thus, the computation of $i(n)$ can be shown in Table 3.5.

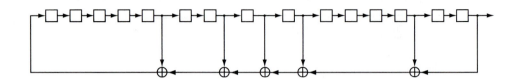

Figure 3.13 A 15-stage LFSR based on the reciprocal tap polynomial
$$i(x) = 1 + x^2 + x^6 + x^7 + x^8 + x^{10} + x^{15}$$

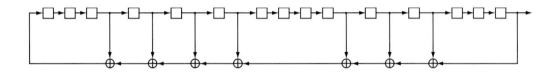

Figure 3.14 A 15-stage LFSR based on the reciprocal tap polynomial
$$q(x) = 1 + x^3 + x^4 + x^5 + x^9 + x^{10} + x^{11} + x^{12} + x^{15}$$

Table 3.5   Generation of $i(n)$ for the In-Phase PN Sequence $\mathcal{P}_I$

n	$\displaystyle\sum_{k=15,10,8,7,6,2} i(n-k),\ 1 \le n \le 32767$	$i(n)$
1		0
2		0
.		.
.		.
.		.
15	$i(n)=i(n-15)=0,\ 1 \le n \le 14$	1
16		0
17		1
.		.
.		.
.		.
21	$0 \oplus 0 \oplus 0 \oplus 0 \oplus 1 \oplus 1$	0
22	$0 \oplus 0 \oplus 0 \oplus 1 \oplus 0 \oplus 0$	1
.		.
.		.
.		.

Table 3.5 Generation of $i(n)$ for the In-Phase PN Sequence $\mathcal{P}_I$ *(continued)*

n	$\displaystyle\sum_{k=15,10,8,7,6,2} i(n-k),\ 1 \le n \le 32767$	$i(n)$
27	$0 \oplus 1 \oplus 1 \oplus 0 \oplus 0 \oplus 1$	1
28	$0 \oplus 0 \oplus 0 \oplus 0 \oplus 1 \oplus 1$	0
.	.	.
.	.	.
.	.	.
34	$1 \oplus 0 \oplus 1 \oplus 1 \oplus 0 \oplus 0$	1
35	$0 \oplus 1 \oplus 1 \oplus 0 \oplus 1 \oplus 1$	0
.	.	.
.	.	.
.	.	.
45	$0 \oplus 0 \oplus 1 \oplus 1 \oplus 1 \oplus 1$	0
46	$0 \oplus 1 \oplus 1 \oplus 1 \oplus 0 \oplus 0$	1
.	.	.
.	.	.
.	.	.
55	$0 \oplus 0 \oplus 0 \oplus 0 \oplus 0 \oplus 1$	1
56	$0 \oplus 1 \oplus 0 \oplus 0 \oplus 0 \oplus 1$	0
.	.	.
.	.	.
.	.	.
74	$0 \oplus 1 \oplus 0 \oplus 1 \oplus 0 \oplus 1$	1
75	$1 \oplus 1 \oplus 1 \oplus 0 \oplus 0 \oplus 0$	1
.	.	.
.	.	.
.	.	.
86	$0 \oplus 0 \oplus 0 \oplus 1 \oplus 0 \oplus 0$	1
87	$1 \oplus 1 \oplus 1 \oplus 0 \oplus 1 \oplus 1$	1
.	.	.
.	.	.
.	.	.
97	$0 \oplus 1 \oplus 1 \oplus 0 \oplus 0 \oplus 1$	1
98	$1 \oplus 0 \oplus 0 \oplus 0 \oplus 1 \oplus 1$	1
99	$0 \oplus 1 \oplus 0 \oplus 1 \oplus 0 \oplus 1$	1
.	.	.
.	.	.
.	.	.

Finally, the in-phase pilot PN sequence $\mathcal{P}_I$ is obtained by inserting a 0 (zero) in $i(n)$ after a 14 consecutive 0 outputs (see arrow ↓), as follows:

> Insert 0
> ↓
> P_I : 0000000000000001010100100111010001101111100110011000
>   0011110000100011010010101101010101011101001001111

Similarly, the quadrature-phase pilot PN sequence P_Q is computed as shown in Table 3.6.

Table 3.6   Generation of *q(n)* for the Quadrature-Phase PN Sequence P_Q

n	$\sum_{k=15, 12, 11, 10, 9, 5, 4, 3} q(n-k), 1 \le n \le 32767$	q(n)
1		0
2		0
.	$q(n)=q(n-15)=0, \ 1 \le n \le 14$	.
.		.
.		.
14		0
15		1
16	$0 \oplus 0 \oplus 0 \oplus 0 \oplus 0 \oplus 0 \oplus 0 \oplus 0$	0
17	$0 \oplus 0 \oplus 0 \oplus 0 \oplus 0 \oplus 0 \oplus 0 \oplus 0$	0
18	$0 \oplus 0 \oplus 0 \oplus 0 \oplus 0 \oplus 0 \oplus 0 \oplus 1$	1
.		.
.		.
.		.
25	$0 \oplus 0 \oplus 0 \oplus 1 \oplus 0 \oplus 1 \oplus 1 \oplus 0$	1
26	$0 \oplus 0 \oplus 1 \oplus 0 \oplus 0 \oplus 1 \oplus 0 \oplus 1$	1
27	$0 \oplus 1 \oplus 0 \oplus 0 \oplus 1 \oplus 0 \oplus 1 \oplus 0$	1
.		.
.		.
.		.
35	$1 \oplus 1 \oplus 0 \oplus 1 \oplus 1 \oplus 0 \oplus 1 \oplus 1$	0
36	$1 \oplus 0 \oplus 1 \oplus 1 \oplus 1 \oplus 1 \oplus 1 \oplus 0$	0
37	$0 \oplus 1 \oplus 1 \oplus 1 \oplus 0 \oplus 1 \oplus 0 \oplus 1$	1
.		.
.		.
.		.
46	$1 \oplus 1 \oplus 0 \oplus 0 \oplus 1 \oplus 0 \oplus 0 \oplus 1$	0
47	$1 \oplus 0 \oplus 0 \oplus 1 \oplus 1 \oplus 0 \oplus 1 \oplus 0$	0
.		.
.		.
.		.

Table 3.6   Generation of $q(n)$ for the Quadrature-Phase PN Sequence P_Q (continued)

n	$\displaystyle\sum_{k=15,\,12,\,11,\,10,\,9,\,5,\,4,\,3} q(n-k),\ 1 \le n \le 32767$	$q(n)$
55	$0 \oplus 1 \oplus 0 \oplus 1 \oplus 0 \oplus 1 \oplus 0 \oplus 0$	1
56	$0 \oplus 0 \oplus 1 \oplus 0 \oplus 0 \oplus 0 \oplus 0 \oplus 1$	0
.		.
.		.
.		.
71	$0 \oplus 1 \oplus 1 \oplus 0 \oplus 1 \oplus 0 \oplus 1 \oplus 1$	1
72	$0 \oplus 1 \oplus 0 \oplus 1 \oplus 1 \oplus 1 \oplus 1 \oplus 1$	0
.		.
.		.
.		.
80	$0 \oplus 1 \oplus 1 \oplus 0 \oplus 1 \oplus 1 \oplus 0 \oplus 0$	0
81	$0 \oplus 1 \oplus 0 \oplus 1 \oplus 0 \oplus 0 \oplus 0 \oplus 0$	0
.		.
.		.
.		.
88	$0 \oplus 0 \oplus 0 \oplus 1 \oplus 0 \oplus 0 \oplus 0 \oplus 1$	0
89	$1 \oplus 0 \oplus 0 \oplus 1 \oplus 0 \oplus 0 \oplus 1 \oplus 0$	1
.		.
.		.
.		.
97	$1 \oplus 1 \oplus 0 \oplus 0 \oplus 0 \oplus 0 \oplus 1 \oplus 1$	0
98	$0 \oplus 0 \oplus 0 \oplus 1 \oplus 1 \oplus 1 \oplus 0$	1
99	$0 \oplus 0 \oplus 1 \oplus 1 \oplus 1 \oplus 0 \oplus 0$	1
.		.
.		.
.		.

Using Table 3.6, we can obtain the quadratic-phase pilot PN sequence by inserting a 0 in $q(n)$ after 14 consecutive 0s, as shown below:

```
                  Insert 0
                     ↓
P_Q  :00000000000000001001111010111010110100111000101 0011
        1001110001101100011101001100010010010 0011001100011
```

As discussed at the end of Example 3.6, the in-phase sequence I and the quadrature-phase sequence Q are easily computed using

$I = \text{DSS} \oplus P_I$ and $Q = \text{DSS} \oplus P_Q$, respectively.

Both I and Q data are applied to the I and Q baseband filters. But the data impulse Q should be delayed by half a PN chip time (406.9 ns) so that the offset QPSK can be applied.

3.1.8   Baseband Filtering

Following the quadrature spreading operation, that is, I=DSS$\oplus P_I$ and Q=DSS$\oplus P_Q$, the I and Q data are applied to the inputs of the I and Q baseband filters. Let $S(f)$ be a frequency spectrum of the baseband filters. The limits of the normalized frequency response of the filter are confined within $\pm\delta_1$ in the passband $0 \leq f \leq f_P$ and the normalized response should be less than or equal to $-\delta_2$ in the stopband $f \geq f_S$ as shown in Fig. 3.15. Specifically, the numerical values for the limit parameters are $\delta_1 = 1.5$ dB, $\delta_2 = 40$ dB, $f_P = 590$ kHz, and f_S=740 kHz.

Let $S(t)$ be the impulse response of the baseband filter. Then $S(t)$ satisfies the following equation:

$$\text{Mean Squared Error} = \sum_{k=0}^{47} [\alpha S(kTs - \tau) - h(k)]^2 \leq 0.03 \qquad (3.9)$$

where α and τ are the constants for minimizing the mean squared error. The constant T_S equal to 1/4 (1.2288 x 10^6) = 203.451 ns, which is just one quarter of a PN chip time.

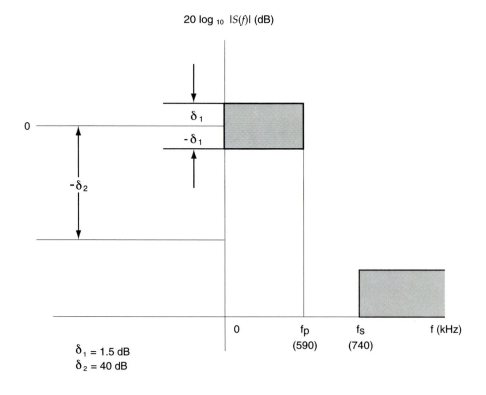

Figure 3.15 Frequency response limits of baseband filters (After IS-95)

The values of the coefficients $h(k)=h(47-k)=0$ for $k \geq 48$; while $h(k)$ for $k < 48$ are given in Table 3.7.

Table 3.7   Coefficients $h\ (k)$

k	$h(k)$
0, 47	−0.025288315
1, 46	−0.034167931
2, 45	−0.035752323
3, 44	−0.016733702
4, 43	0.021602514
5, 42	0.064938487
6, 41	0.091002137
7, 40	0.081894974
8, 39	0.037071157
9, 38	−0.021998074
10, 37	−0.060716277
11, 36	−0.051178658
12, 35	0.007874526
13, 34	0.084368728
14, 33	0.126869306
15, 32	0.094528345
16, 31	−0.012839661
17, 30	−0.143477028
18, 29	−0.211829088
19, 28	−0.140513128
20, 27	0.094601918
21, 26	0.441387140
22, 25	0.785875640
23, 24	1.0

(After TIA/EIA/IS-95)

3.1.9   Offset Quadrature Phase Shift Keying (OQPSK)

The pilot PN sequences repeat every $2^{15}/1.2288 \times 10^6 = 26.666$ ms. Hence, there are exactly 75 repetitions in every 2 seconds (i.e., $2/26.666 \times 10^{-3}=75$). Since the data spread by the Q pilot PN sequence is delayed by half a chip time, offset quadrature phase shift keying is used for the spreading modulation which is advantageous for transmitting simultaneously on two carriers in phase quadrature.

OQPSK is different from standard nonoffset QPSK in the alignment of the two baseband wareforms. Let $I(t)$ and $Q(t)$ be two output streams coming out from the baseband filters. The timing of the two pulse streams $I(t)$ and $Q(t)$ is offset by $T_b/2$ seconds due to delaying Q by half a chip time. In standard nonoffset QPSK, the two baseband streams coincide in time and the carrier phase can change only once every T_b.

An orthogonal QPSK waveform s(t) is obtained by amplitude modulation $I(t)$ and $Q(t)$ each onto the cosine and sine functions of a carrier wave, as shown in Fig. 3.16.

The in-phase stream $I(t)$ amplitude-modulates the cosine function with an amplitude of +1 or −1, which produces a BPSK waveform. Whereas the quadrature-phase stream $Q(t)$ modulates the sine function, resulting in a BPSK waveform orthogonal to the cosine function. Thus, the summation of these two orthogonal BPSK waveforms will yield the QPSK waveform.

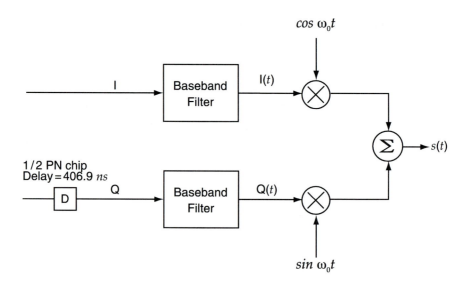

Figure 3.16 OQPSK waveform generation by amplitude modulation

Let $s(t)$ be the QPSK waveform expressed as

$$s(t) = I(t) \cos \omega_0 t + Q(t) \sin \omega_0 t \qquad (3.10)$$

Putting $I(t) = \sqrt{2} \cos \theta (t)$ and $Q(t) = \sqrt{2} \sin \theta (t)$, respectively, $s(t)$ becomes

$$s(t) = \sqrt{2} \cos (\omega_0 t - \theta (t)) \qquad (3.11)$$

where $\theta (t) = \tan^{-1} \dfrac{Q(t)}{I(t)}$

The value of $\theta (t)$ will correspond to one of the four possible combinations of $I(t)$ and $Q(t)$. The QPSK stream $s(t)$ corresponding to the specific values of $I(t)$ and $Q(t)$ can be determined according to the value of $\theta (t)$ as follows:

1. For $\theta (t) = \dfrac{\pi}{4}$, $I(t)$ and $Q(t)$ become $I(t) = \sqrt{2} \cos (\dfrac{\pi}{4}) = 1$,
   $Q(t) = \sqrt{2} \sin(\dfrac{\pi}{4}) = 1$, and $s(t) = \sqrt{2} \cos (\omega_0 t - \dfrac{\pi}{4})$.

2. For $\theta (t) = \dfrac{3\pi}{4}$, it gives $I(t) = \sqrt{2} \cos (\dfrac{3\pi}{4}) = -1$, $Q(t) = \sqrt{2} \sin (\dfrac{3\pi}{4}) = 1$,
   and $s(t) = \sqrt{2} \cos (\omega_0 t - \dfrac{3\pi}{4})$

3. For $\theta (t) = -\dfrac{3\pi}{4}$, we have $I(t) = \sqrt{2} \cos (-\dfrac{3\pi}{4}) = -1$,
   $Q(t) = \sqrt{2} \sin (-\dfrac{3\pi}{4}) = -1$, and $s(t) = \sqrt{2} \cos (\omega_0 t + \dfrac{3\pi}{4})$

4. For $\theta (t) = -\dfrac{\pi}{4}$, it follows $I(t) = \sqrt{2} \cos (-\dfrac{\pi}{4}) = 1$,
   $Q(t) = \sqrt{2} \sin (-\dfrac{\pi}{4}) = -1$, and $s(t) = \sqrt{2} \cos (\omega_0 t + \dfrac{\pi}{4})$

Based on the above analysis, $I(t)$ and $Q(t)$ mapping for the reverse CDMA channel (either access channel or reverse traffic channel) can be shown in Table 3.8.

Table 3.8 $I(t)$ and $Q(t)$ Mapping for the Access Channel (also for the Reverse Traffic Channel)

$\theta (t)$	$I(t)$	$Q(t)$		$I(t)$	$Q(t)$
$\pi /4$	1	1		0	0
$3\pi /4$	−1	1		1	0
$-3\pi /4$	−1	−1	⇨	1	1
$-\pi /4$	1	−1		0	1
	(NRZ Value)			(Binary Value)	

Using Table 3.8, the signal constellation mapping and phase transition are shown in Fig. 3.17.

The QPSK pulse stream $s(t) = \sqrt{2} \cos(\omega_0 t - \theta (t))$ can be sketched using $(I(t), Q(t), \theta (t))$ mapping which is shown in Table 3.9.

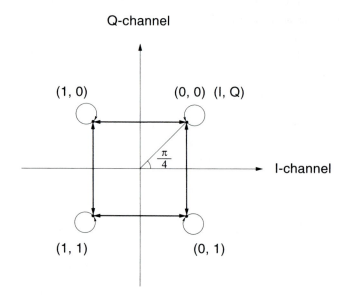

Figure 3.17 Signal constellation and phase transition applicable to
either access channel or reverse traffic channel

Table 3.9   Data for the QPSK Waveform Plot

$\omega_0 t$	$s(t)= \sqrt{2} \cos (\omega_0 t - \theta (t))$			
	$(1,1),\ \theta(t)=\pi/4$	$(-1,1),\ \theta(t)=3\pi/4$	$(-1,-1),\ \theta(t)=-3\pi/4$	$(1,-1),\ \theta(t)=-\pi/4$
0	1	-1	-1	1
$\pi/4$	$\sqrt{2}$	0	$-\sqrt{2}$	0
$\pi/2$	1	1	-1	-1
$3\pi/4$	0	$\sqrt{2}$	0	$-\sqrt{2}$
π	-1	1	1	-1
$-3\pi/4$	$-\sqrt{2}$	0	$\sqrt{2}$	0
$-\pi/2$	-1	-1	1	1
$-\pi/4$	0	$-\sqrt{2}$	0	$\sqrt{2}$
2π	1	-1	1	1

Utilizing Table 3.9 as a tool, the QPSK pulses, which are important elements for plotting $s(t)$, can be sketched as illustrated in Fig. 3.18.

Example 3.10 As calculated previously, the in-phase sequence I and the quadrature-phase sequence Q are, respectively,

I:	10010011	00101101	11101100	11100000	11100111	. . .
Q:	10010011	00101101	10000011	11100001	00101111	. . .

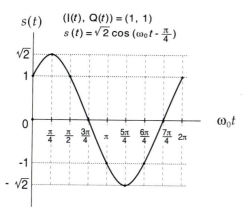

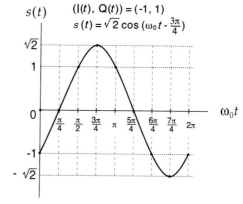

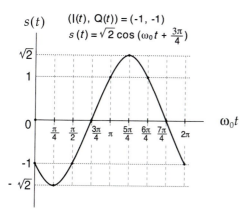

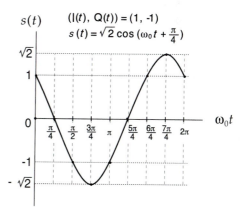

Figure 3.18 Illustration of QPSK pulses according to $(I(t), Q(t), \theta(t)$ mapping over $T_b/2$

Consider only the partial sequences $I(t)$=(11101100) and $Q(t)$=(10000011) for demonstrative purposes. These individual pulses $I(t)$ and $Q(t)$, as well as the QPSK waveform, are shown as illustrated in Fig. 3.19.

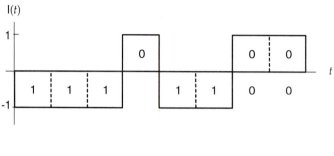

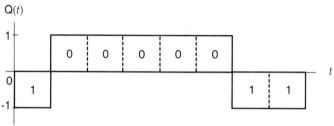

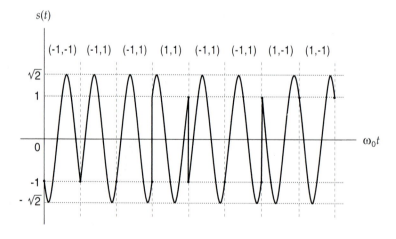

Figure 3.19 The QPSK waveform corresponding to the given
$I(t) = (11101100)$ and, $Q(t) = (10000011)$

Similarly, the offset QPSK pulse stream caused by delaying Q by half a chip time, $T_b/2$, can be shown as illustrated in Fig. 3.20.

3.1.10 Access Channel Preamble and Message Capsule

The reverse CDMA channel may contain up to 32 access channels numbered 0 through 31 per supported paging channel. Each access channel is associated with a single paging channel. The access channel is used by the mobile station to initiate communication with the base station and to respond to paging channel messages. The frame of an access channel at a fixed data rate of 4.8 kbps contains 96 bits in 20 ms time duration and begins only when system time is an integral multiple of 20 ms.

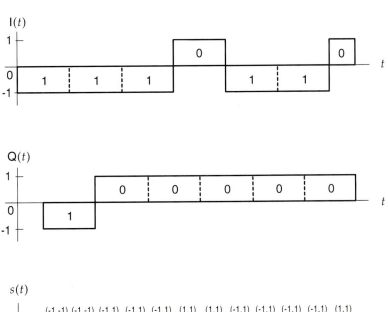

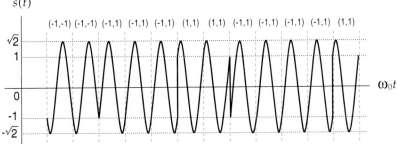

Figure 3.20 The OQPSK waveform $s(t)$ corresponding to the given

$I(t) = (11101100)$ and $Q(t) = (10000011)$

An access channel transmission consists of the access channel preamble and the access channel message capsule. The access channel preamble is transmitted to aid the base station in acquiring an access channel transmission. An access channel message capsule consists of an access channel message and padding as shown in Fig. 3.21(c). The access channel message includes the message length field (8 bits), the message body (2 to 842 bits), and the CRC (30 bits) in that order, but not including the preamble or padding.

The length of the access channel message capsule is an integer number of access channel frames which is given by

$$\text{Capsule Size} = \left\lceil \frac{\text{Message Length} + \text{Message Body Length} + \text{CRC}}{\text{Message Frame without Encoder Tail}} \right\rceil$$

$$= \left\lceil \frac{8 + (2 \text{ to } 842) + 30}{88} \right\rceil \tag{3.12}$$

The message body size is selected so that the capsule size does not exceed 3+MAX-CAP-SZ (the maximum number of access channel frames in an access channel message capsule).

The mobile station transmits padding consisting of zero or more 0 bits immediately following the access channel message. The length of padding will be such that

$$8 + \text{Message Body Length} + 30 + \text{Padding Length} = 88 \times \text{Capsule Size} \tag{3.13}$$

Example 3.11 Determine the capsule size and padding length when the message body length is assumed to be 460 bits.

$$\text{Capsule Size} = \left\lceil \frac{8 + 460 + 30}{88} \right\rceil = \left\lceil \frac{498}{88} \right\rceil = 6$$

Padding Length = (88x6)–498 = 528–498 = 30 bits (all zeros).

Since the capsule size cannot exceed $3 + N_{mf}$, the maximum number of access channel frames N_{mf} is at least 3 due to $6 \le 3 + N_{mf}$.

Using Eqs. 3.12 and 3.13, we can determine the capsule size and the number of access channel frames versus the message body length as shown in Table 3.10

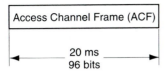

(a) Access Channel Frame

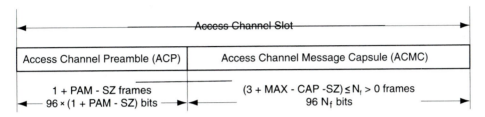

(b) Access Channel Slot

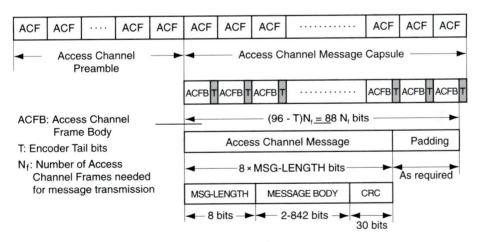

(c) Access Channel Message Capsule = Access Channel Message + Padding

Figure 3.21 Access Channel Structure: (a) Access Channel Frame
(b) Access Channel Slot (c) Access Channel Message Capsule = Access
Channel Message + Padding

Table 3.10   Determination of Capsule Size and Number of Access Channel Frames

Message Body Length (MBL)	Capsule Size (CS)	Number of Access Channel Frame (N_f)
$227 \leq MBL \leq 314$	4	1
$315 \leq MBL \leq 402$	5	2
$403 \leq MBL \leq 490$	6	3
$491 \leq MBL \leq 578$	7	4
$579 \leq MBL \leq 666$	8	5
$667 \leq MBL \leq 754$	9	6
$755 \leq MBL \leq 842$	10	7

Using Table 3.10, plots of the capsule size against the message body length, and the number of access channel frames against the capsule size, are shown in Fig. 3.22.

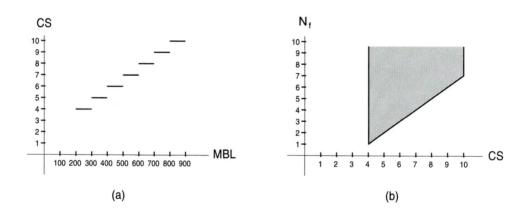

(a)   (b)

Figure 3.22 Plots of capsule size versus message body length (a); number of access channel frames versus capsule size (b)

3.1.11 Access Channel CRC Computation

A 30-bit CRC is computed for each access channel signaling message. The CRC sequence is generated from the message length field plus the message body. The generator polynomial for the CRC is given as shown below:

$$g(x) = 1 + x + x^2 + x^6 + x^7 + x^8 + x^{11} + x^{12} + x^{13} + x^{15} + x^{20} + x^{21} + x^{29} + x^{30} \qquad (3.14)$$

whose logic diagram is shown in Fig. 3.23.

The CRC field can be computed by the following procedure and the corresponding CRC encoder is illustrated in Fig. 3.24.

1. Initial contents of the shift register is set to binary one in order to make the CRC field to nonzero values even for all-zero data.
2. Set the switch in the down position and close the gate in the feedback path of the CRC encoder.
3. Then the k input bits start to transmit into the encoder output as well as the feedback path of the shift register. The k-bit input represents 8 + message body length in bits.
4. The register shall be clocked k times with the k-bit input.
5. Set the switch to the up position and open the gate in the feedback path.
6. Clocking the register by an additional 30 times, the CRC field will be produced.
7. Thus, the bits will be transmitted in the order of the message length field, the message body length, and the CRC field at the CRC encoder output.

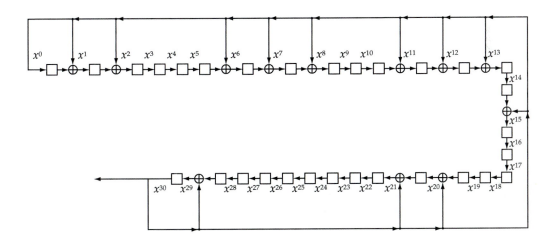

Figure 3.23 LFSR circuit representing $g(x)$ of Eq. 3.14

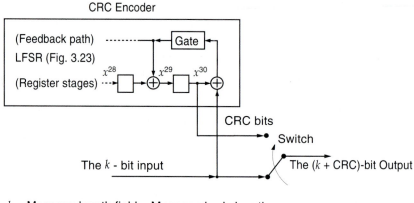

k = Message length field + Message body length
= 8+(2 to 842) in bits

Figure 3.24 Encoder block diagram for generation of CRC field

The mobile station may consider any message with a CRC that checks to be valid. The mobile station will ignore any message which is not valid.

3.1.12 Modulation Parameters for the Access Channel

So far, we have covered the access channel in great detail. In summary, all data transmitted on the access channel are processed by taking the following steps:

1. The 8-bit encoder tail is added to the information bits in order to group into 20 ms frames.
2. Data is convolutionally encoded.
3. Code symbols reading out from the convolutional encoder are repeated before being interleaved.
4. The purpose of using block interleaving is to protect from burst errors which occur while sending the data prior to modulation.
5. In 64-ary orthogonal modulation, the modulation symbol is one of 64 mutually orthogonal waveforms generated by using Walsh function.
   One of 64 possible modulation symbols is transmitted for each six code symbols.
6. The direct sequence spread by the long code is applied before transmission. This spreading operation involves modulo-2 addition of the 64-ary orthogonal output and the long code.
7. Following direct sequence spreading, the access channel is spread in quadrature. The quadrature data sequences I and Q are the resulting sequences by EX-ORing the zero-off-

set I and Q pilot PN sequences and the input data resulted from the direct sequence spreading by the long code.

8. The quadrature-sequence data Q spread by the Q-pilot PN sequence is delayed by half a PN chip time (406.9 ns).

   These I and Q impulses are applied to the I and Q baseband filters. After baseband filters, the data impulses of I and Q are mapped into phase transitions.

9. An orthogonal QPSK waveform $s(t)$ is obtained by amplitude modulation $I(t)$ and $Q(t)$ each onto the cosine and sine functions of a carrier wave. Since the data spread by the Q pilot PN sequence is delayed by half a chip time, offset quadrature-phase shift keying (OQPSK) is used for the spreading modulation.

The modulation parameters for the access channel are illustrated in Table 3.11.

Table 3.11   Access Channel Modulation Parameters

Parameter	Data Rate (4800 bps)	Units
PN chip rate	1.2288	Mcps
Code rate	1/3	bits/code sym
Code symbol repetition	2	Symbols/code sym
Transmit duty cycle	100.0	%
Code symbol rate	28,800	sps
Modulation	6	Code sym/mod symbol
Modulation symbol rate	4800	sps
Walsh chip rate	307.20	kcps
Mod symbol duration	208.33	μ s
PN chips/code symbol	42.67	PN chips/code sym
PN chips/mod symbol	256	PN chips/mod sym
PN chips/Walsh chip	4	PN chips/Walsh chip

3.2 REVERSE TRAFFIC CHANNEL

The reverse traffic channel (RTC) is used for the transmission of user and signaling information to the base station during a call. The reverse traffic channel may use variable transmission rates of 9.6, 4.8, 2.4 or 1.2 kbps.

The RTC data rate is selected on a frame-by-frame basis. The frame denotes a basic timing interval in the CDMA system. For the reverse traffic channel, a frame is 20 ms long.

The transmission duty cycle on the reverse traffic channel varies with the transmission data rate. Specifically, the duty cycles for transmission rate frames of 9.6 kbps, 4.8 kbps, 2.4 kbps, and 1.2 kbps are 100%, 50%, 25%, and 12.5%, respectively.

Each reverse traffic channel is identified by a distinct user long code sequence. The long code uniquely identifies a mobile station on both the reverse traffic channel and the forward traffic channel.

The reverse traffic channel has the overall structure shown in Fig. 3.25. All data transmitted on the reverse traffic channel is convolutionally encoded, block interleaved, modulated by the 64-ary orthogonal modulation, and direct-sequence spread prior to transmission.

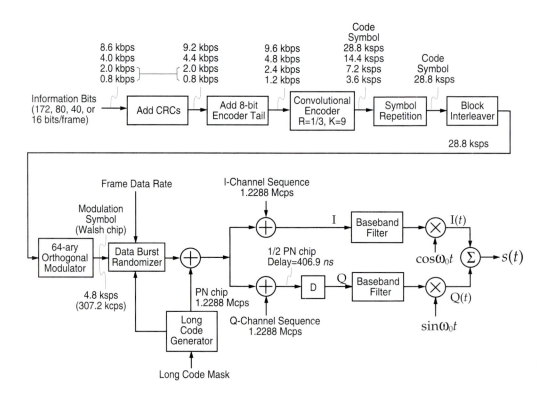

Figure 3.25 Reverse traffic channel structure

Referring to Fig. 3.25, computation of transmission data rates of 9.6 kbps and 4.8 kbps are obtained by taking the following two steps:

1. Data rates of 9.2 kbps and 4.4 kbps are computed from information rates of 8.6 kbps and 4.0 kbps by adding the frame quality indicator (i.e., add the 12-bit CRC for the 9.2 kbps rate and the 8-bit CRC for the 4.4 kbps rate).
2. By adding the additional 8-bit encoder tail, transmission data rates of 9.6 kbps and 4.8 kbps are computed from the 9.2 kbps and 4.4 kbps rates. Referring to Table 3.12, the frame length at the 9.6 kbps rate is 192 bits. The transmission rate from 8.6 kbps to 9.6 kbps is computed as 192 bits/20 ms = 9.6 kbps. Similarly, the frame length at the 4.8 kbps rate is 96 bits. The transmission rate from the 4.0 kbps to the 4.8 kbps rate is found as 96 bits/20 ms = 4.8 kbps.

Next, computation of transmission rates of 2.4 kbps and 1.2 kbps are obtained from information data rates of 2.0 kbps and 0.8 kbps by adding the 8-bit encoder tail each. Using Table 3.12, the frame length at 2.4 kbps rate is 48 bits. Therefore, the 2.4 kbps transmission data rate from the 2.0 kbps information data rate can be computed as 48 bits/20 ms = 2.4 kbps. Similarly, for the 1.2 kbps transmission rate, it can be found as 24 bits/20 ms = 1.2 kbps.

3.2.1   RTC Frame Structure

Each RTC data frame is 20 ms in duration and consists of either information-CRC-tail bits or information-tail bits depending on transmission rates, as shown below.

1. RTC frames sent at the 9.6 kbps and 4.8 kbps transmission rates consist of 192 bits and 96 bits, respectively.
   For 9600 bps,
   Frame bits = 172 information bits+12 CRC bits+8 encoder tail bits = 192 bits
   For 4800 bps,
   Frame bits = 80 information bits+8 CRC bits+8 encoder tail bits = 96 bits
2. RTC frames sent at the 2.4 kbps and 1.2 kbps transmission rates consist of 48 bits and 24 bits, respectively.
   For 2400 bps,
   Frame bits = 40 information bits + 8 encoder tail bits = 48 bits
   For 1200 bps,
   Frame bits = 16 information bits + 8 encoder tail bits = 24 bits

Summarizing these RTC frame bits, we have Table 3.12.

Table 3.12   RTC Frame Structure

Data rates, bps	Information bits	CRC bits	Encoder tail bits	Frame length,bits
9600	172	12	8	192
4800	80	8	8	96
2400	40	–	8	48
1200	16	–	8	24

3.2.2   Modulation Parameters and Characteristics

The reverse traffic channel may use any of four data rates for transmission, but the transmission duty cycle varies with the transmission data rate, as shown in Table 3.13. Specifically, the transmission duty cycle is 100% for 9.6 kbps frames, 50% for 4.8 kbps frames, 25% for 2.4 kbps frames, and 12.5% for 1.2 kbps frames, respectively. However, the actual code symbol rate is fixed at 28.8 ksps. The modulation of the reverse traffic channel is 64-ary orthogonal modulation at a transmission rate of 9.6, 4.8, 2.4, or 1.2 kbps. One of 64 possible modulation symbols is transmitted for each six code symbols. The modulation symbol will be one of 64 mutually orthogonal waveforms generated using the Walsh function. Since the code symbol rate at the block interleaver output is fixed at 28.8 ksps, a fixed Walsh chip rate becomes $28.8/6=4.8$ ksps or $4.8 \times 64=307.2$ kcps. Each Walsh chip is spread by four PN chips, because the ratio of PN chips to Walsh chip is $1228.8/307.20=4$. The rate of the spreading PN sequence is fixed at 1.2288 Mcps. The modulation symbol duration is $1/4800=208.33\,\mu s$. Thus, Table 3.13 defines the signal rates and their relationships for the various transmission rates on the reverse traffic channel.

Table 3.13   Reverse Traffic Channel Modulation Parameters

Parameter	Data Rate, bps				Units
	9600	**4800**	**2400**	**1200**	
PN chip rate	1.2288	1.2288	1.2288	1.2288	Mcps
Code rate	1/3	1/3	1/3	1/3	bits/code symbols
Transmit duty cycle	100	50	25	12.5	%
Code symbol rate	28,800	28,800	28,800	28,800	sps
Modulation	6	6	6	6	code symbol/mod symbol
Modulation symbol rate	4800	4800	4800	4800	sps
Walsh chip rate	307.20	307.20	307.20	307.20	kcps
Mod symbol duration	208.33	208.33	208.33	208.33	μ s
PN chips/code symbol	42.67	42.67	42.67	42.67	PN chip/code symbol
PN chips/mod symbol	256	256	256	256	PN chip/mod symbol
PN chips/Walsh chip	4	4	4	4	

The overall structure of the reverse traffic channel is shown in Fig. 3.25.

3.2.3   RTC Frame Quality Indicator

Among four different data frames for transmission, each 9600 bps and 4800 bps frame will include a frame quality indicator (CRC). No frame quality indicator is used for the 2400 bps and 1200 bps transmission rates.

For both the 9600 bps and 4800 bps rates, the CRC will be calculated on information bits within the frame, except the CRC itself and the encoder tail bits. The generator polynomial for computing the 9600 bps CRC is given as

$$g(x) = 1 + x + x^4 + x^8 + x^9 + x^{10} + x^{11} + x^{12} \tag{3.15}$$

whose logic diagram is shown in Fig. 3.26.

The generator polynomial for calculating the 4800 bps CRC is given as

$$g(x) = 1 + x + x^3 + x^4 + x^7 + x^8 \tag{3.16}$$

whose LFSR circuit is shown in Fig. 3.27.

The frame quality indicators (CRCs) are calculated according to the following procedure using Figs. 3.26 and 3.27 as follows:

1. All shift register stages are initially set to logical one, the switch sets in the down position and Gates 1 and 2 on.
2. The register is clocked 172 times for the 192-bit frame (20 ms) or 80 times for the 96-bit frame (20 ms).
3. After all the information bits are read in, set Gate 1 off.
4. After setting the switch in the up position, the register is then clocked an additional 12 times for the 192-bit frame or 8 times for the 96-bit frame. The 12 or 8 additional output bits will be the CRC bits.
5. These CRC bits will be transmitted in the order computed.

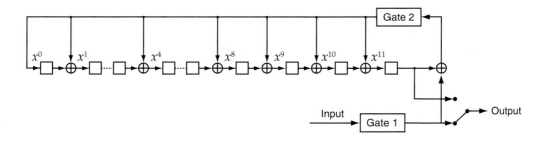

Figure 3.26 CRC computation circuit at the 9600 bps data rate

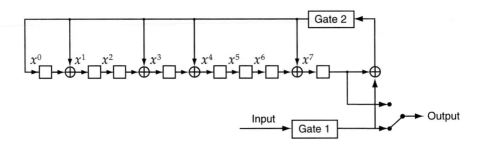

Figure 3.27 CRC computation circuit at the 4800 bps data rate

The last eight bits of each RTC frame, called the encoder tail bits, will be set to logical zero (0).

Example 3.12 Consider the 80-bit information data sequence for the 4800 bps frame (20 ms) as follows:

> 1010110011 1001101111 0010010100 0100110011
> 0011001000 1110010110 1001100110 0110011110

The register is clocked 80 times for the 96-bit frame. Using Fig. 3.27, the CRC bits can be computed as shown in Table 3.14.

Table 3.14   CRC Computation for the 4800 bps Frame by Setting Register Initial Contents as (1 1 1 1 1 1 1 1)

Shift No. n	Information Bits	Feedback Bits	Register Contents
1	1	$1 \oplus 1 = 0$	01111111
2	0	$0 \oplus 1 = 1$	11100110
3	1	$1 \oplus 0 = 1$	10101010
4	0	$0 \oplus 0 = 0$	01010101
5	1	$1 \oplus 1 = 0$	00101010
6	1	$1 \oplus 0 = 1$	11001100

Table 3.14 CRC Computation for the 4800 bps Frame by Setting Register Initial
Contents as (1 1 1 1 1 1 1 1) *(continued)*

.	.	.	.	.	.	.
21	0	$0 \oplus 1 = 1$	10000011			
22	0	$0 \oplus 1 = 1$	10011000			
23	1	$1 \oplus 0 = 1$	10010101			
.	.	.	.	.	.	.
35	1	$1 \oplus 1 = 0$	00000000			
36	1	$1 \oplus 0 = 1$	11011001			
37	0	$0 \oplus 1 = 1$	10110101			
.	.	.	.	.	.	.
50	0	$0 \oplus 0 = 0$	00010111			
51	1	$1 \oplus 1 = 0$	00001011			
52	1	$1 \oplus 1 = 0$	00000101			
.	.	.	.	.	.	.
78	1	$1 \oplus 1 = 0$	00110000			
79	1	$1 \oplus 0 = 1$	11000001			
80	0	$0 \oplus 1 = 1$	10111001			

Referring to Table 3.14, the register content for $n = 80$ is 1 0 1 1 1 0 0 1.

When the register is now clocked an additional 8 times, the register output will be the CRC bits. The information sequence after the CRC computation becomes

```
1010110011 1001101111 0010010100 0100110011 0011001000
1110010110 1001100110 0110011110 10011101
```

Finally, the RTC frame sequence for the 4800 bps rate will be computed by simply adding the encoder tail bits (extra all-zero eight bits), as shown below.

```
1010110011 1001101111 0010010100 0100110011 0011001000
1110010110 1001100110 0110011110 1001110100 000000
```                     (I)

This frame sequence will be the input to the (3,1,8) convolutional encoder.

Example 3.13 Suppose the 172-bit information sequence for the 9600 bps frame (20 ms) is as shown below:

```
1100100011 0111001111 0011010110 0100110011 0011001111 0011010110
1110110011 0011001111 0011010110 1001100110 0110011001 1110011011
0001010010 0110011001 1011001010 1001100110 0110011110 01
```

Using Fig. 3.26, the CRC computation can be achieved by clocking the shift register 172 times for the 192-bit frame as shown in Table. 3.15

Table 3.15   CRC Computation for the 9600 bps Frame by Setting the Sequence (11111111111) as the Initial Contents

Shift No. n	Information Bits	Feedback Bits	Register Contents
1	1	$1 \oplus 1 = 0$	01111111111
2	1	$1 \oplus 1 = 0$	00111111111
3	0	$0 \oplus 1 = 1$	110101110000
.	.	.	.
80	1	$1 \oplus 0 = 1$	100000010110
81	0	$0 \oplus 0 = 0$	010000001011
82	0	$0 \oplus 1 = 1$	111010001010
.	.	.	.
125	0	$0 \oplus 1 = 1$	111101001100
126	1	$1 \oplus 0 = 1$	101100101001
127	0	$0 \oplus 1 = 1$	100100011011
.	.	.	.
170	0	$0 \oplus 0 = 0$	000011111100
171	0	$0 \oplus 0 = 0$	000001111110
172	1	$1 \oplus 0 = 1$	110010110000

The register content for $n=172$ in Table 3.15 will be the CRC bits to be produced by clocking an additional 12 times.

Thus, the information plus CRC bits becomes

```
1100100011 0111001111 0011010110 0100110011 0011001111 0011010110 1110110011
0011001111 0011010110 1001100110 0110011001 1110011011 0001010010 0110011001
1011001010 1001100110 0110011110 0100001101 0011 . . .   . . .
```

Finally, the RTC frame sequence for the 9600 bps rate is computed by simply adding the encoder tail bits as follows:

1100100011 0111001111 0011010110 0100110011 0011001111 0011010110
1110110011 0011001111 0011010110 1001100110 0110011001 1110011011   (II)
0001010010 0110011001 1011001010 1001100110 0110011110 0100001101
0011000000 000000

This frame sequence becomes the input to the (3,1,8) convolutional encoder.

3.2.4   Convolutional Encoder with R=1/3 and K=9

The mobile station convolutionally encodes the frame data sequence transmitted on the reverse traffic channel prior to block interleaving. The same structure of the (3, 1, 8) convolutional encoder (Fig. 3.4) is equally applicable to either an access channel or a reverse traffic channel. The generator sequences for this encoder are $g_1^{(1)}$=(101101111), $g_1^{(2)}$= (110110011), $g_1^{(3)}$ =(111001001).

The initial state of the convolutional encoder is assumed to be all-zero. The encoder generates three code symbols (c_0, c_1, c_2) for each bit of the encoder input sequence.

Convolutional encoding involves the modulo-2 addition of selected taps of a serially time-delayed data sequence. The length of the data sequence delay is equal to m=K–1=8, where K denotes the code constraint length. Figure 3.4 illustrates the convolutional encoder for the reverse traffic channel.

Example 3.14   The RTC frame at 1200 bps transmission rate consists of 24 bits which is composed of 16 information bits followed by 8 encoder tail bits. The encoder tail bits are set to logical 0. Suppose the 16 information bits are 1110001011010011 and the 8 encoder tail bits are 00000000.

The input frame sequence to the convolutional encoder then becomes 111000101101001100000000. Since the code rate is R=1/3, the encoder generates three code symbols (c_0, c_1, c_2) for each input frame bit.

Using Fig. 3.4, three code symbols can be computed as shown in Table 3.16.

Table 3.16   Generation of Three Code Symbols (c_0, c_1, c_2) by Using Fig. 3.4

Shift No. n	Input Bit	Register Contents	Output Code Symbols (c_0, c_1, c_2)
1	1	00000000	111
2	1	10000000	100
3	1	11000000	001
4	0	11100000	000

Table 3.16   Generation of Three Code Symbols (c_0, c_1, c_2) by
Using Fig. 3.4 *(continued)*

Shift No. n	Input Bit	Register Contents	Output Code Symbols (c_0, c_1, c_2)
5	0	01110000	001
6	0	00111000	001
7	1	00011100	100
8	0	10001110	100
9	1	01000111	111
10	1	10100011	011
11	0	11010001	011
12	1	01101000	001
13	0	10110100	011
14	0	01011010	100
15	1	00101101	111
16	1	10010110	100
17	0	11001011	010
18	0	01100101	000
19	0	00110010	010
20	0	00011001	000
21	0	00001100	001
22	0	00000110	010
23	0	00000011	001
24	0	00000001	111

Thus, the generated code symbols at each modulo-2 adder are, respectively

$$c_0 = (1\,1\,0\,0\,0\,0\,1\,1\,1\,0\,0\,0\,0\,1\,1\,1\,0\,0\,0\,0\,0\,0\,0\,1)$$
$$c_1 = (1\,0\,0\,0\,0\,0\,0\,0\,1\,1\,1\,0\,1\,0\,1\,0\,1\,0\,1\,0\,0\,1\,0\,1)$$
$$c_2 = (1\,0\,1\,0\,1\,1\,0\,0\,1\,1\,1\,1\,1\,0\,1\,0\,0\,0\,0\,0\,1\,0\,1\,1)$$

Finally, the code symbol output sequence can be produced by column-by-column shifting of c_0, c_1, c_2 through the commutating switch.

1111000010 0000100110 0100111011 0110010111	(III)
0011110001 0000010000 0010100011 11	

Example 3.15 Consider again computation of the output sequence of the (3, 1, 8) convolutional encoder at the RTC 1200 bps transmission rate. We will seek alternative solutions for the encoder output sequence as shown below.

Let d be the information sequence with tail bits:

$$d = (1110\ 0010\ 1101\ 0011\ 0000\ 0000)$$

Generator sequences:

$$g_1^{(1)} = (557) = (101\ 101\ 111\ ) = D^8 + D^6 + D^5 + D^3 + D^2 + D + 1$$
$$g_1^{(2)} = (663) = (110\ 110\ 011\ ) = D^8 + D^7 + D^5 + D^4 + D + 1$$
$$g_1^{(3)} = (711) = (111\ 001\ 001\ ) = D^8 + D^7 + D^6 + D^3 + 1$$

1. Solution by Discrete Convolution

$$c_0 = d * g_1^{(1)} = (\ 11100010110100110000000\ ) * (\ 101101111\ )$$
$$= (1100\ 0011\ 1000\ 0111\ 0000\ 0001)$$
$$c_1 = d * g_1^{(2)} = (11100010110100110000000) * (110110011)$$
$$= (1000\ 0000\ 1110\ 1010\ 1010\ 0101)$$
$$c_2 = d * g_1^{(3)} = (11100010110100110000000) * (111001001)$$
$$= (1010\ 1100\ 1111\ 1010\ 0000\ 1011)$$

The code symbol sequence c is then computed by concatenating three symbols at a time, coming out of c_0, c_1, and c_2.

$$c = (111\ 100\ 001\ 000\ 001\ 001\ 100\ 100\ 111\ 011\ 011\ 001$$
$$011\ 100\ 111\ 100\ 010\ 000\ 010\ 000\ 001\ 010\ 001\ 111)$$

2. Solution by Scalar Matrix Product

$$c = d \bullet G$$

where G is a semi-infinite generator matrix:

$$G = \begin{bmatrix} G_0 & G_1 & G_2 & . & . & . & G_m & & & \\ & G_0 & G_1 & G_2 & . & . & . & G_m & & \\ & & G_0 & G_1 & G_2 & . & . & . & G_m & \\ & & & \vdots & & & & & \vdots & \\ & & & & \vdots & & & & & \vdots \end{bmatrix}$$

Component matrices G_i, $0 \le i \le 8$, is easily obtained from $g_1^{(1)}$, $g_1^{(2)}$, and $g_1^{(3)}$ as follows:

$G_0 = 111$, $G_1 = 011$, $G_2 = 101$, $\cdots$, $G_7 = 110$, $G_8 = 111$.

Thus, the code symbol word c is computed as

$$c = d \bullet G$$
$$= (11100010110100110000000)$$

```
111 011 101 110 010 101 100 110 111
  111 011 101 110 010 101 100 110 111
    111 011 101 110 010 101 100 110 111
      111 011 101 110 010 101 100 110 111
        111 011 101 110 010 101 100 110 111
          111 011 101 110 010 101 100 110 111
            111 011 101 110 010 101 100 110 111
              111 011 101 110 010 101 100 110 111
                111 011 101 110 010 101 100 110 111
                  111 011 101 110 010 101 100 110 111
                    111 011 101 110 010 101 100 110 111
                      111 011 101 110 010 101 100 110 111
                        111 011 101 110 010 101 100 110 111
                          111 011 101 110 010 101 100 110 111
                            111 011 101 110 010 101 100 110 111
                              111 011 101 110 010 101 100 110 111
                                111 011 101 110 010 101 100 110 111
                                  111 011 101 110 010 101 100 110 111
```

$$= (111\ 100\ 001\ 000\ 001\ 001\ 100\ 100\ 111\ 011\ 011\ 001$$
$$011\ 100\ 111\ 100\ 010\ 000\ 010\ 000\ 001\ 010\ 001\ 111)$$

This is the code symbol word c as expected.

3. Solution by Polynomial Matrix

The information polynomial represents

$$d(D) = D^{23} + D^{22} + D^{21} + D^{17} + D^{15} + D^{14} + D^{12} + D^9 + D^8$$

The generator polynomial matrix is expressed as

$$G(D) = \left[\ g_1^{(1)}(D),\ g_1^{(2)}(D),\ g_1^{(3)}(D)\ \right]$$

$$= \left[D^8 + D^6 + D^5 + D^3 + D^2 + D + 1,\ D^8 + D^7 + D^5 + D^4 + D + 1,\ D^8 + D^7 + D^6 + D^3 + 1 \right]$$

The code symbol polynomial matrix at the modulo-2 address is obtained as

$$C(D) = d(D)G(D) = \left[\ c_0(D),\ c_1(D),\ c_2(D)\ \right]$$

$$= \big[\ D^{31} + D^{30} + D^{25} + D^{24} + D^{23} + D^{18} + D^{17} + D^{16} + D^8,\ D^{31} + D^{23} + D^{22} + D^{21} +$$
$$D^{19} + D^{17} + D^{15} + D^{13} + D^{10} + D^8,\ D^{31} + D^{29} + D^{27} + D^{26} + D^{23} + D^{22} + D^{21} + D^{20}$$
$$+ D^{19} + D^{17} + D^{11} + D^9 + D^8\ \big]$$

Finally, the delay operation by the commutating switch will produce the encoder output sequence as follows:

$$C(D) = D^2 c_0(D^3) + D c_1(D^3) + c_2(D^3)$$
$$= D^2\,(D^{93} + D^{90} + D^{75} + D^{72} + D^{69} + D^{54} + D^{51} + D^{48} + D^{24}\,) +$$
$$D\,(D^{93} + D^{69} + D^{66} + D^{63} + D^{57} + D^{51} + D^{45} + D^{39} + D^{30} + D^{24}) +$$
$$(D^{93} + D^{87} + D^{81} + D^{78} + D^{69} + D^{66} + D^{63} + D^{60} + D^{57} + D^{51} + D^{33} +$$
$$D^{27} + D^{24})$$
$$= D^{95} + D^{94} + D^{93} + D^{92} + D^{87} + D^{81} + D^{78} + D^{77} + D^{74} + D^{71} + D^{70} + D^{69} +$$
$$D^{67} + D^{66} + D^{64} + D^{63} + D^{60} + D^{58} + D^{57} + D^{56} + D^{53} + D^{52} + D^{51} + D^{50} + D^{46} +$$
$$D^{40} + D^{33} + D^{31} + D^{27} + D^{26} + D^{25} + D^{24}$$
$$= (111\ 100\ 001\ 000\ 001\ 001\ 100\ 100\ 111\ 011\ 011\ 001$$
$$011\ 100\ 111\ 100\ 010\ 000\ 010\ 000\ 001\ 010\ 001\ 111)$$

It is thus proved that alternative solutions through three different approaches presented above are identical with the output symbol sequence (III).

Example 3.16 The 4800 bps frame consists of 96 bits which is composed of 80 information bits followed by 8 CRC bits and 8 encoder tail bits. As shown in Example 3.12, the input to the (3,1,8) convolutional encoder was computed as the sequence (I).

Utilizing Fig. 3.4, the encoder output symbols at three modulo-2 adders can be computed as follows:

c_0= (1001010001 1000100001 0111110000 1001100111 1001100011
         1111110111 1011100111 0011001010 1100010011 011011)

c_1= (1110010000 1100100001 0001000000 0001101100 1010101100
         0101110010 1101010001 0101010001 0100101001 010111)

c_2 = (1101010001 1001111110 0110100011 0100111110 1100110100
         0010000111 0101000101 1001100011 0001111010 110101)

Thus, the serialized encoder output sequence can be computed by using the switch in Fig. 3.4 as follows:

> 1110110101010001110000000000101 1101000000111100100100100110
> 0001011011101011000000000001001 1000010001101110010111111101100
> 1110010101001110010100111001001 001101011101101100001011111101   (IV)
> 1100111001111000100001011001111 0010101001100010101000000101011
> 10011000000101110101100010111 0001111100011110111

Example 3.17   The 9600 bps frame consists of 192 bits (20 ms).

Referring to Example 3.13, the RTC frame sequence for the 9600 bps transmission rate was computed as the sequence (II) which is the input to the (3,1,8) convolutional encoder. Using Fig. 3.4, the encoder output symbols at three modulo-2 adders are, respectively:

c_0 = (1110011111 0011001010 0000101001 1000011111 1001100101
           0000101001 0001110110 0001100101 0000101001 0111100111
           0011001100 1010000100 1010101011 1001111100 1110100111
           0100110111 0011001010 0011001000 1001000000 01)

c_1 = (1011100001 1111011011 1110101111 1011110100 1010101000
           1110101111 0101001011 0010101000 1110101111 0001010001
           0101010101 0001110100 1110000000 1100100101 0110110001
           0111000001 0101010001 1110110001 1100101001 01)

c_2 = (1001100000 0010111000 1101010010 1001110110 1100110001
           1101010010 0100000000 0100110001 1101010010 0001000101
           1001100110 0011101011 0001100001 0000110110 0101001101
           1110110101 1001100011 1001111001 1000000000 11)

Finally, the serialized encoder output can be obtained by column-by-column interleaving of c_0, c_1, c_2 through the switch in Fig. 3.4 as follows:

$$
\begin{array}{l}
11110011001101110010010010011001010111110001011111000110010 \\
01101101000111000111001001111011100001001101111110011110110\,0 \\
11001010100110010101010000010101101101000111000111001001111\,0 \\
00001100011010010001010011001000001010100111001010100000010\,1 \\
01101101000111000111001001111000010010010011110001000010110011\,1 \\
001010100111001010100111001010\,10000010101101101000111000100\,1 \\
11001011000110100010000010010111001000010011111011001110010\,10 \\
10011111000111001000110110011100111101101010110100010110011\,1 \\
001010100111001010100000101011\,011010110101011011101000000011 \\
11101000010001000001000000101000\,1111
\end{array}
\quad (V)
$$

3.2.5   Code symbol Repetition

The convolutional encoder output symbols must be repeated before being interleaved when the data rate is lower than 9600 bps. The RTC symbol repetition varies with the data rate. There is no symbol repetition for the 9600 bps data rate. For the 4800 bps data rate, each code symbol is repeated 1 time. For the 2400 bps data rate, each code symbol is repeated 3 times. For the 1200 bps, each code symbol is repeated 7 times.

For all of the data rates, each code symbol is thus repeated as follows:

Data rate, bps	Repetition times/Symbol	Consecutive occurring times/Symbol
9600	0	1
4800	1	2
2400	3	4
1200	7	8

The repeated code symbol will be the input to the block interleaver, and all but one of the code symbol repetitions will be deleted before transmission due to the variable transmission duty cycle.

Example 3.17   Consider again the RTC frame at 1200 bps transmission rate.

Referring to Example 3.14, the symbol output sequence of the convolutional encoder is obtained as

$$
\begin{array}{l}
1111000010\ 0000100110\ 0100111011\ 0110010111 \\
0011110001\ 0000010000\ 0010100011\ 11
\end{array}
\quad (III)
$$

Each code symbol at the 1200 bps data rate will be repeated 7 times (each symbol occurs 8 consecutive times). Therefore, repeated code symbols are shown as follows:

```
11111111 11111111 11111111 11111111 00000000 00000000 00000000
00000000 11111111 00000000 00000000 00000000 00000000 00000000
11111111 00000000 00000000 11111111 11111111 00000000 00000000
11111111 00000000 00000000 11111111 11111111 11111111 00000000
11111111 11111111 00000000 11111111 11111111 00000000 00000000
11111111 00000000 11111111 11111111 11111111 00000000 00000000
11111111 11111111 11111111 11111111 00000000 00000000 00000000
11111111 00000000 00000000 00000000 00000000 00000000 11111111
00000000 00000000 00000000 00000000 00000000 00000000 11111111
00000000 11111111 00000000 00000000 00000000 11111111 11111111
11111111 11111111
```

These repeated code symbols will be input to the block interleaver, but they will not be transmitted multiple times. Rather, all but one of the code symbol repetitions will be deleted prior to actual transmission.

3.2.6   Block Interleaving

A block interleaver spanning 20 ms interleaves all code symbols on the reverse traffic channel prior to orthogonal modulation.

The interleaver forms a 576-cell array with 32 rows and 18 columns.

Repeated code symbols at data rates lower than 9600 bps are written into the interleaver by columns filling the complete 32 x 18 matrix.

Tables 3.2, 3.17, 3.18, and 3.19 illustrate the ordering of writing operations of repeated code symbols into the interleaver array for transmission rates of 4800, 9600, 2400, and 1200 bps, respectively.

RTC code symbols will be output from the interleaver by rows. The interleaver rows will be output in the following order depending on the transmission rate.

For the 9600 bps data rate (Table 3.17):

1	2	3	4	5	6	7	8	9	10	11	12	13	14	15	16
17	18	19	20	21	22	23	24	25	26	27	28	29	30	31	32

The interleaver output sequence sends row by row up to row 32 in a regular order.
For the 4800 bps data rate (Table 3.2):

1	3	2	4	5	7	6	8	9	11	10	12	13	15	14	16
17	19	18	20	21	23	22	24	25	27	26	28	29	31	30	32

The interleaver output sequence sends by the following order of rows:
$$1+4i, 3+4i, 2+4i, 4+4i, 0 \leq i \leq 7$$
For the 2400 bps data rate (Table 3.18):

1	5	2	6	3	7	4	8	9	13	10	14	11	15	12	16
17	21	18	22	19	23	20	24	25	29	26	30	27	31	28	32

The interleaver output sequence sends by the following order of rows:
$$1+8i, 5+8i, 2+8i, 6+8i, 3+8i, 7+8i, 4+8i, 8+8i, 0 \leq i \leq 3$$
For the 1200 bps data rate (Table 3.19):

1	9	2	10	3	11	4	12	5	13	6	14	7	15	8	16
17	25	18	26	19	27	20	28	21	29	22	30	23	31	24	32

The interleaver output sequence sends by the following order of rows:
$$1+16i, 9+16i, 2+16i, 10+16i, 3+16i, 11+16i, 4+16i, 12+16i, 5+16i,$$
$$13+16i, 6+16i, 14+16i, 7+16i, 15+16i, 8+16i, 16+16i, i=0,1$$

Table 3.17   Reverse Traffic Channel Interleaver Memory (9600 bps) (TIA/EIA/IS-95)

1	33	65	97	129	161	193	225	257	289	321	353	385	417	449	481	513	545
2	34	66	98	130	162	194	226	258	290	322	354	386	418	450	482	514	546
3	35	67	99	131	163	195	227	259	291	323	355	387	419	451	483	515	547
4	36	68	100	132	164	196	228	260	292	324	356	388	420	452	484	516	548
5	37	69	101	133	165	197	229	261	293	325	357	389	421	453	485	517	549
6	38	70	102	134	166	198	230	262	294	326	358	390	422	454	486	518	550
7	39	71	103	135	167	199	231	263	295	327	359	391	423	455	487	519	551
8	40	72	104	136	168	200	232	264	296	328	360	392	424	456	488	520	552
9	41	73	105	137	169	201	233	265	297	329	361	393	425	457	489	521	553
10	42	74	106	138	170	202	234	266	298	330	362	394	426	458	490	522	554
11	43	75	107	139	171	203	235	267	299	331	363	395	427	459	491	523	555
12	44	76	108	140	172	204	236	268	300	332	364	396	428	460	492	524	556
13	45	77	109	141	173	205	237	269	301	333	365	397	429	461	493	525	557
14	46	78	110	142	174	206	238	270	302	334	366	398	430	462	494	526	558
15	47	79	111	143	175	207	239	271	303	335	367	399	431	463	495	527	559
16	48	80	112	144	176	208	240	272	304	336	368	400	432	464	496	528	560
17	49	81	113	145	177	209	241	273	305	337	369	401	433	465	497	529	561
18	50	82	114	146	178	210	242	274	306	338	370	402	434	466	498	530	562
19	51	83	115	147	179	211	243	275	307	339	371	403	435	467	499	531	563
20	52	84	116	148	180	212	244	276	308	340	372	404	436	468	500	532	564
21	53	85	117	149	181	213	245	277	309	341	373	405	437	469	501	533	565
22	54	86	118	150	182	214	246	278	310	342	374	406	438	470	502	534	566
23	55	87	119	151	183	215	247	279	311	343	375	407	439	471	503	535	567
24	56	88	120	152	184	216	248	280	312	344	376	408	440	472	504	536	568
25	57	89	121	153	185	217	249	281	313	345	377	409	441	473	505	537	569
26	58	90	122	154	186	218	250	282	314	346	378	410	442	474	506	538	570
27	59	91	123	155	187	219	251	283	315	347	379	411	443	475	507	539	571
28	60	92	124	156	188	220	252	284	316	348	380	412	444	476	508	540	572
29	61	93	125	157	189	221	253	285	317	349	381	413	445	477	509	541	753
30	62	94	126	158	190	222	254	286	318	350	382	414	446	478	510	542	754
31	63	95	127	159	191	223	255	287	319	351	383	415	447	479	511	543	575
32	64	96	128	160	192	224	256	288	320	352	384	416	448	480	512	544	576

Table 3.18 Reverse Traffic Channel Interleaver Memory (2400 bps)
(TIA/EIA/IS-95)

1	9	17	25	33	41	49	57	65	73	81	89	97	105	113	121	129	137
1	9	17	25	33	41	49	57	65	73	81	89	97	105	113	121	129	137
1	9	17	25	33	41	49	57	65	73	81	89	97	105	113	121	129	137
1	9	17	25	33	41	49	57	65	73	81	89	97	105	113	121	129	137
2	10	18	26	34	42	50	58	66	74	82	90	98	106	114	122	130	138
2	10	18	26	34	42	50	58	66	74	82	90	98	106	114	122	130	138
2	10	18	26	34	42	50	58	66	74	82	90	98	106	114	122	130	138
2	10	18	26	34	42	50	58	66	74	82	90	98	106	114	122	130	138
3	11	19	27	35	43	51	59	67	75	83	91	99	107	115	123	131	139
3	11	19	27	35	43	51	59	67	75	83	91	99	107	115	123	131	139
3	11	19	27	35	43	51	59	67	75	83	91	99	107	115	123	131	139
3	11	19	27	35	43	51	59	67	75	83	91	99	107	115	123	131	139
4	12	20	28	36	44	52	60	68	76	84	92	100	108	116	124	132	140
4	12	20	28	36	44	52	60	68	76	84	92	100	108	116	124	132	140
4	12	20	28	36	44	52	60	68	76	84	92	100	108	116	124	132	140
4	12	20	28	36	44	52	60	68	76	84	92	100	108	116	124	132	140
5	13	21	29	37	45	53	61	69	77	85	93	101	109	117	125	133	141
5	13	21	29	37	45	53	61	69	77	85	93	101	109	117	125	133	141
5	13	21	29	37	45	53	61	69	77	85	93	101	109	117	125	133	141
5	13	21	29	37	45	53	61	69	77	85	93	101	109	117	125	133	141
6	14	22	30	38	46	54	62	70	78	86	94	102	110	118	126	134	142
6	14	22	30	38	46	54	62	70	78	86	94	102	110	118	126	134	142
6	14	22	30	38	46	54	62	70	78	86	94	102	110	118	126	134	142
6	14	22	30	38	46	54	62	70	78	86	94	102	110	118	126	134	142
7	15	23	31	39	47	55	63	71	79	87	95	103	111	119	127	135	143
7	15	23	31	39	47	55	63	71	79	87	95	103	111	119	127	135	143
7	15	23	31	39	47	55	63	71	79	87	95	103	111	119	127	135	143
7	15	23	31	39	47	55	63	71	79	87	95	103	111	119	127	135	143
8	16	24	32	40	48	56	64	72	80	88	96	104	112	120	128	136	144
8	16	24	32	40	48	56	64	72	80	88	96	104	112	120	128	136	144
8	16	24	32	40	48	56	64	72	80	88	96	104	112	120	128	136	144
8	16	24	32	40	48	56	64	72	80	88	96	104	112	120	128	136	144

Table 3.19 Reverse Traffic Channel Interleaver Memory (1200 bps) (TIA/EIA/IS-95)

1	5	9	13	17	21	25	29	33	37	41	45	49	53	57	61	65	69
1	5	9	13	17	21	25	29	33	37	41	45	49	53	57	61	65	69
1	5	9	13	17	21	25	29	33	37	41	45	49	53	57	61	65	69
1	5	9	13	17	21	25	29	33	37	41	45	49	53	57	61	65	69
1	5	9	13	17	21	25	29	33	37	41	45	49	53	57	61	65	69
1	5	9	13	17	21	25	29	33	37	41	45	49	53	57	61	65	69
1	5	9	13	17	21	25	29	33	37	41	45	49	53	57	61	65	69
1	5	9	13	17	21	25	29	33	37	41	45	49	53	57	61	65	69
2	6	10	14	18	22	26	30	34	38	42	46	50	54	58	62	66	70
2	6	10	14	18	22	26	30	34	38	42	46	50	54	58	62	66	70
2	6	10	14	18	22	26	30	34	38	42	46	50	54	58	62	66	70
2	6	10	14	18	22	26	30	34	38	42	46	50	54	58	62	66	70
2	6	10	14	18	22	26	30	34	38	42	46	50	54	58	62	66	70
2	6	10	14	18	22	26	30	34	38	42	46	50	54	58	62	66	70
2	6	10	14	18	22	26	30	34	38	42	46	50	54	58	62	66	70
2	6	10	14	18	22	26	30	34	38	42	46	50	54	58	62	66	70
3	7	11	15	19	23	27	31	35	39	43	47	51	55	59	63	67	71
3	7	11	15	19	23	27	31	35	39	43	47	51	55	59	63	67	71
3	7	11	15	19	23	27	31	35	39	43	47	51	55	59	63	67	71
3	7	11	15	19	23	27	31	35	39	43	47	51	55	59	63	67	71
3	7	11	15	19	23	27	31	35	39	43	47	51	55	59	63	67	71
3	7	11	15	19	23	27	31	35	39	43	47	51	55	59	63	67	71
3	7	11	15	19	23	27	31	35	39	43	47	51	55	59	63	67	71
3	7	11	15	19	23	27	31	35	39	43	47	51	55	59	63	67	71
4	8	12	16	20	24	28	32	36	40	44	48	52	56	60	64	68	72
4	8	12	16	20	24	28	32	36	40	44	48	52	56	60	64	68	72
4	8	12	16	20	24	28	32	36	40	44	48	52	56	60	64	68	72
4	8	12	16	20	24	28	32	36	40	44	48	52	56	60	64	68	72
4	8	12	16	20	24	28	32	36	40	44	48	52	56	60	64	68	72
4	8	12	16	20	24	28	32	36	40	44	48	52	56	60	64	68	72
4	8	12	16	20	24	28	32	36	40	44	48	52	56	60	64	68	72
4	8	12	16	20	24	28	32	36	40	44	48	52	56	60	64	68	72

Example 3.18   For the 1200 bps data rate, we compute the interleaver output by making use of rows in Table 3.19. Using the encoder output sequence (III) and ordering number, the interleaver output can be obtained as shown in Table 3.20.

Table 3.20   Interleaver Output Symbols corresponding to the 1200 bps Transmission Rate

Row Number	Interleaver Output	Row Number	Interleaver Output
[1]	101000111001000011	[17]	100110100110000101
[9]	100011110101100001	[25]	100000011110010001
[2]	101000111001000011	[18]	100110100110000101
[10]	100011110101100001	[26]	100000011110010001
[3]	101000111001000011	[19]	100110100110000101
[11]	100011110101100001	[27]	100000011110010001
[4]	101000111001000011	[20]	100110100110000101
[12]	100011110101100001	[28]	100000011110010001
[5]	101000111001000011	[21]	100110100110000101
[13]	100011110101100001	[29]	100000011110010001
[6]	101000111001000011	[22]	100110100110000101
[14]	100011110101100001	[30]	100000011110010001
[7]	101000111001000011	[23]	100110100110000101
[15]	100011110101100001	[31]	100000011110010001
[8]	101000111001000011	[24]	100110100110000101
[16]	100011110101100001	[32]	100000011110010001

Example 3.19   For the 9600 bps transmission rate, the interleaver output can be computed by making use of rows from Table 3.17. Since the convolutional encoder output represents the sequence (V), the interleaver output is then computed according to the following ordering number as shown in Table 3.21.

Table 3.21   Interleaver Output Symbols Corresponding to the 9600 bps
                  Transmission Rate

Row Number	Interleaver Output	Row Number	Interleaver Output
[1]	1010001100011111011	[17]	0101011001100111110
[2]	1011110010100001000	[18]	0101010110000011000
[3]	1100010010100101110	[19]	1111010101101001110
[4]	1010011111101000000	[20]	0000100000011011000
[5]	0101100000010111100	[21]	0011000110111100001
[6]	0101100101001010110	[22]	1011100101101000000
[7]	1100100011101011001	[23]	0110000001010000001
[8]	1111011100011110100	[24]	0110101111101010001
[9]	0111010000110100001	[25]	1011110010101101011
[10]	0011111011010111110	[26]	0001011010000110001
[11]	1001000011101111001	[27]	0111100011110011101
[12]	1001100011000101100	[28]	1010101100001100101
[13]	0101011010010010101	[29]	1010000101001101110
[14]	1010101100000000001	[30]	0101110100001011111
[15]	1110000001010011100	[31]	0100011011100010111
[16]	1111010100011100111	[32]	1001001000100111011

Example 3.20   For the 4800 bps transmission rate, the interleaver output can be computed using the code symbol sequence (IV) and numbering rows of Table 3.2 as shown in Table 3.22.

Table 3.22   Interleaver Output Symbols Corresponding to the 4800 bps
                  Transmission Rate

Row Number	Interleaver Output	Row Number	Interleaver Output
[1]	1110000101111111101	[17]	0001101100010100011
[3]	1100100110010100011	[19]	1011010111011000001
[2]	1110000101111111101	[18]	0001101100010100011
[4]	1100100110010100011	[20]	1011010111011000001
[5]	1011100100100010010	[21]	0011110101010001011

Table 3.22 Interleaver Output Symbols Corresponding to the 4800 bps Transmission Rate *(continued)*

[7]	000000100111000111	[23]	111010101010110111
[6]	101110010010001001	[22]	001111010101001011
[8]	000000100111000111	[24]	111010101010110111
[9]	100010111001111111	[25]	001000000100111010
[11]	100110011011000000	[27]	010000110011010101
[10]	100010111001111111	[26]	001000000100111010
[12]	100110011011000000	[28]	010000110011010101
[13]	000011101110010000	[29]	010000101111001101
[15]	100000100000000000	[31]	111101111111101101
[14]	000011101110010000	[30]	010000101111001101
[16]	100000100000000000	[32]	111101111111101101

3.2.7 Orthogonal Modulation for RTC

Modulation for RTC is 64-ary orthogonal modulation. One of 64 possible modulation symbols generated using Walsh functions is transmitted for each six code symbols. These modulation symbols numbering 0 through 63 are given in Table 2.4. Since the code symbol rate at the block interleaver output is 28.8 ksps, the symbol rate at the orthogonal modulator output becomes 4.8 ksps or 307.2 kcps.

Modulation symbols are selected according to the following modulation symbol index (MSI).

$$MSI = c_0 + 2c_1 + 4c_2 + 8c_3 + 16c_4 + 32c_5$$

where each c_i represents the binary valued code symbol.

Consider the case of I/O relationship of 64-ary orthogonal modulator through the following examples.

Example 3.21 For the 9600 bps transmission rate, the interleaver output of the first row is 1 0 1 0 0 0 1 1 0 0 0 1 1 1 1 0 1 1 (see Example 3.19).

Since one of 64 modulation symbols is per each 6 code symbols, it is better to divide the interleaver output into 6-symbol groups of equal length.

For the 6-symbol input 101000, the modulation symbol index is 5 because $c_0 = 1$, $c_1 = 0$, $c_2 = 1$, $c_3 = 0$, $c_4 = 0$, $c_5 = 0$. Using Table 2.4, the modulation symbol corresponding to MSI=5 is computed as

010110100101101001011010001011010

010110100101101001011010001011010

Similarly, for the 6-symbol input 110001, we have MSI=35.

Then the modulation symbol corresponding to MSI = 35 is

011001100110011001100110011001100110

100110011001100110011001100110011001

For the 6-symbol input 111011, MSI=55. The modulation symbol corresponding to MSI = 55 is then

011010010110100110010110100010110

100101101001011001101001011011001

Thus, we have computed the three 192-chip modulation symbols corresponding to the 18 code symbols. This output corresponds to just one row out of 32 rows.

Example 3.22   For the 1200 bps transmission rate, the interleaver rows are output in the following order:

1	9	2	10	3	11	4	12	5	13	6	14	7	15	8	16
17	25	18	26	19	27	20	28	21	29	22	30	23	31	24	32

For demonstrative purposes, let us consider the first two rows, 1 and 9. The interleaver output corresponding to these two rows was computed in Example 3.18.

Row 1: 1 0 1 0 0 0 1 1 1 0 0 1 0 0 0 0 1 1

Row 9: 1 0 0 0 1 1 1 1 0 1 0 1 1 0 0 0 0 1

[Computation for Row 1]

1. For the 6-symbol input 1 0 1 0 0 0 ($c_0 = 1.$ $c_1 = 0.$ $c_2 = 1.$ $c_3 = 0.$ $c_4 = 0.$ $c_5 = 0.$), MSI = 5. Using Table 2.4, the modulation symbol is obtained as

   010110100101101001011010010110100101011010

   010110100101101001011010010110100101011010

2. For the input symbol 1 1 1 0 0 1 ($c_0 = 1.$ $c_1 = 1.$ $c_2 = 1.$ $c_3 = 0.$ $c_4 = 0.$ $c_5 = 1.$), MSI = 39. The modulation symbol is

   011010010110100101101001011011001

   100101101001011010010110100101101010

3. For the input symbol 0 0 0 0 1 1 ($c_0 = 0.$ $c_1 = 0.$ $c_2 = 0.$ $c_3 = 0.$ $c_4 = 1.$ $c_5 = 1.$) MSI = 48. The corresponding modulation symbol is

   000000000000000011111111111111111

   11111111111111110000000000000000

[Computation for Row 9]

4. Input symbol = 1 0 0 0 1 1 , MSI = 49
   Modulation symbol = 0101010101010101101010101010101010

   101010101010101001010101010101010101

5. Input symbol = 1 1 0 1 0 1 , MSI = 43

   Modulation symbol = 0110011010011001011001101001 1001
   10011001011001101001100101100110

6. Input symbol = 1 0 0 0 0 1 , MSI = 33

   Modulation symbol = 010101010101010101010101010101
   101010101010101010101010101010

As a result from I/O computation, we summarize modulation symbols as follows:

Input (Interleaver output symbols relating to Rows 1 and 9)

101000111001000011
100011110101100001

(VI)

Output (Modulation symbols corresponding to Rows 1 and 9)

010110100101101001011010010110 10
010110100101101001011010010110 10
011010010110100101101001011010 01
100101101001011010010110100101 10
000000000000000111111111111111 11
111111111111111110000000000000 000
010101010101010110101010101010 10
101010101010101001010101010101 01
011001101001100101100110100110 01
100110010110011010011001011001 10
010101010101010101010101010101 01
101010101010101010101010101010 10

3.2.8   Data Burst Randomizer

The RTC interleaver output is gated with a time filter that permits transmission of certain interleaver output symbols and deletion of others. As illustrated in Fig. 3.28, the transmission gating process varies with the transmit data rate. For the 9600 bps data rate, the transmission gate allows 100% of all interleaver output symbols to be transmitted. When the data rate is 4800 bps, the gate allows one-half (50%) of the interleaver output symbols to be transmitted. With the 2400 transmission rate, the gate permits a quarter (25%) of the interleaver output symbols to be transmitted. Finally, for the 1200 bps transmission rate, the gate permits 12.5% of the interleaver output symbols to be transmitted.

The gating process operates by dividing the 20 ms frame into 16 power control groups numbered from 0 to 15 as shown in Fig. 3.28. Each power control group (PCG) possesses exactly the 1 . 2 5 ms length. The gated-on and gated-off assignment is called data burst randomizing. The data burst randomizer ensures that every code symbol input to the repetition process is transmitted exactly once.

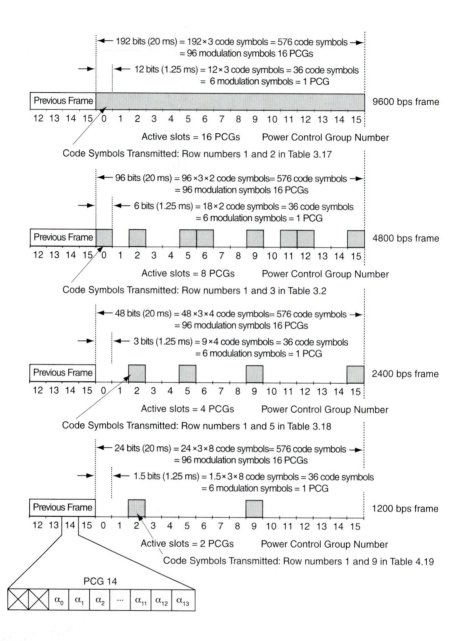

Figure 3.28 The data burst randomizer pulse trains
varying with the transmission rate

The gated-on power groups are pseudorandomized in their positions in the frame; and the mobile station transmits at normal controlled power levels only during gated-on periods. During the gated-off periods, the mobile station will reduce its mean output power either by at least 20 dB with respect to the mean output power of the most recent power control group, or to the transmitter noise, whichever is the greater power.

The data burst randomizer generates a masking binary pattern that randomly masks out the redundant data generated by the code repetition. The masking pattern is determined by the data rate of the frame and by a block of 14 bits taken from the long code. This 14-bit PN sequence consists of the last 14 bits of the long code used for spreading in the previous power control group (PCG 14) to the last power control group (PCG 15) of the previous frame.

Let this 14-bit stream be denoted as α_0, α_1, α_2, α_3, α_4, α_5, α_6, α_7, α_8, α_9, α_{10}, α_{11}, α_{12}, α_{13} where α_0 represents the oldest bit, and α_{13} denotes the latest bit. These 14 bits occur exactly one power control group (1.25 ms) before each RTC frame boundary.

Data Burst Randomizer Algorithm

Transmission will occur on PCGs numbered between 0 and 15 according to the data rate selected. The randomizing algorithm will be as follows:

Data Rate, bps	PCGs Numbered
9600	0, 1, 2, 3, 4, 5, 6, 7, 8, 9, 10, 11, 12, 13, 14, 15
4800	α_0, $2+\alpha_1$, $4+\alpha_2$, $6+\alpha_3$, $8+\alpha_4$, $10+\alpha_5$, $12+\alpha_6$, $14+\alpha_7$
2400	α_0 if $\alpha_8 = 0$, $2+\alpha_1$ if $\alpha_8 = 1$ $4+\alpha_2$ if $\alpha_9 = 0$, $6+\alpha_3$ if $\alpha_9 = 1$ $8+\alpha_4$ if $\alpha_{10} = 0$, $10+\alpha_5$ if $\alpha_{10} = 1$ $12+\alpha_6$ if $\alpha_{11} = 0$, $14+\alpha_7$ if $\alpha_{11} = 1$
1200	α_0 if $\alpha_8=0$ and $\alpha_{12}=0$, $2+\alpha_1$ if $\alpha_8=1$ and $\alpha_{12}=0$ $4+\alpha_2$ if $\alpha_9=0$ and $\alpha_{12}=1$, $6+\alpha_3$ if $\alpha_9=1$ and $\alpha_{12}=1$ $8+\alpha_4$ if $\alpha_{10}=0$ and $\alpha_{13}=0$, $10+\alpha_5$ if $\alpha_{10}=1$ and $\alpha_{13}=0$ $12+\alpha_6$ if $\alpha_{11}=0$ and $\alpha_{13}=1$, $14+\alpha_8$ if $\alpha_{11}=1$ and $\alpha_{13}=1$

The 20 ms frame is equally divided into 16 power control groups numbered from 0 to 15, but the gated-on power groups are different in accordance with variable data rate transmission. Figure 3.28 based on Table 3.23 illustrates one example of RTC variable data rate transmission.

Table 3.23   RTC Variable Data Rate Transmission

Data Rate bps	Information Bits per Frame	Encoder Output Symbols	Consecutive Occurring Times	Code Symbols after Interleaving	Modulation Symbols	PCGs	Gated-on Power Groups
9600	192 (20ms)	576	1	576	96	16	16 (100%)
	12 (1.25ms)	36		36	6	1	One PCG
4800	96 (20ms)	288	2	576	96	16	8 (50%)
	6 (1.25ms)	18		36	6	1	One PCG
2400	48 (20ms)	144	4	576	96	16	4 (25%)
	3 (1.25ms)	9		36	6	1	One PCG
1200	24 (20ms)	72	8	576	96	16	2 (12.5%)
	1.5 (1.25ms)	4.5		36	6	1	One PCG

Example 3.23   One frame (20 ms) consists of 16 slots (PCGs).

For the 1200 bps transmission rate, the information bits equal 24 bits per frame, which is the input to the convolutional encoder. Since the encoder code rate is R=1/3, the code symbols of encoder output is 72 bits. By the repetition process, we obtain 576 (=72 x 8) code symbols. These code symbols are interleaved prior to modulation and transmission. Consequently, the modulation symbols are 96 (=576/6).

For a single PCG, Time duration = 1.25 (= 20/16) ms, Information bits = 1.5 (=24/16) bits, Encoder code symbols = 4.5 (= 1.5 x 3) bits, Repetition symbols = 36 (=4.5 x 8) bits, and Walsh modulation = 6 (= 36/6) modulation symbols.

Suppose the scrambling PN sequence is $(\alpha_0, \alpha_1, \cdots, \alpha_{13})$ = (00101101100100) whence $\alpha_0 = 0$, $\alpha_1 = 0$, $\alpha_2 = 1$, $\alpha_3 = 0$, $\alpha_4 = 1$, $\alpha_5 = 1$, $\alpha_6 = 0$, $\alpha_7 = 1$, $\alpha_8 = 1$, $\alpha_9 = 0$, $\alpha_{10} = 0$, $\alpha_{11} = 1$, $\alpha_{12} = 0$, $\alpha_{13} = 0$.

This scrambling pattern represents the 14-bit PN sequence.

The power control group numbers are determined following the data burst randomizer algorithm for the 1200 bps data rate as follows:

$\alpha_0 = 0$ if $\alpha_8 = 0$ and $\alpha_{12} = 0$ which corresponds to PCG No. 0.

$8 + \alpha_4 = 9$ if $\alpha_{10} = 0$ and $\alpha_{13} = 0$ which corresponds to PCG No. 9.

It is clear from Table 3.20 that the interleaver output symbols corresponding to rows 1 and 9 represent PCG No. 0 and that the interleaver output symbols corresponding to rows 18 and 26 represent PCG No. 9. Hence the code symbols for PCG No. 0 are 101000111001000011100011110101100001 and the code symbols for PCG No. 9 are 100110100110000101100000011110010001, respectively. Actually the code symbol sequence corresponding to PCG No. 0 is the symbol sequence (VI) and the modulation symbols corresponding to the symbol sequence (VI) represent the symbol sequence (VII). Thus, PCG No. 0 (or Slot 0) consists of the 36-bit code symbol

$$101000 \ 111001 \ 000011 \ 100011 \ 110101 \ 100001 \qquad (VI)$$

and the 6 modulation symbols (6 x 64=384 chips)

```
010110100101101001011010010110100101101001011010010110100101011010
011010010110100101101001011010011001011010010110100101011010010110
000000000000000011111111111111111111111111111111110000000000000000    (VII)
010101010101010110101010101010101010101010101010100101010101010101
011001101001100101100110100110011001100101100110100110010110011 0
010101010101010101010101010101011010101010101010101010101010101010
```

Next, consider the case for PCG No. 9. Each 6-bit code symbols versus the 64-chip modulation symbols are listed in the following table.

$$100110 \ 100110 \ 000101 \ 100000 \ 011110 \ 010001 \qquad (VIII)$$

```
010101011010101010101010010101010101010101101010101010101001010101
010101011010101010101010010101010101010101101010101010101001010101
000000001111111100000000111111111111111100000000111111100000000    (IX)
010101010101010101010101010101010101010101010101010101010101010101
001110011000011110000110011100001111001100001111000011001111100
001100110011001100110011001100111100110011001100110011001100110 0
```

For the 1200 bps transmission rate, the modulation symbol sequences (VII) and (IX) represent nothing but the data burst randomizer output streams for PCG No. 0 & No. 9, respectively.

3.2.9   Direct Sequence Spreading

Consider direct sequence spreading by the long code for the reverse traffic channel. For the RTC, this spreading operation involves modulo-2 addition of the data burst randomizer output stream and the long code.

The long code is periodic with period $2^{42}-1$ chips specified by the LFSR tap polynomial $p(x)$ of the code generator:

$$p(x) = 1 + x + x^2 + x^3 + x^5 + x^6 + x^7 + x^{10} + x^{16} + x^{17} + x^{18} + x^{19} + x^{21} + x^{22} + x^{25} + x^{26} +$$
$$x^{27} + x^{31} + x^{33} + x^{35} + x^{42} \qquad (3.17)$$

Each PN chip of the long code is generated by the modulo-2 sum of the results through the AND gates inputting a 42-bit mask with the 42-stage LFSR of the sequence generator as shown in Fig. 3.8. The long code provides limited privacy. The long code mask varies depending on the channel type on which the mobile station is transmitting. The long code mask consists of a 42-bit binary sequence that creates the unique identity of the long code.

When transmitting on the RTC, the mobile station uses one of two long code masks, that is, a public long code mask unique to the mobile station's ESN or a private long code mask. The public long code mask is created as follows:

$M_{41}-M_{32}$: Set to 1100011000.

$M_{31}-M_{0}$: Set to a permutation of the mobile station's ESN bits.

Defining the mobile station's ESN bits as

ESN = (E_{31} , E_{30} , E_{29} , E_{28} , E_{27} , E_{26} , $\cdots$, E_{3} , E_{2} , E_{1} , E_{0}),

a permutation of the ESN bits should be specified in such a way that this permutation prevents high correlation between long codes as shown below:

Permuted ESN = (E_{0}, E_{31}, E_{22}, E_{13}, E_{4}, E_{26}, E_{17}, E_{8}, E_{30}, E_{21}, E_{12}, E_{3}, E_{25}, E_{16}, E_{7}, E_{29},
$\quad$ E_{20}, E_{11}, E_{2}, E_{24}, E_{15}, E_{6}, E_{28}, E_{19}, E_{10}, E_{1}, E_{23}, E_{14}, E_{5}, E_{27}, E_{18}, E_{9})

The public long code mask format is illustrated in Fig. 3.29.

The private long code mask is not available at this moment because distribution for its foreign use is controlled by TIA.

Example 3.24   Consider again the case for the 1200 bps transmission rate.

If the mobile station's ESN bits are chosen as $0100110010110011101001110$ 1100111, then the permuted ESN becomes $100101111100010010101100111$ 01101. The public long code mask format (see Fig. 3.29) is represented as $110001100010010111100010010101100111011101101$ (MSB→LSB). If this public long code mask is arranged in the reverse order, it becomes 1011011 $100110101001000111110100100011100011$ (LSB→MSB). Referring to Fig. 3.8, this reversed 42-bit long code mask and the 42-stage outputs of LFSR are subjected to input to the AND gates. The resulted AND gate outputs are to be modulo-2 added. This output is the binary value '0' if an integral number from the modulo-2 adder is even and is the binary value '1' if an integral number is odd.

From Eq. 3.17, the generator sequence can be expressed as 11110111001000 $0011110110011100010101000000$. Assume that the initial contents of LFSR is $1000000000000000000000000000000000000000000$.

Applying these sequences to Fig. 3.8, we generate the long code PN chips as follows:

$$
\begin{array}{l}
\text{1011 0111 0011 0101 0010 0011 1110 1001 0001 1000} \\
\text{1111 1011 1010 1000 0010 0011 1000 0001 1000 1100} \quad\text{(X)}\\
\text{0000 0001 0100 0100 0011}
\end{array}
$$

This sequence represents the long code generated partially by the first 100 shifting operations.

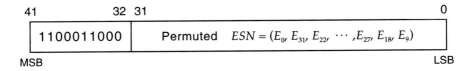

Figure 3.29 Public long code mask format

Finally, consider direct sequence spreading (DSS) by EX-ORing the orthogonal modulation symbols and the long code PN chips. For the 1200 bps data rate, the modulation symbols represent the data burst randomizer output stream. For simplicity, consider only the first row of the long code sequence (X).

Recall that PN chips/Walsh chip is 4 so that computation would be as follows:

Modulation symbols:	0	1	0	1	1	0	1	0	0	0
Long code chips :	1011	0111	0011	0101	0010	0011	1110	1001	0001	1000
EX-OR ($\oplus$) :	1011	1000	0011	1010	1101	0011	0001	1001	0001	0111

Thus, DSS by the long code PN chips is computed as

$$\boxed{1011\ 1000\ 0011\ 1010\ 1101\ 0011\ 0001\ 1001\ 0001\ 0111} \qquad (XI)$$

Example 3.25 For the 4800 bps transmission rate, the interleaver output will result in the following order of rows in Table 3.2: 1, 3, 2, 4, 5, 7, 6, 8, 9, 11, 10, 12, 13, 15, 14, 16, 17, 19, 18, 20, 21, 23, 22, 24, 25, 27, 26, 28, 29, 31, 30, 32. The interleaver output symbols were already computed as shown in Table 3.22.

Let us consider only the first two rows 1 and 3, which are the input to the 64-ary orthogonal modulator. Thus, the interleaver output for rows 1 and 3 is

111000 010111 111101
110010 011001 010011

The 64-ary orthogonal modulator output stream (or modulation symbols) is computed using the modulation symbol index (MSI) as shown below:

Interleaver Output	MSI	Modulator Output Stream (see Table 2.10)
111000	7	011010010110100101101001
		011010010110100101010100101101001
010111	58	001100111100110011001100110000110011
		110011000011001100110011111001100
111101	47	011010011001011001101001100100110
		100101100110100110010110011001101001
110010	19	011001100110011010011001100110011001
		011001100110011010011001100110011001
011001	38	001111000011110000111110000111100
		110000111100001111100001111000011
010011	50	001100110011001111100110011001100
		110011001100110000011001100110011

This modulator output stream is equivalent to PCG No. 0.

For the 4800 bps data rate, the transmission gate allows one-half of the interleaver output symbols to be transmitted. If the 14-bit PN sequence used for scrambling is assumed to be

$(\alpha_0, \alpha_1, \alpha_2, \alpha_3, \alpha_4, \alpha_5, \alpha_6, \alpha_7, \alpha_8, \alpha_9, \alpha_{10}, \alpha_{11}, \alpha_{12}, \alpha_{13})$
$= (00101101100100)$, transmission occurs on PCGs numbered:

$\alpha_0, 2 + \alpha_1, 4 + \alpha_2, 6 + \alpha_3, 8 + \alpha_4, 10 + \alpha_5, 12 + \alpha_6, 14 + \alpha_7$.

Since $\alpha_0 = 0$, PCG No. 0	Since $2 + \alpha_1 = 2$, PCG No. 2
Since $4 + \alpha_2 = 5$, PCG No. 5	Since $6 + \alpha_3 = 6$, PCG No. 6
Since $8 + \alpha_4 = 9$, PCG No. 9	Since $10 + \alpha_5 = 11$, PCG No. 11
Since $12 + \alpha_6 = 12$, PCG No. 12	Since $14 + \alpha_7 = 15$, PCG No. 15

Thus, we see that the assignment of gated-on PCGs is just 1/2 within the 20 ms frame, as illustrated in Fig. 3.30. The gated-on PCGs are actually the data burst randomizer output consisting of slots 0, 2, 5, 6, 9, 11, 12, and 15.

According to Table 3.22, row numbers 1 and 3 correspond to PCG No. 0, row numbers 5 and 7 correspond to PCG No. 2, row numbers 10 and 12 correspond to PCG 5, and so on. For the 4800 bps transmission rate, the RTC interleaver output stream is gated allowing transmission of one-half of the interleaver output symbols (i.e., active slots) and deletion of others. Table 3.24 illustrates the 36-bit interleaver output symbols corresponding to the gated-on groups (PCG Nos.)

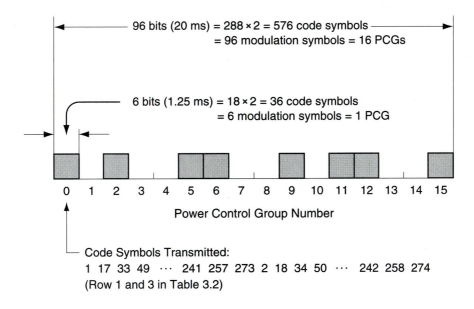

Figure 3.30 The assignment of gated-on power control groups
(1/2 of 16 PCGs) for the RTC 4800 bps frame (20 ms)

Table 3.24 Interleaver Output Symbols Corresponding to Active PCG Numbers

Row No.	PCG No.	Interleaver Output Symbols (36 bits)
1, 3	0	111000010111111101 110010011001010011
5, 7	2	101110010010001001 000000100111000111
10, 12	5	100010111001111111 100110011011000000
13, 15	6	000011101110010000 100000100000000000
18, 20	9	000110110001010011 101101011101100001
22, 24	11	001111010101001011 111010101010110111
25, 27	12	001000000100111010 010000110011010101
30, 32	15	010000101111001101 111101111111101101

Consider PCG No. 2 and PCG No. 5. The modulator output stream (modulation symbols) corresponding to the modulator input stream (interleaver output symbols) is computed as follows:

PCG No.	MSI	Modulation symbols (Data burst randomizer output)
	29	0101101010100101101001010101101001011010101001011010010101011010
	18	0011001100110011110011001100110000110011001100111100110011001100
	36	0000111100001111000011110000111111110000111100001111000011110000
2	0	0000000000000000000000000000000000000000000000000000000000000000
	57	0101010110101010101010100101010110101010010101010101010110101010
	56	0000000011111111111111110000000011111111000000000000000011111111
	17	0101010101010101101010101010101001010101010101011010101010101010
	39	0110100101101001011010010110100110010110100101101001011010010110
	63	0110100110010110100101100110100110010110011010010110100110010110
5	25	0101010110101010101010100101010101010101101010101010101001010101
	54	0011110000111100110000111100001111000011110000110011110000111100
	0	0000000000000000000000000000000000000000000000000000000000000000

Computation for PCG Nos. 6, 9, 11, 12, and 15 will be left to the reader as an exercise.

3.2.10 Quadrature Spreading

After the direct sequence spreading (DSS), the reverse traffic channel (RTC) is subject to spread the DSS stream in quadrature. The I and Q pilot PN sequences are generated based on the following LFSR tap polynomials with period 2^{15} chips, respectively. They are the same as used for the access channel:

$$P_I(x) = 1 + x^5 + x^7 + x^8 + x^9 + x^{13} + x^{15}$$
$$P_I = (100001011000101) \qquad \text{(vector form)}$$
and $\quad P_Q(x) = 1 + x^3 + x^4 + x^5 + x^6 + x^{10} + x^{11} + x^{12} + x^{15}$
$$P_Q = (100111100011001) \qquad \text{(vector form)}$$

The quadrature spreading sequences I and Q are readily computed from the results of EX-ORing either P_I or P_Q pilot sequence with the DSS stream, respectively.

The I-channel stream I and Q-channel stream Q are the output sequences resulting from EX-ORing the I and Q pilot chips and the DSS stream input.

Let us consider first the generation of pilot PN chips. Using the given tap sequences P_I and P_Q, the I and Q pilot PN chips stream can be computed under the initial contents (100000000000000) of the LFSRs, as shown in Table 3.4.

The in-phase and quadrature-phase pilot PN chips are, respectively

$$\downarrow$$

$\mathcal{P}_I =$ 000000000000000010101001001110100011011110011001000
00111100001000110100101011010101010111010010011111

$$\downarrow$$

and $\mathcal{P}_Q =$ 0000000000000001001111010111010110100111000101001
10011100011011000111010011000100100100011001100111

Note that a binary value '0' is inserted in $\mathcal{P}_I$ and $\mathcal{P}_Q$ after 14 consecutive 0 bits as seen from the arrow mark ($\downarrow$).

Example 3.26 Using $\mathcal{P}_I$ and $\mathcal{P}_Q$ pilot PN sequences, the I-channel stream I is computed by EX-ORing the $\mathcal{P}_I$ and DSS stream; while the Q-channel stream Q is computed by EX-ORing the $\mathcal{P}_Q$ and DSS stream.

For the 1200 bps transmission rate, the in-phase channel sequence can be computed as

DSS: 10111000 00111010 11010011 00011001 00010111

$\oplus$

$\mathcal{P}_I$:   00000000 00000001 01010010 01110100 01101111

I: 10111000 00111011 10000001 01101101 01111000

While the quadrature channel sequence Q is computed as

DSS: 10111000 00111010 11010011 00011001 00010111

$\oplus$

$\mathcal{P}_Q$:   00000000 00000001 00111101 01110101 10100111

Q: 10111000 00111011 11101110 01101100 10110000

Example 3.27 For the 4800 bps data transmission rate, the DSS stream can be produced by EX-ORing the data burst randomizer output (PCGs) and the long code sequence (X). The long code PN chips for the 1200 bps transmission rate is exactly identical with that of the 4800 bps data rate. That is,

1011 0111 0011 0101 0010 0011 1110 1001 0001 1000

which represents only the first portion of the entire long code PN chips.

The data burst randomizer output corresponding to PCG No. 0 with respect to MSI = 7 is

<u>0110100101</u>101001011010010110100101101001
0110100101101001011010010101101001

Let us consider only the first 10 symbols 0110100101 which will correspond to the following 40 chips 0000111111110000111100000000111100001111 because the ratio of PN chips/Walsh chip is 4.

Now, we can compute the DSS stream as follows:

Randomizer output (PCG No.0): ⊕	0	1	1	0	1	0	0	1	0	1
Long code chips:	1011	0111	0011	0101	0010	0011	1110	1001	0001	1000
DSS stream:	1011	1000	1100	0101	1101	0011	1110	0110	0001	0111

The I-channel pilot PN chips P_I and the Q-channel pilot PN chips P_Q for the 4800 bps data rate are exactly the same as that for the 1200 bps data rate. They are, respectively

P_I : 00000000 00000001 01010010 01110100 01101111

P_Q : 00000000 00000001 00111101 01110101 10100111

Thus, the in-phase channel sequence I and the quadrature-phase channel sequence Q are computed by EX-ORing the DSS sequence and P_I or P_Q, respectively, as shown below:

The in-phase sequence I is obtained as

DSS stream:	10111000	11000101	11010011	11100110	00010111
⊕					
P_I :	00000000	00000001	01010010	01110100	01101111
In-phase I:	10111000	11000100	10000001	10010010	01111000

while the quadrature-phase sequence Q is computed as

DSS stream:	10111000	11000101	11010011	11100110	00010111
⊕					
P_Q :	00000000	00000001	00111101	01110101	10100111
Quadrature-phase Q:	10111000	11000100	11101110	10010011	10110000

3.2.11 RTC Quadrature Phase Shift Keying

The I and Q data stream obtained in the previous section will be applied to the I and Q baseband filters. The data stream Q spread by the Q-channel pilot PN chips is delayed by half a chip time. Let $I(t)$ and $Q(t)$ be two output streams coming out of the baseband filters. The timing of the two pulse streams $I(t)$ and $Q(t)$ is offset by $T_b/2$ seconds due to delaying Q by half a chip time.

The offset QPSK waveform s(t) is obtained by amplitude modulation $I(t)$ and $Q(t)$ each onto the cosine and sine functions of a carrier wave as shown in Fig. 3.16. The in-phase stream $I(t)$ is amplitude-modulated the cosine function with an amplitude of +1 or –1, which produces a BPSK waveform. The quadrature-phase stream $Q(t)$ modulates the sine function, resulting in a BPSK waveform orthogonal to the cosine function. Thus, the summation of these two orthogonal BPSK waveforms will yield the QPSK waveforms s(t). The offset QPSK pulse stream caused by delaying Q by half a chip time $T_b/2$ is illustrated in the following examples.

Example 3.28 Sketch the offset QPSK pulse stream for the 1200 bps transmission rate.
As computed previously, the in-phase channel sequence I is
$$I = (10111000\ 00111011\ \underline{10000001}\ 01101101\ 01111000)$$
and the quadrature channel sequence Q is
$$Q = (10111000\ 00111011\ \underline{11101110}\ 01101100\ 100110000)$$
For demonstrative purposes, pick the third elements, $I_3 = (10000001)$ and $Q_3 = (11101110)$, respectively, from I and Q streams. The offset QPSK pulse stream caused by Q_3 by half a chip time, $T_b/2$, can be illustrated as shown in Fig. 3.31.

Example 3.29 Compute the offset QPSK pulse trains for the 2400 bps transmission rate. The in-phase sequence I and quadrature-phase sequence Q were already computed in Example 3.27. They are, respectively
$$I = (10111000\ 11000100\ 10000001\ \underline{10010010}\ 01111000)$$
$$Q = (10111000\ 11000100\ 11101110\ \underline{10010011}\ 101100000)$$
Consider the pair of fourth elements $I_4 = (10010010)$ and $Q_4 = (10010011)$ from the I and Q streams. The offset QPSK pulse trains caused by Q_4 by half a chip time, $T_b/2$, can be illustrated as sketched in Fig. 3.32.

Example 3.30 For the 9600 bps transmission rate, referring to Table 3.21, the entire interleaver outputs corresponding to the row numbers from 1 to 32 were computed. The 192-chip modulation symbol sequence corresponding to the first row 101000110001111011 has been shown in Example 3.21. The 192-chip modulation symbols represent nothing but the orthogonal modulator output as shown below.

```
0101101001011010010110100101101001011010 0101101001011010010110100101101001011010
0110011001100110011001100110011001100110 0110011001100110011001100110011001100110
0110100101101001100101101001010110 1001011010010110011010010110101001
```

1200 bps: I-channel output: 10000001 (-1, 1, 1, 1, 1, 1, 1,-1)
Q-channel output: 11101110 (-1,-1,-1, 1,-1,-1,-1, 1)

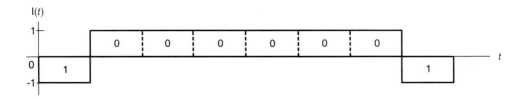

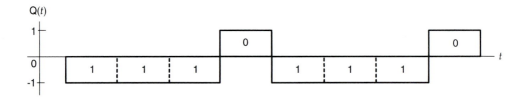

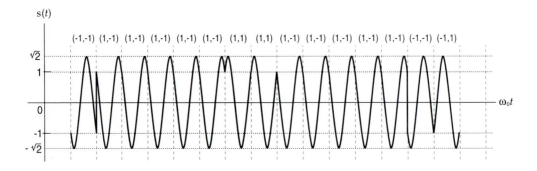

Figure 3.31 The offset QPSK waveform for I_3 and Q_3
sequences at the 1200 bps transmission rate

4800 bps: I-channel output: 10010010 (-1, 1, 1,-1, 1, 1,-1, 1)

Q-channel output: 10010011 (-1, 1, 1,-1, 1, 1,-1,-1)

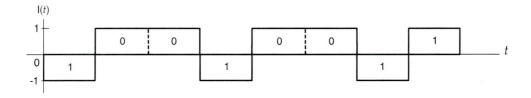

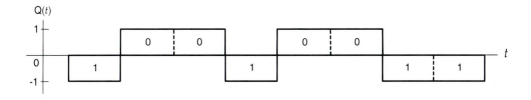

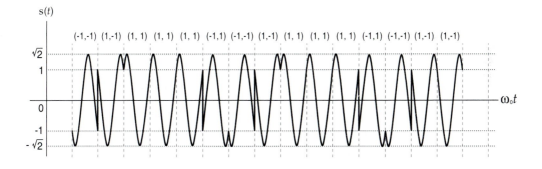

Figure 3.32 The offset QPSK waveform for I_4 and Q_4

sequences at the 4800 bps transmission rate

Since there is no repetition required for the 9600 bps transmission rate, the orthogonal modulator output is the same chip stream as the data burst randomizer output. In other words, the data burst randomizer is not needed to provide for the 9600 bps RTC data rate.

Since the permuted ESN is 1 0 0 1 0 1 1 1 1 1 0 0 0 1 0 0 1 0 1 0 1 1 0 0 1 1 1 0 1 1 0 1, the 42-bit public long code mask sequence is represented as 1 1 0 0 0 1 1 0 0 0 1 0 0 1 0 1 1 1 1 1 0 0 0 1 0 0 1 0 1 0 1 1 0 0 1 1 1 0 1 1 0 1 (MSB-LSB).

Thus, the long code PN chips are generated as follows:

LFSR initial contents: 10000000000000000000000000000000000000000

Generator tap sequence: 11110111001000001111011001110001010100000

Public long code mask bits: 11000110001001011111000100101011001110110 1 (LSB-MSB)

Using Fig. 3.8, the long code PN chips are computed from the modulo-2 sum of AND gate outputs. These PN chips for the first 100 clocking times are obtained as

1011 0111 0011 0101 0010 0011 1110 1001 0001 1000
1111 1011 1010 1000 0010 0011 1000 0001 1000 1100
0000 0001 0100 0011

Considering the first 40 chips only, the DSS data sequence is computed through the EX-ORing operation with the orthogonal modulator output and the long code PN chips.

Modulator symbols:	0	1	0	1	1	0	1	0	0	1
$\oplus$										
Long code PN chips:	1011	0111	0011	0101	0010	0011	1110	1001	0001	1000
DSS sequence:	1011	1000	0011	1010	1101	0011	0001	1001	0001	0111

The I-channel pilot PN chips P_I and the Q-channel pilot chips P_Q for the 9600 bps transmission rate are computed as, respectively

Insert zero '0'

↓

P_I: 0000000000000001010100100111010001101111001100100
00011110000100011010010101101010101011010010011111

Insert zero '0'

↓

P_Q: 00000000000000010011110101110101101001110001010011
10011100011011000111010011000100100100011001100011

Thus, the in-phase channel stream I and the quadrature-channel stream Q are computed by EX-ORing the DSS sequence and P_I or P_Q, respectively, as shown below:

DSS sequence:	10111000	00111010	11010011	00011001	00010111
⊕					
P_I:	00000000	00000001	01010010	01110100	01101111
In-phase *I*:	10111000	00111011	10000001	<u>01101101</u>	01111000

and

DSS sequence:	10111000	00111010	11010011	00011001	00010111
⊕					
P_Q:	00000000	00000001	00111101	01110101	10100111
Quadrature-phase *Q*:	10111000	00111011	11101110	<u>01101100</u>	10110000

Considering the pair of fourth elements $I_4 = (01101101)$ and $Q_4 = (01101100)$ in both I and Q streams, the OQPSK pulse trains caused by Q_4 by half a chip time, $T_b/2$, is illustrated as shown in Fig. 3.33.

9600 bps: I-channel output: 01101101 (1,-1,-1, 1,-1,-1, 1,-1)
                 Q-channel output: 01101100 (1,-1,-1, 1,-1,-1, 1, 1)

I(t)

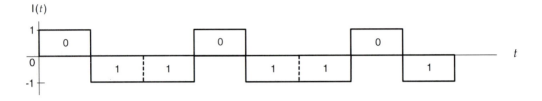

Q(t)

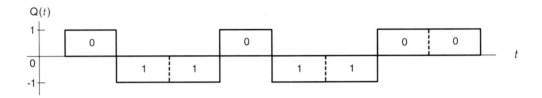

s(t)

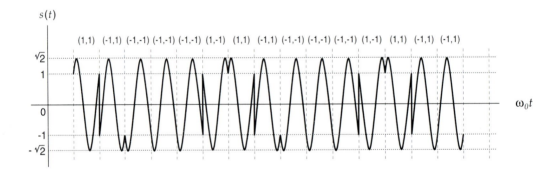

Figure 3.33 The offset QPSK waveform for I_4 and Q_4
                 sequences at the 9600 bps transmission rate

CHAPTER 4

Forward CDMA Channel

The forward CDMA channel consists of one or more code channels that are transmitted on a CDMA frequency assignment using a particular pilot PN offset. The code channels are associated with the pilot channel, sync channel, paging channels, and forward traffic channels. Each base station uses a time offset of the pilot PN sequence to identify a forward CDMA channel. Time offset can be reused within a CDMA cellular system.

The overall structures of the forward CDMA channel are shown in Figs. 4.2, 4.10, 4.18, and 4.26. Requirements that are specific to CDMA base station operation will be discussed in the following.

4.1 FORWARD CODE CHANNELS

The forward CDMA channel consists of the following code channels: The pilot channel, up to one sync channel, up to seven paging channels, and a number of forward traffic channels. Each of these code channels is orthogonally spread by the appropriate Walsh function and is then spread by a quadrature pair of pilot PN sequences at a fixed chip rate of 1.2288 Mcps. Multiple forward CDMA channels may be used within a base station in a frequency division multiplexed manner.

A typical example of the channels transmitted by a base station is shown in Fig. 4.1. Out of the 64 code channels available for use, the example depicts the pilot channel, one sync channel, seven paging channels, and 55 traffic channels.

Code channel number zero (W_0) is always assigned to the pilot channel.

If the sync channel is present, it will be assigned to code channel 32 (W_{32}). If paging channels are present, they should be assigned to code channel numbers one through seven (W_1 to W_7) in sequence. The remaining code channels are available to the forward traffic channels (W_8, W_{31}, W_{33}, $\cdots$, W_{63}).

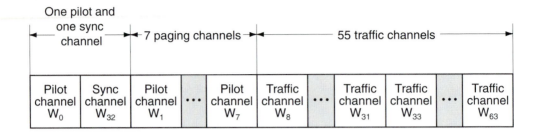

Figure 4.1 A forward CDMA channel transmitted by a base station

Each code channel transmitted on the forward CDMA link is spread with a Walsh function at a fixed chip rate of 1.2288 Mcps to provide orthogonal channelization among all code channels on a given forward CDMA channel. Following the orthogonal spreading, each code channel is spread in quadrature. After the spreading operation, the I and Q impulses are applied to the inputs of the I and Q baseband filters. Following baseband filtering, the binary I and Q at the output of the quadrature spreading is mapped into specified phase transitions.

4.2 PILOT CHANNEL

The pilot channel is an unmodulated spread spectrum signal which is transmitted at all times by the base station on each active forward CDMA channel. The pilot channel is illustrated in Fig. 4.2. The base station continually transmits a pilot channel for every CDMA channel supported by the base station. The mobile station monitors the pilot channel at all times except when not receiving in the slotted mode. A mobile station within the coverage area of the base station uses the pilot channel for synchronization. The pilot channel is a reference channel which allows a mobile station to acquire the timing of the forward CDMA channel and thus provides a phase reference for coherent demodulation. During pilot and sync channel processing, the mobile station acquires and synchronizes to the CDMA system while it is in the mobile station initialization state.

In the pilot channel acquisition substate, the mobile station acquires the pilot channel of the selected CDMA system. Upon entering this substate, the mobile station tunes to the CDMA channel number equal to CDMACHs, sets its code channel for the pilot channel, and searches for the pilot channel. If the mobile station acquires the pilot channel within $T_{20m} = 15$ seconds, the mobile station enters the sync channel acquisition substate. If the mobile station does not acquire the pilot channel within T_{20m} seconds, the mobile station enters the system determination substate (as will be discussed in Chapter 5). T_{20m} denotes the maximum time to remain in the pilot channel acquisition substate of the mobile station initialization state.

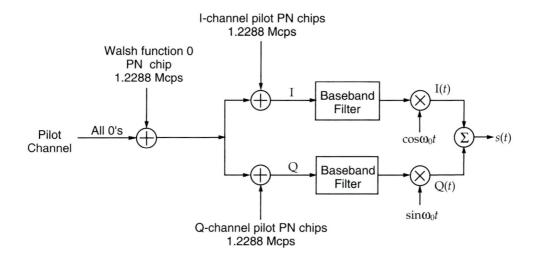

Figure 4.2 Pilot channel in the forward CDMA channel

In the CDMA cellular telephone system, each cell site transmits a pilot carrier signal at each link frequency. This pilot carrier is used by the mobile station to obtain initial system synchronization and to provide robust time, frequency, and phase tracking of the signals from the cell site. This signal is tracked continuously by each mobile station. Variations in the transmitted power level of the pilot signal control the coverage area of the cell.

The pilot carriers are transmitted by each cell site using the same code but with different spread spectrum code phase offsets, allowing them to be distinguished. The fact that the pilots all use the same code allows the mobile station to find system timing synchronization by a single search through all code phases. The strongest signal found corresponds to the code phase of the cell site.

4.2.1 Orthogonal Spreading with a Walsh Function

Since the pilot bits are all zero (0s) as seen from Fig. 4.2, the Walsh function (or modulation symbol index) is 0 and the corresponding Walsh chip generated by the Walsh function 0 is represented by $0000\cdots000$ (64 zeros).

This 64-zero Walsh chip is input to the quadrature spreading.

4.2.2 Quadrature Spreading

Following the orthogonal spreading, the pilot channel is spread in quadrature as shown in Fig. 4.2. The spreading sequence, called the pilot PN sequence, is a quadrature sequence of length $2^{15} = 32768$ PN chips. In order to obtain the I and Q pilot PN sequence of period 2^{15}, a '0' is inserted in the LFSR sequence $i(n)$ and $q(n)$ after 14 consecutive '0' outputs, as shown in Example 4.1.

These spreading sequences are based on the following LFSR tap polynomials. For the in-phase sequence, the tap polynomial is given as

$$P_I(x) = 1 + x^5 + x^7 + x^8 + x^9 + x^{13} + x^{15} \tag{4.1}$$

or $P_I = (1000010111000101)$

For the quadrature-phase sequence, it gives

$$P_{Q(x)} = 1 + x^3 + x^4 + x^5 + x^6 + x^{10} + x^{11} + x^{12} + x^{15} \tag{4.2}$$

or $P_Q = (1001111000111001)$

Equations 4.1 and 4.2 are the same as Eqs. 3.3 and 3.4.

The I and Q pilot PN sequences are generated from the maximum length LFSRs based on the tap polynomials $P_I(x)$ and $P_Q(x)$, respectively.

The LFSRs representing $P_I(x)$ and $P_Q(x)$ are shown in Figs 4.3 and 4.4. The chip rate for the pilot sequence is 1.2288 Mcps. The same I and Q pilot PN sequences are also generated by the reciprocal polynomials $i(x) = x^n P_I(x^{-1})$ and $q(x) = x^n P_Q(x^{-1})$. They are, respectively

$$i(x) = 1 + x^2 + x^6 + x^7 + x^8 + x^{10} + x^{15} \tag{4.3}$$

and $$q(x) = 1 + x^3 + x^4 + x^5 + x^9 + x^{10} + x^{11} + x^{12} + x^{15} \tag{4.4}$$

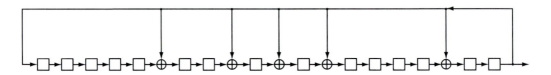

Figure 4.3 A 15-stage LFSR based on the tap polynomial
$$P_I(x) = 1 + x^5 + x^7 + x^8 + x^9 + x^{13} + x^{15}$$

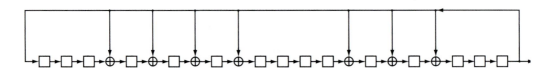

Figure 4.4 A 15-stage LFSR based on the tap polynomial
$$P_Q(x) = 1 + x^3 + x^4 + x^5 + x^6 + x^{10} + x^{11} + x^{12} + x^{15}$$

Using these tap polynomials $i(x)$ and $q(x)$, the LFSRs based on Eqs. 4.3 and 4.4 can be sketched as shown in Figs. 4.5 and 4.6.

The binary sequences $i(n)$ and $q(n)$ are called the LFSR sequences which are generated by using the following linear recursive operations:

$$i(n) = i(n-15) \oplus i(n-10) \oplus i(n-8) \oplus i(n-7) \oplus i(n-6) \oplus i(n-2) \tag{4.5}$$

and $\quad q(n)=q(n-15) \oplus q(n-12) \oplus q(n-11) \oplus q(n-10) \oplus q(n-9) \oplus q(n-5)$

$$\oplus\, q(n-4) \oplus q(n-3) \tag{4.6}$$

where $i(n)$ and $q(n)$ for $1 \le n \le 32767$ represent binary-valued 0 or 1, $i(15)=1$ and $q(15)=1$, and $\oplus$ denotes the modulo-2 addition.

The binary sequences I and Q at the output of the quadrature spreading are computed by EX-ORing the I and Q pilot sequences, $\mathcal{P}_I$ and $\mathcal{P}_Q$, and the all-zero input sequence to the modulo-2 adders.

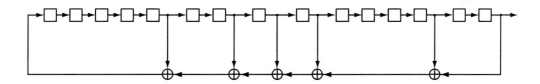

Figure 4.5 A 15-stage LFSR based on the reciprocal tap polynomial
$$i(x) = 1 + x^2 + x^6 + x^7 + x^8 + x^{10} + x^{15}$$

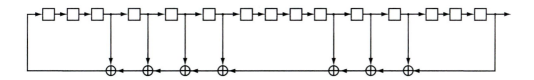

Figure 4.6 A 15-stage LFSR based on the reciprocal tap polynomial
$$q(x) = 1 + x^3 + x^4 + x^5 + x^9 + x^{10} + x^{11} + x^{12} + x^{15}$$

Example 4.1   Consider first the generation of pilot PN chips. The in-phase tap sequence is P_I = (1000010111000101) and the quadrature-phase tap sequence is P_Q = (100111000111001). Assuming the initial contents of LFSR as (100000000000000), the shifted $\mathcal{P}_I$ and $\mathcal{P}_Q$ sequences are computed as shown below:

Insert 0
↓

$\mathcal{P}_I$ = (000000000000000101010010011101000110111100110001000
00111100001000110100101011010101010111010010011111)
      This is the in-phase pilot PN chips for $1 \leq n \leq 99$, where n = shifting number.

Insert 0
↓

$\mathcal{P}_Q$ = (0000000000000001001111010111010110100111000101 0011
10011100011011000111010011000100100100011001 1000111)
      This is the quadrature-phase pilot PN chips for $1 \leq n \leq 99$, where n = shifting number. Notice that a zero (0) is inserted in $\mathcal{P}_I$ and $\mathcal{P}_Q$ after 14 consecutive 0 outputs as shown by the arrow (↓).

The I-channel sequence I and the Q-channel sequence Q at the output of the quadrature spreading are computed by EX-ORing $\mathcal{P}_I$ and $\mathcal{P}_Q$ and the all-zero input sequence to the modulo-2 adders.

All zero input	00000000 00000000 00000000 00000000 00000000
⊕	
$\mathcal{P}_I$	00000000 00000001 01010010 01110100 01101111
I	00000000 00000001 01010010 01110100 01101111
All zero input	00000000 00000000 00000000 00000000 00000000
⊕	
$\mathcal{P}_Q$	00000000 00000001 00111101 01110101 10100111
Q	00000000 00000001 00111101 01110101 10100111

In the pilot channel, we can now see that the I-channel stream I is nothing but the in-phase pilot PN sequence$\mathcal{P}_I$; while the Q-channel stream Q is exactly identical to the quadrature-phase pilot PN sequence $\mathcal{P}_Q$.

Example 4.2   Consider the binary sequences $i(n)$ and $q(n)$ which are induced from the tap polynomials $i(x)$ and $q(x)$. Since $i(n)=0$ for $1 \leq n \leq 14$ and $i(15)=1$, $i(n)$ for $16 \leq n \leq 32767$ can be computed as follows:

Using $i(n) = \displaystyle\sum_{k=15,10,8,7,6,2} i(n-k)$, $1 \leq n \leq 32767$, it generates

$i(16) = i(1) \oplus i(6) \oplus i(8) \oplus i(9) \oplus i(10) \oplus i(14)$
$= 0 \oplus 0 \oplus 0 \oplus 0 \oplus 0 \oplus 0 = 0$, for $n = 16$
$i(17) = i(2) \oplus i(7) \oplus i(9) \oplus i(10) \oplus i(11) \oplus i(15)$
$= 0 \oplus 0 \oplus 0 \oplus 0 \oplus 0 \oplus 1 = 1$, for $n = 17$
$i(18) = i(3) \oplus i(8) \oplus i(10) \oplus i(11) \oplus i(12) \oplus i(16)$
$= 0 \oplus 0 \oplus 0 \oplus 0 \oplus 0 \oplus 0 = 0$, for $n = 18$
$i(19) = i(4) \oplus i(9) \oplus i(11) \oplus i(12) \oplus i(13) \oplus i(17)$
$= 0 \oplus 0 \oplus 0 \oplus 0 \oplus 0 \oplus 1 = 1$, for $n = 19$

. .
. .
. .

Thus, the in-phase pilot PN sequence P_I is obtained by inserting a 0 (zero) in $i(n)$ after 14 consecutive 0 outputs (see arrow $\downarrow$) as follows:

Insert 0
$\downarrow$

$P_I = (0000000000000000101010010011101000110111100110011000$
$0011110000100011010010101101010101011101001001111111)$

Likewise, the quadrature-phase pilot PN sequence P_Q can be computed using $q(n)=0$ for $1 \leq n \leq 14$, $q(15)=1$, and $q(n) = \displaystyle\sum_{K=15,10,8,7,6,2} i(n-k)$, $1 \leq n \leq 32767$.

The quadrature-phase pilot PN sequence is obtained by inserting a 0 in $q(n)$ after 14 consecutive 0s as follows:

Insert 0
$\downarrow$

$P_Q = (0000000000000001001111010111010110100111000101 00 11$
$1001110001101100011101001100010010010001100110000111)$

The I-channel sequence I and the Q-channel sequence Q at the output of quadrature spreading results in the same data sequences as shown in Example 4.1.

4.2.3 Baseband Filtering

Following the quadrature spreading, the I and Q data streams are subject to apply to the inputs of the I and Q baseband filters. Let $S(f)$ be a frequency spectrum of the baseband filters. The limits of the normalized frequency response of the filter are confined within $\pm \delta_1$ in the passband $0 \leq f \leq f_P$ and the normalized response should be less than or equal to $-\delta_2$ in the stopband $f \leq f_S$ as shown in Fig. 3.15. Specifically, the numerical values for the limit parameters are $\delta_1 = 1.5$dB, $\delta_2 = 40$ dB, $f_P = 590$ kHz, and $f_S = 740$ kHz.

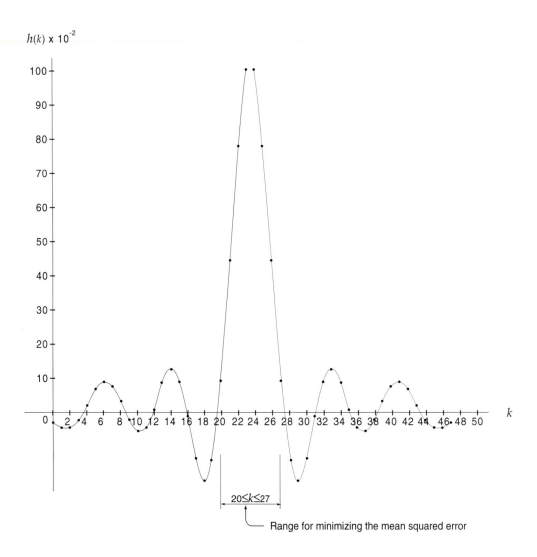

Figure 4.7 The values of coefficients $h(k)$ confined within $0 \leq k \leq 47$

Let $S(t)$ be the impulse response of the baseband filter. Then $S(t)$ satisfies the following equation:

$$\text{Mean Squared Error} = \sum_{k=0}^{47} [aS(kT_s - \tau) - h(k)]^2 \leq 0.03 \tag{4.7}$$

where a and τ are the constants for minimizing the mean squared error. The constant T_S is equal to 1/4 (1.2288 x 10^6) = 203.451 ns, which is just one quarter of a PN chip time.

According to Table 4.1, the values of the coefficients $h(k)$ confined within $20 \leq k \leq 27$ are also contributable to minimize the mean square error, as shown in Fig. 4.7.

Table 4.1   Coefficients $h(k)$

k	$h(k)$
0, 47	−0.025288315
1, 46	−0.034167931
2, 45	−0.035752323
3, 44	−0.016733702
4, 43	0.021602514
5, 42	0.064938487
6, 41	0.091002137
7, 40	0.081894974
8, 39	0.037071157
9, 38	−0.021998074
10, 37	−0.060716277
11, 36	−0.051178658
12, 35	0.007874526
13, 34	0.084368728
14, 33	0.126869306
15, 32	0.094528345
16, 31	−0.012839661
17, 30	−0.143477028
18, 29	−0.211829088
19, 28	−0.140513128
20, 27	0.094601918
21, 26	0.441387140
22, 25	0.785875640
23, 24	1.0

(After TIA/EIA/IS-95)

4.2.4   Quadrature Phase Shift Keying (QPSK)

Let $I(t)$ and $Q(t)$ be two output streams coming out from the baseband filters. In non-offset QPSK, the two baseband streams coincide in time so that the carrier phase could be changed once every T_b seconds.

An orthogonal QPSK waveform $s(t)$ is obtained by amplitude modulation of $I(t)$ and $Q(t)$ each onto the cosine and sine functions of a carrier wave, as shown in Fig. 4.2. The in-phase stream $I(t)$ amplitude-modulates the cosine function with an amplitude of $+1$ or -1, which produces a BPSK wave form; whereas the quadrature-phase stream $Q(t)$ modulates the sine function, resulting in a BPSK waveform orthogonal to the cosine function. Thus, the sum of these two orthogonal BPSK waveforms yields the QPSK waveform.

Let $s(t)$ be the QPSK waveform as shown by

$$
\begin{aligned}
s(t) &= I(t)\cos\omega_o t + Q(t)\sin\omega_o t \\
&= \sqrt{2}\,\cos(\omega_o t - \theta(t))
\end{aligned}
\tag{4.8}
$$

where $I(t) = \sqrt{2}\,\cos\theta(t)$, $Q(t) = \sqrt{2}\,\sin\theta(t)$, and $\theta(t) = \tan^{-1}\dfrac{Q(t)}{I(t)}$

The phase $\theta(t)$ is an offset from a reference version of the periodic sequence.

The QPSK stream $s(t)$ with respect to the specific values of $I(t)$ and $Q(t)$ can be determined according to the chosen values of $\theta(t)$ as follows:

1. Since $I(t) = 1$ and $Q(t) = 1$ for $\theta(t) = \dfrac{\pi}{4}$, $s(t) = \sqrt{2}\cos\left(\omega_0 t - \dfrac{\pi}{4}\right)$.

2. Since $I(t) = -1$ and $Q(t) = 1$ for $\theta(t) = \dfrac{3\pi}{4}$, $s(t) = \sqrt{2}\cos\left(\omega_0 t - \dfrac{3\pi}{4}\right)$.   (4.9)

3. Since $I(t) = -1$ and $Q(t) = -1$ for $\theta(t) = -\dfrac{3\pi}{4}$, $s(t) = \sqrt{2}\cos\left(\omega_0 t + \dfrac{3\pi}{4}\right)$.

4. Since $I(t) = 1$ and $Q(t) = -1$ for $\theta(t) = -\dfrac{\pi}{4}$, $s(t) = \sqrt{2}\cos\left(\omega_0 t + \dfrac{\pi}{4}\right)$.

Based on the analysis given by Eq. (4.9), $I(t)$ and $Q(t)$ mapping for the forward CDMA channel (Pilot, Sync, Paging, and Forward traffic channels) are summarized in Table 4.2.

Table 4.2   I(t) and Q(t) Mapping for Forward CDMA channel

$\theta(t)$	I(t)		Q(t)	
	NRZ	**(Binary)**	**NRZ**	**(Binary)**
$\pi/4$	1	(0)	1	(0)
$3\pi/4$	-1	(1)	1	(0)
$-3\pi/4$	-1	(1)	-1	(1)
$-\pi/4$	1	(0)	-1	(1)

Using Table 4.2, the resulting signal constellation and phase transition are shown in Fig. 4.8. After baseband filtering, the binary (0s and 1s) $I(t)$ and $Q(t)$ at the output of the quadrature spreading is mapped into phase according to Table 4.2.

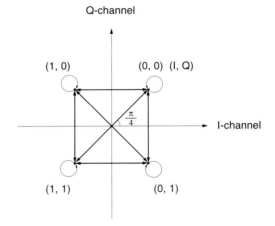

Figure 4.8 Forward CDMA channel signal constellation and phase transition

Data for the QPSK waveform plot is tabulated in Table 4.3.

Table 4.3   Data Tabulation for the Plot of QPSK Waveform $s(t)$

$\omega_0 t$	$s(t) = \sqrt{2} \cos(\omega_0 t - \theta(t))$			
	(1,1), $\theta(t)=\pi/4$	(-1,1), $\theta(t)=3\pi/4$	(-1,-1), $\theta(t)=-3\pi/4$	(1,-1), $\theta(t)=-\pi/4$
0	1	-1	-1	1
$\pi/4$	$\sqrt{2}$	0	$-\sqrt{2}$	0
$\pi/2$	1	1	-1	-1
$3\pi/4$	0	$\sqrt{2}$	0	$-\sqrt{2}$
π	-1	1	1	-1
$-3\pi/4$	$-\sqrt{2}$	0	$\sqrt{2}$	0
$-\pi/2$	-1	-1	1	1
$-\pi/4$	0	$-\sqrt{2}$	0	$\sqrt{2}$
2π	1	-1	1	1

Using Table 4.3, the QPSK pulse train based on $I(t)$ and $Q(t)$ can be drawn as shown in the following example.

Example 4.3   The in-phase sequence I and the quadrature-phase sequence Q were calculated in Example 4.1. They are, respectively

I :	00000000	00000001	01010010	01110100	01101111
Q :	00000000	00000001	00111101	01110101	10100111

If we draw the (I, Q) pair of pulses inside the square bracket, the QPSK waveform $s(t)$ can be shown as illustrated in Fig. 4.9.

Pilot channel
I-channel output: 01010010 (1, -1, 1, -1, 1, 1,-1, 1)
Q-channel output: 00111101 (1, 1, -1, -1 -1,-1, 1,-1)

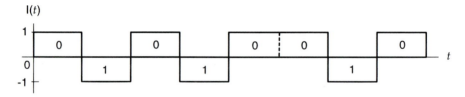

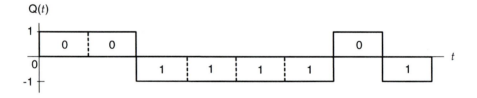

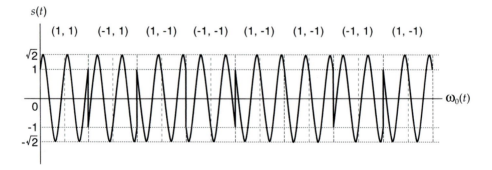

Figure 4 .9 The pilot channel QPSK waveform plot

4.3 SYNC CHANNEL

The sync channel is defined as the code channel 32 (W_{32}) in the forward CDMA channel which transports the synchronization message to the mobile station. The sync channel will operate at a fixed rate of 1200 bps if it is present. The sync channel is convolutionally encoded before transmission and each convolutionally encoded symbol is repeated one time, i.e., occurring 2 consecutive times prior to block interleaving. All symbols after repetition on the sync channel are block interleaved. The sync channel uses a block interleaver spanning 26.667 (=128/4800) msec which is equivalent to 128 modulation symbols at the symbol rate of 4800 sps. A sync channel frame is 26.667 ms in duration. The sync channel transmitted on the forward CDMA channel is spread with a Walsh function at a fixed chip rate of 1.2288 Mcps to provide orthogonal channelization among all code channels on a given forward CDMA channel.

In summary, the sync channel is an encoded, interleaved, spread, and modulated spread spectrum signal that is used by mobile stations to acquire initial time synchronization. The sync channel structure (1200 bps) is drawn as shown in Fig. 4.10.

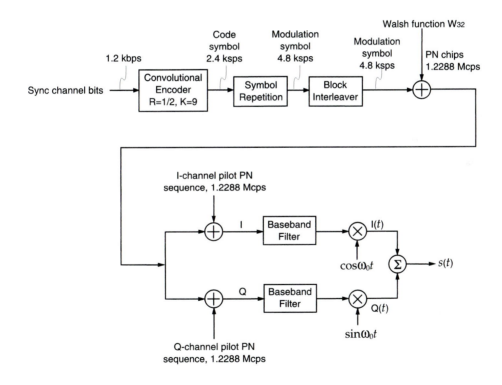

Figure 4.10 Sync channel structure (1200 bps)

The I and Q channel pilot sequences for the sync channel use the same pilot PN sequence offset as the pilot channel for a given base station. The sync channel is used during the system acquisition stage. Once the mobile station acquires the system, it will not normally reuse the sync channel until it powers on again. Once the mobile station achieves pilot PN sequence synchronization by acquiring the pilot channel, the synchronization for the sync channel is immediately known. This is because the sync channel is spread with the same pilot PN sequence, and because the frame and interleaver timing on the sync channel are aligned with the pilot PN sequence.

A sync channel superframe is formed by three sync channel frames (26.667 x 3 = 80ms). Messages transmitted on the sync channel begin only at the start of a sync channel superframe as shown in Fig. 4.11.

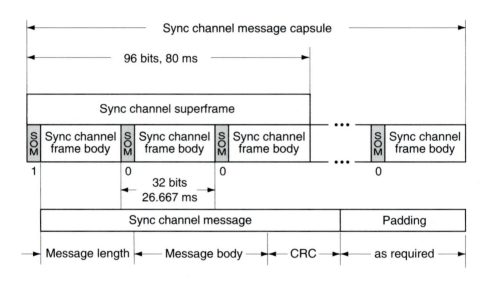

SOM: Start of Message bit

Figure 4.11 Sync channel superframe structure

Finally, the modulation parameters of a sync channel are listed in Table 4.4.

Table 4.4   Sync Channel Modulation Parameters

Parameters	Data Rate 1200 bps	Units
PN chip rate	1.2288	Mcps
Code rate	1/2	bits/code symbol
Code repetition	2	mod symbol/code symbol
Modulation symbol rate	4800	sps
PN chips/Modulation symbol	456	PN chips/mod.symbol
PN chips/bit	1024	PN chips/bit

4.3.1   Sync Channel Encoding

An (n, k, m) convolutional code designates the code rate $R=k/n$ with encoder stages of $m=K-1$, where K is the constraint length of the code. The sync channel is encoded by the (2, 1, 8) convolutional encoder with the code rate $R=1/2$ and a constraint length of K=9. Convolutional encoding involves the modulo-2 addition of selected taps of a serially time-delayed data sequence. Hence, the length of the data sequence delay is equal to K-1. The generator sequences of encoder are given by $g_1^{(1)} = 753$ (octal) = (111101011) (binary) and $g_1^{(2)} = 561$ (octal) = (101110001) (binary), respectively.

On the other hand, the generator polynomials also can be written as

$$g_1^{(1)}(x) = 1 + x + x^2 + x^3 + x^5 + x^7 + x^8 \qquad (4.10)$$
$$g_1^{(2)}(x) = 1 + x^2 + x^3 + x^4 + x^8 \qquad (4.11)$$

Since $k=1$, $n=2$, and $m=8$, this convolutional encoder consists of a single input terminal, an 8-stage shift register coupled with two modulo-2 adders, as illustrated in Fig. 4.12 (a) & (b). Since the code rate is $R=1/2$, two code symbols are generated at the output modulo-2 adders for each information bit to the input. These two code symbols will be output so that the code symbol c_0 encoded with the generator sequence $g_1^{(1)}$ is the first output and the code symbol c_1 encoded with the generator sequence $g_1^{(2)}$ is the second output. The initial state of the convolutional encoder is assumed to be all zero. The first code symbol output after initialization will be a code symbol encoded with the generator sequence $g_1^{(1)}$.

Example 4.4 Referring to the convolutional encoder shown in Fig. 4.12, two generator sequences are given as $g_1^{(1)} = (111101011)$ and $g_1^{(2)} = (101110001)$, respectively. If the 24-bit information sequence is assumed as $d = (1001110100\ 11100010100101)$, the 32-bit encoder output symbols c_0 and c_1 are computed from

$$c_0 = d_0 + d_1 + d_2 + d_3 + d_5 + d_7 + d_8 \qquad (4.12)$$
$$c_1 = d_0 + d_2 + d_3 + d_4 + d_8$$

Assuming the initial contents of the convolutional encoder is all zero, the code symbol c_0 encoded with $g_1^{(1)} = 753$ (octal) is the first output symbol; and the code symbol c_1 encoded with $g_1^{(2)} = 561$ (octal) is the second symbol.

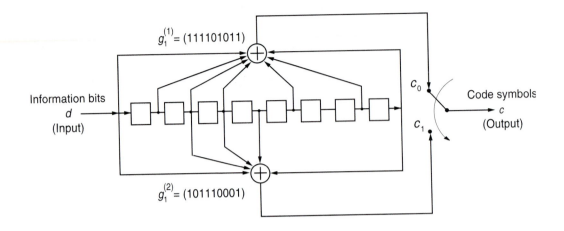

(a) Encoder based on the generator vectors

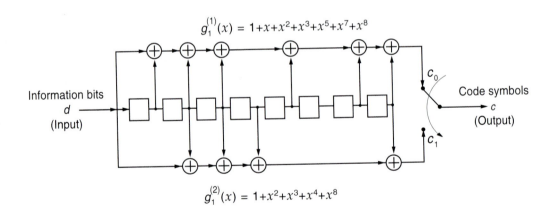

(b) Encoder based on the generator polynomials

Figure 4.12 The (2, 1, 8) convolutional encoder (a) based on $g_1^{(1)}$ and $g_1^{(2)}$;

(b) based on $g_1^{(1)}(x)$ and $g_1^{(2)}(x)$

Each code symbol sequence consists of 32 bits due to 48 bits by R=1/2 and additional 8 bits stored in the encoder memory.

When the information sequence reads in the encoder bit by bit, c_0 and c_1 can be computed as follows:

$c_0 = (1110001000\ 0000110100\ 0110101001\ 11)$

$c_1 = (1010000010\ 0111110010\ 1010000101\ 01)$

Multiplexing and pairwise concatenating $(c_0,\ c_1)$ results in a single output symbol sequence as shown below:

$c = (1110110000001000010000010101111$

$0010010001101100100010010011011)$

Alternative solutions by several analytical methods will be demonstrated in the following.

Example 4.5 Consider again the $(2, 1, 8)$ convolutional code with generator sequences $g_1^{(1)} = (111\ 101\ 011)$ and $g_1^{(2)} = (101\ 110\ 001)$.

Given the information sequence $d = (1001\ 1101\ 0011\ 1000\ 1010\ 0101)$, the identical output code sequence can be computed as follows:

1. Discrete Convolution Method

$c_0 = d * g_1^{(1)} = (1001\ 1101\ 0011\ 1000\ 1010\ 0101) * (111101011)$

$= (1110\ 0010\ 0000\ 0011\ 0100\ 0110\ 1010\ 0111)$

$c_1 = d * g_1^{(2)} = (1001\ 1101\ 0011\ 1000\ 1010\ 0101) * (101110001)$

$= (1010\ 0000\ 1001\ 1111\ 0010\ 1010\ 0001\ 0101)$

By multiplexing and pairwise concatenating (c_0, c_1), the code symbol word c becomes

$c = (11\ 10\ 11\ 00\ 00\ 00\ 10\ 00\ 01\ 00\ 00\ 01\ 01\ 01\ 11\ 11$

$00\ 10\ 01\ 00\ 01\ 10\ 11\ 00\ 10\ 00\ 10\ 01\ 00\ 11\ 10\ 11)$

2. Scalar Matrix Method

$$c = d \cdot G$$

where G is a semi-infinite generator matrix.

$$G = \begin{bmatrix} G_0\ G_1\ G_2\ \cdots\ G_m & & \\ & G_0\ G_1\ G_2\ \cdots\ G_m & \\ & & G_0\ G_1\ G_2\ \cdots\ G_m \\ & \cdot & \\ & \cdot & \cdot \\ \cdot & & \cdot \end{bmatrix}$$

Component matrix G_i, $0 \leq i \leq 8$, is obtained from $g_1^{(1)}$ and $g^{1(2)}$ as follows:

$G_0 = 11,\ G_1 = 10,\ G_2 = 11,\ G_3 = 11,\ G_4 = 01,\ G_5 = 10,\ G_6 = 00,\ G_7 = 10,\ G_8 = 11$

Thus, the code symbol word c is now computed as

c = $d \cdot$ G

   = (1001 1101 0011 1000 1010 0101)

$$
\begin{bmatrix}
11\ 10\ 11\ 11\ 01\ 10\ 00\ 10\ 11 \\
\ \ 11\ 10\ 11\ 11\ 01\ 10\ 00\ 10\ 11 \\
\ \ \ \ 11\ 10\ 11\ 11\ 01\ 10\ 00\ 10\ 11 \\
\ \ \ \ \ \ 11\ 10\ 11\ 11\ 01\ 10\ 00\ 10\ 11 \\
\ \ \ \ \ \ \ \ 11\ 10\ 11\ 11\ 01\ 10\ 00\ 10\ 11 \\
\ \ \ \ \ \ \ \ \ \ 11\ 10\ 11\ 11\ 01\ 10\ 00\ 10\ 11 \\
\ \ \ \ \ \ \ \ \ \ \ \ 11\ 10\ 11\ 11\ 01\ 10\ 00\ 10\ 11 \\
\ \ \ \ \ \ \ \ \ \ \ \ \ \ 11\ 10\ 11\ 11\ 01\ 10\ 00\ 11\ 11 \\
\ \ \ \ \ \ \ \ \ \ \ \ \ \ \ \ 11\ 10\ 11\ 11\ 01\ 10\ 00\ 11\ 11 \\
\ \ \ \ \ \ \ \ \ \ \ \ \ \ \ \ \ \ 11\ 10\ 11\ 11\ 01\ 10\ 00\ 11\ 11 \\
\ 11\ 10\ 11\ 11\ 01\ 10\ 00\ 11\ 11 \\
\ 11\ 10\ 11\ 11\ 01\ 10\ 00\ 11\ 11 \\
\ 11\ 10\ 11\ 11\ 01\ 10\ 00\ 11\ 11 \\
\ 11\ 10\ 11\ 11\ 01\ 10\ 00\ 11\ 11 \\
\ 11\ 10\ 11\ 11\ 01\ 10\ 00\ 11\ 11 \\
\ 11\ 10\ 11\ 11\ 01\ 10\ 00\ 11\ 11 \\
\ 11\ 10\ 11\ 11\ 01\ 10\ 00\ 11\ 11 \\
\ 11\ 10\ 11\ 11\ 01\ 10\ 00\ 11\ 11 \\
\ 11\ 10\ 11\ 11\ 01\ 10\ 00\ 11\ 11 \\
\ 11\ 10\ 11\ 11\ 01\ 10\ 00\ 11\ 11 \\
\ 11\ 10\ 11\ 11\ 01\ 10\ 00\ 11\ 11 \\
\ 11\ 10\ 11\ 11\ 01\ 10\ 00\ 11\ 11
\end{bmatrix}
$$

   = (11 10 11 00 00 00 10 00 01 00 00 01 01 01 11 11 00 10 01 00 01 10)

This is the code symbol word c as expected.

3. Polynomial Matrix Method

The information polynomial is represented by

$d(D) = D^{23} + D^{20} + D^{19} + D^{18} + D^{16} + D^{13} + D^{12} + D^{11} + D^{7} + D^{5} + D^{2} + 1$

The generator polynomial matrix is expressed as

$$G(d) = [g_1^{(1)}(D), g_1^{(2)}(D)]$$

$$= [D^8 + D^7 + D^6 + D^5 + D^3 + D + 1, D^8 + D^6 + D^5 + D^4 + 1]$$

The code symbol polynomial matrix at the modulo-2 adders is obtained as

$$c'(D) = \quad d(D)G(D) = \quad [C_0(D), C_1(D)]$$

$$= [D^{31} + D^{30} + D^{29} + D^{25} + D^{17} + D^{16} + D^{14} + D^{10} + D^9 + D^7 + D^5 + D^2 + D + 1,$$

$$D^{31} + D^{29} + D^{23} + D^{20} + D^{19} + D^{18} + D^{17} + D^{16} + D^{13} + D^{11} + D^9 + D^4 + D^2 + 1]$$

Thus, the delay operation by the commutating switch will produce the encoder output symbol sequence as follows:

$$c(D) = Dc_0(D^2) + c_1(D^2)$$
$$= D(D^{62} + D^{60} + D^{58} + D^{50} + D^{34} + D^{32} + D^{28} + D^{20} + D^{18} + D^{14} + D^{10} + D^4 + D^2 + 1)$$
$$+ (D^{62} + D^{58} + D^{46} + D^{40} + D^{38} + D^{36} + D^{34} + D^{32} + D^{26} + D^{22} + D^{18} + D^8 + D^4 + 1)$$
$$= D^{63} + D^{62} + D^{61} + D^{59} + D^{58} + D^{51} + D^{46} + D^{40} + D^{38} + D^{36} + D^{35} + D^{34} + D^{33}$$
$$+ D^{32} + D^{29} + D^{26} + D^{22} + D^{21} + D^{19} + D^{18} + D^{15} + D^{11} + D^8 + D^5 + D^4 + D^3 + D + 1$$
$$c = (11\ 10\ 11\ 00\ 00\ 00\ 10\ 00\ 01\ 00\ 00\ 01\ 01\ 01\ 11\ 11$$
$$00\ 10\ 01\ 00\ 01\ 10\ 11\ 00\ 10\ 00\ 10\ 01\ 00\ 11\ 10\ 11)$$

It is enough to prove that alternative solutions presented previously resulted in the identical sequences. It will be convenient for the readers to understand these alternative approaches.

4.3.2   Code Symbol Repetition

The transmission rate supporting the sync channel is fixed at 1200 bps. For the sync channel, each convolutionally encoded symbol will be repeated 1 time (each symbol occurs 2 consecutive times) before being interleaved, as shown in Table 4.5.

4.3.3   Block Interleaving Operation

All code symbols after repetition on the sync channel will be block interleaved. The purpose of using block interleaving is to avoid burst errors while sending the data through a multipath fading environment, and to provide the access redundancy for achieving performance improvement.

The sync channel is divided into 80 ms superframes. Since the sync channel frame consists of 32 bits and 26.667 (=32/1200) ms, each superframe is divided into three 26.667 ms frames. The sync channel will use a block interleaver spanning 26.667 ms which is equivalent to 128 (=16 x 8) modulation symbols at the symbol rate of 4800 sps. Notice that each repetition of a code symbol is a modulation symbol.

The input and output symbol sequences to the sync channel interleaver are given in Tables 4.5 and 4.6. The input symbol table (Table 4.5) is read down by columns from the left to the right. The first input symbol (1) is at the top left, the second input symbol (1) is just below the first input symbol, the 17th input symbol (9) is just to the right of the first input symbol, and the 18th input symbol (9) is just to the right of the second input symbol. In this input symbol table, symbols with the same number denote repeated code symbols. The output symbol table (Table

4.6) is read the same way as the input symbol table (Table 4.5). That is, the first output symbol (1) is at the top left, the second output symbol (33) is just below the first output symbol, and the 17th output symbol (3) is just to the right of the first output symbol.

Table 4.5 Input Symbol Array to the Sync Channel Interleaver (TIA/EIA/IS-95)

1	9	17	25	33	41	49	57
1	9	17	25	33	41	49	57
2	10	18	26	34	42	50	58
2	10	18	26	34	42	50	58
3	11	19	27	35	43	51	59
3	11	19	27	35	43	51	59
4	12	20	28	36	44	52	60
4	12	20	28	36	44	52	60
5	13	21	29	37	45	53	61
5	13	21	29	37	45	53	61
6	14	22	30	38	46	54	62
6	14	22	30	38	46	54	62
7	15	23	31	39	47	55	63
7	15	23	31	39	47	55	63
8	16	24	32	40	48	56	64
8	16	24	32	40	48	56	64

Table 4.6 Output Symbol Array from the Sync Channel Interleaver (TIA/EIA/IS-95)

1	3	2	4	1	3	2	4
33	35	34	36	33	35	34	36
17	19	18	20	17	19	18	20
49	51	50	52	49	51	50	52
9	11	10	12	9	11	10	12

Table 4.6 Output Symbol Array from the Sync Channel Interleaver
(TIA/EIA/IS-95) *(continued)*

41	43	42	44	41	43	42	44
25	27	26	28	25	27	26	28
57	59	58	60	57	59	58	60
5	7	6	8	5	7	6	8
37	39	38	40	37	39	38	40
21	23	22	24	21	23	22	24
53	55	54	56	53	55	54	56
13	15	14	16	13	15	14	16
45	47	46	48	45	47	46	48
29	31	30	32	29	31	30	32
61	63	62	64	61	63	62	64

Example 4.6 The input sequence to the sync channel interleaver consists of the convolutionally encoded repeated symbols. Using Table 4.5, all code symbols after repetition will be the 128-bit symbol sequence shown below:

11111100111100000000000011000000001100000000000110011001111111111
00001100001100000011110011110000110000001100011000011111001111

Example 4.7 Consider the computation of the output symbol sequence from the sync channel interleaver. For convenience, the convolutionally encoded symbols are arranged as follows:

1110110000001000
0100000101011111
0010010001101100
1000100100111011

The output symbols are found by reading down by columns from the left to the right, as demonstrated in the following. As an example, let us show only the first and second columns in Table 4.6.

Interleaver Output Order	Corresponding Symbols
1, 33, 17, 49, 9, 41, 25, 57, 5, 37, 21, 53, 13, 45, 29, 61	1001000010011111
3, 35, 19, 51, 11, 43, 27, 59 7, 39, 23, 55, 15, 47, 31, 63	1100010100000011

Using Table 4.6, the interleaver output symbols corresponding to the third column through the eighth column can be similarly computed. Thus, the complete list of the interleaver output symbol sequence can be shown as follows:

Column No.	Interleaver Output
1	1001000010011111
2	1100010100000011
3	1010011011000110
4	0000001100110011
5	1001000010011111
6	1100010100000011
7	1010011011000110
8	0000001100110011

4.3.4   Orthogonal Spreading by Walsh Function

The sync channel transmitted on the forward CDMA channel is spread with a Walsh function at a fixed chip rate of 1.2288 Mcps.

The sync channel that is spread using the Walsh function 32 (which is one of sixty-four time-orthogonal Walsh functions) will be assigned to code channel number 32 (W_{32}).

The Walsh function index n for the sync channel is designated by $n=32$ either in the column heading or in the row heading of Table 4.7. The Walsh function spreading sequence repeats with a period of 52.083 μs (= 64/1.2288 Mcps).

The Walsh chip within a Walsh index 32 is
00000000 00000000 00000000 00000000 11111111 11111111 11111111 11111111

Since the interleaver output consists of 128 bits, the Walsh chip within a Walsh function index 32 will be used twice. Due to the fact that the modulation symbol rate is 4800 sps and PN chip rate is 1.2288 Mcps, the ratio of these two rates is 256, which corresponds to 256 Walsh chips with respect to 1 interleaver output symbol.

Thus, EX-ORing with the Walsh function index 32 will produce the orthogonal modulator output as follows:

    (a) 11 ⋯ 1 11 ⋯ 1 11 ⋯ 1 11 ⋯ 1 11 ⋯ 1 11 ⋯ 1 11 ⋯ 1 11 ⋯ 1
    (b) 00 ⋯ 0 11 ⋯ 1 00 ⋯ 0 11 ⋯ 1 00 ⋯ 0 11 ⋯ 1 00 ⋯ 0 11 ⋯ 1
    (c) 11 ⋯ 1 00 ⋯ 0 11 ⋯ 1 00 ⋯ 0 11 ⋯ 1 00 ⋯ 0 11 ⋯ 1 00 ⋯ 0

where(a): The first symbol 1 of interleaver output, repeated 256 times, for EX-ORing with 256 Walsh chips.

(b): Walsh chips (64 x 4 = 256 bits) corresponding to 4 times of Walsh index 32.

(c): Modulator output by EX-ORing (a) with (b), i.e., (c) = (a) $\oplus$ (b).

Table 4.7   64-ary Walsh Functions

Walsh Chip within a Walsh Index
1111111111222222222233333333334444444444555555555566666
0123456789012345678901234567890123456789012345678901234567890123

0	0000000000000000000000000000000000000000000000000000000000000000
1	0101010101010101010101010101010101010101010101010101010101010101
2	0011001100110011001100110011001100110011001100110011001100110011
3	0110011001100110011001100110011001100110011001100110011001100110
4	0000111100001111000011110000111100001111000011110000111100001111
5	0101101001011010010110100101101001011010010110100101101001011010
6	0011110000111100001111000011110000111100001111000011110000111100
7	0110100101101001011010010110100101101001011010010110100101101001
8	0000000011111111000000001111111100000000111111110000000011111111
9	0101010110101010010101011010101001010101101010100101010110101010
10	0011001111001100001100111100110000110011110011000011001111001100
11	0110011010011001011001101001100101100110100110010110011010011001
12	0000111111110000000011111111000000001111111100000000111111110000
13	0101101010100101010110101010010101011010101001010101101010100101
14	0011110011000011001111001100001100111100110000110011110011000011
15	0110100110010110011010011001011001101001100101100110100110010110
16	0000000000000000111111111111111100000000000000001111111111111111
17	0101010101010101101010101010101001010101010101011010101010101010
18	0011001100110011110011001100110000110011001100111100110011001100
19	0110011001100110100110011001100101100110011001101001100110011001
20	0000111100001111111100001111000000001111000011111111000011110000
21	0101101001011010100101101001010101011010010110101010101001011010...
22	0011110000111100110000111100001100111100001111001100001111000011
23	0110100101101001100101101001010110011010010110100110010110010110
24	0000000011111111111111110000000000000000111111111111111100000000
25	0101010110101010101010100101010101010101101010101010101001010101
26	0011001111001100110011000011001100110011001111001100110010000110011
27	0110011010011001100110010110011001100110100110011001100100101100110
28	0000111111110000111100000000111100001111111100001111000000001111
29	0101101010100101101001010101101001011010101001010110010101011010

Table 4.7   64-ary Walsh Functions *(continued)*

Walsh Chip within a Walsh Index
30 0011110011000011110000110011110000111100110000111100001100111100
31 0110100110010110100101100110100101101001100101101001011001101001
32 0000000000000000000000000000000011111111111111111111111111111111
33 0101010101010101010101010101010110101010101010101010101010101010
34 0011001100110011001100110011110011001100110011001100110011001100
35 0110011001100110011001100110110100110011001100110011001100110011
36 0000111100001111000011110000111111110000111100001111000011110000
37 0101101001011010010110100101101010100101101001011010010110100101
38 0011110000111100001111000011110011000011110000111100001111000011
39 0110100101101001011010010110100110010110100101101001011010010110
40 0000000011111111000000001111111111111111000000001111111100000000
41 0101010110101010010101011010101010101010010101011010101001010101
42 0011001111001100011001111001100110011000011001111001100001100011
43 0110011010011001011001101001100110011001010110011010011001011001
44 0000111111110000000011111111000011110000000011111111000000001111
45 0101101010010101011010101001011010100101010110101010010101011010
46 0011110011000011001110011000011110000110011100110000110011011100
47 0110100110010110010110011001011010010110100110010110011001101001
48 0000000000000000111111111111111111111111111111110000000000000000
49 0101010101010101101010101010101010101010101010100101010101010101
50 0011001100110011110011001100110011001100110010000110011001100011
51 0110011001101001100110011001100110011001100110011001100110011001
52 0000111100001111111100000111000011100001111000000001111000001111
53 0101101001011010100101101001011010010110100101011010010101011010
54 0011110000111100110000111100001111000011110000110011110000111100
55 0110100101101001100101101001011010010110100101100110100101101001
56 0000000011111111111111110000000011111111000000000000000011111111
57 0101010110101010101010100101010110101010010101011010101101101010
58 0011001111001100110010000110011110011000011001100110011001111001100
59 0110011010011001100110010110011010011001101011001100110011010011001
60 0000111111110000111100000000111111110000000001111000011111110000
61 0101101010010110100101010110101010010101011010010110101010100101
62 0011110011000011110000110011110011000011001111000011110011000011
63 0110100110010110100101100110100110010110011010010110100110010110

Thus, the symbol cover by means of the orthogonal modulator will be computed as

11111111 11111111 11111111 11111111 00000000 00000000 00000000 00000000
11111111 11111111 11111111 11111111 00000000 00000000 00000000 00000000
11111111 11111111 11111111 11111111 00000000 00000000 00000000 00000000
11111111 11111111 11111111 11111111 00000000 00000000 00000000 00000000

Following the orthogonal covering, the sync channel will be spread in quadrature as shown in Fig. 4.10.

4.3.5　Quadrature Spreading

The sync channel is quadrature spread by the I and Q pilot PN sequences as specified in Section 4.2.2. These pilot PN sequences, P_I and P_Q, will be computed from the LFSR binary sequences $i(n)$ and $q(n)$ as shown in the following example.

Example 4.8 Using $i(n)$, $q(n)$ and the initial contents of LFSR, i.e., IC = (1000000000000000), the in-phase pilot PN sequence P_I and the quadrature-phase pilot PN sequence P_Q can be computed as follows:

(I)　Using $i(n) = i(n-15) = 0$ for $1 \leq n \leq 14$, $i(15)=1$, and $i(n) = \sum_{K=15,10,8,7,6,2} i(n-k)$,

16 $\leq n \leq$ 32767, we can compute $i(n)$ from which the in-phase pilot PN sequence P_I is obtained by inserting a 0 (zero) in $i(n)$ after 14 consecutive 0 outputs (see arrow $\downarrow$) as shown below.

Insert 0

$\downarrow$

P_I =　00000000000000010101001001110100011011110011001000
　　　　00111100001000110100101011010101010111010010011111

Notice that the computation is actually executed over $0 \leq n \leq 99$.

(II) Similarly, using $q(n) = q(n-15) = 0$ for $1 \leq n \leq 14$, $q(15)=1$, and $q(n) = \sum_{k=15,12,11,10,9,5,4,3} q(n-k)$,

for 16 $\leq n \leq$ 32767, we can compute $q(n)$ and subsequently the quadrature-phase pilot PN sequence P_Q is obtained by inserting a 0 (zero) at the 15th place in $q(n)$ as seen below.

Insert 0

$\downarrow$

P_Q =　00000000000000010011110101110101101001110001010011
　　　　10011100011011000111010011000100100100011001100011

Example 4.9　The in-phase sequence I and the quadrature-phase sequence Q are computed by the following spreading operation:

　$I = OSC \oplus P_I$ and $Q = OSC \oplus P_Q$

where OSC denotes the orthogonal symbol covering.

```
OSC:    11111111 11111111 11111111 11111111 00000000 · · ·
  ⊕
 𝒫ᵢ     00000000 00000001 01010010 01110100 01101111 · · ·

 I:     11111111 11111110 10101101 10001011 01101111 · · ·
```

and

```
OSC:    11111111 11111111 11111111 11111111 00000000 · · ·
  ⊕
 𝒫_Q    00000000 00000001 00111101 01110101 10100111 · · ·

 Q:     11111111 11111110 11000010 10001010 10100111 · · ·
```

Following the spreading operation, the I and Q impulses are applied to the inputs of the I and Q baseband filters as shown in Fig. 4.10.

4.3.6   Sync Channel Filtering

The sync channel will be filtered as specified in Section 4.2.3. The baseband filters have a frequency response $S(f)$ that satisfies the limit given in Fig. 3.15.

Specifically, the normalized frequency response of the filter is contained within $\pm\delta_1$ in the passband $0 \leq f \leq f_p$ and is less than or equal to $-\delta_2$ in the stopband $f \geq f_S$. The numerical values for these parameters are $\delta_1 = 1.5$ dB, $\delta_2 = 40$ dB, $f_P = 590$ KHz, and $f_S = 740$ KHz.

4.3.7   Quadrature Phase Shift Keying (QPSK)

After baseband filtering, the in-phase and quadrature-phase sequences, $I(t)$ and $Q(t)$, are modulated onto the cosine and sine function of a carrier wave and formed into two respective BPSK waveforms. These two BPSK waveforms will coincide in time and the carrier phase will change only once every T_b.

As seen from Fig. 4.10, the QPSK waveform $s(t)$ will be generated from the summation of these two orthogonal BPSK waveforms $I(t)$ and $Q(t)$. The standard QPSK for any forward code channel was fully discussed in Section 4.2.4.

Example 4.10   As computed previously, the in-phase sequence I and the quadrature-phase sequence Q are, respectively

I:	11111111	11111110	10101101	10001011	01101111
Q:	11111111	11111110	11000010	10001010	10100111

Let us consider a partial pair of two sequences, $I(t) = 10101101$ and $Q(t) = 11000010$. The QPSK waveform $s(t)$ coupled with $I(t)$ and $Q(t)$ are depicted in Fig. 4.13.

4.3.8   Sync Channel Signaling and Message Structure

This section specifies requirements for the signaling message format transmitted on the sync channel.

The sync channel is divided into 80 ms, 96 bits superframes. Each superframe is divided into three 26.667 (=32/1200 bps) ms frames because the sync channel operates at a fixed rate of 1200 bps. The first bit of each frame is a Start-of-Message bit (SOM bit), and the remaining bits in the frame comprise the sync channel frame body.

Sync channel(1200 bps)
          I-channel output: 10101101 (-1, 1,-1, 1,-1,-1, 1,-1)
          Q-channel output: 11000010 (-1,-1, 1, 1, 1, 1,-1, 1)

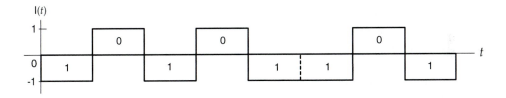

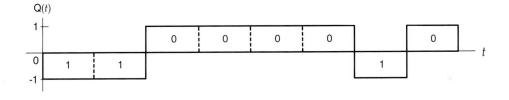

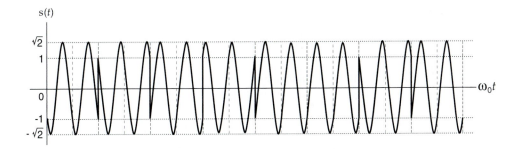

Figure 4 .13 The sync channel QPSK waveform for $s(t)$ plot

A sync channel message capsule is composed of a sync channel message and padding. A sync channel message consists of an 8-bit message length field, a message body, and a 30-bit cyclic redundancy check (CRC) field. Padding bits are set to zero (0) and appended to the end of the sync channel message. A typical example of sync channel structure (1200 bps) is depicted in Fig. 4.14.

Sync channel message capsules begin with the first bit of the first sync channel frame body of a sync channel superframe. The base station will set the SOM bit immediately preceding the beginning of a sync channel message capsule to '1' and set all other SOM bits to '0' as shown in Fig. 4.14.

The base station transmits the sync channel message in consecutive sync channel frame bodies. The base station should include sufficient padding bits in each sync channel message capsule to extend it through the bit preceding the SOM bit at the beginning of the next sync channel superframe. The base station will begin a new sync channel message capsule in the first sync channel frame of that superframe. The base station limits the maximum sync channel message length to 148 octets, that is 148 x 8=1184 bits.

4.3.9   Cyclic Redundancy Check (CRC) for Sync Channel

A 30-bit CRC will be computed for each sync channel signaling message which includes the 8-bit message length field, the message body field, and the 30-bit CRC.

The generator polynomial for the CRC computation is given as

$$g(x) = 1 + x + x^2 + x^6 + x^7 + x^8 + x^{11} + x^{12} + x^{13} + x^{15} + x^{20} + x^{21} + x^{29} + x^{30} \quad (4.13)$$

whose logic diagram is depicted in Fig. 4.15.

The following procedure is used to compute the CRC bits and its encoder logic is shown in Fig. 4.16:

1. Initial contents of the LFSR are set to binary one (1) in order to make the CRC field to nonzero values even for all-zero data.
2. Set the switch in the down position and close the gate in the feedback path of the CRC encoder.
3. Then the k input bits begin to transmit the encoder output as well as the feedback path of the shift register.
   The k-bit input includes the 8-bit message length + the message body length in bits.
4. The register will be clocked k times with the k-bit input.
5. Set the switch to the up position and open the gate in the feedback path.
6. Clocking the register by an additional 30 times, the CRC field will be generated.
7. Thus, the sync channel message is transmitted in the order of the 8-bit message length field, the message body length, and the 30-bit CRC field at the CRC encoder output.

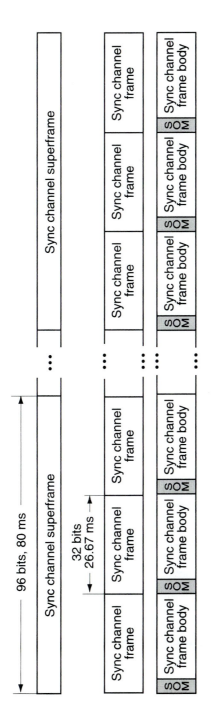

(a) Sync channel message capsule

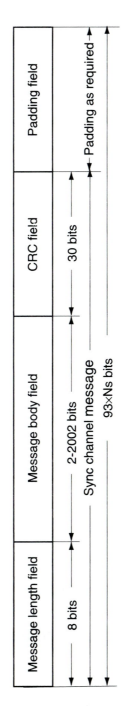

(b) Sync channel message + Padding

SOM: Start of Message bit   Ns: Number of Sync channel superframes needed for message transmission

Figure 4.14 A typical example of the 1200 bps sync channel structure

175

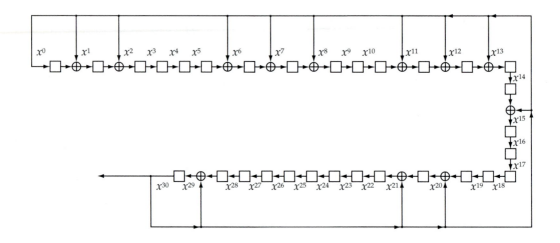

Figure 4.15 LFSR logic diagram representing the generator polynomial
$$g(x) = 1+x+x^2+x^6+x^7+x^8+x^{11}+x^{12}+x^{13}+x^{15}+x^{21}+x^{29}+x^{30}$$

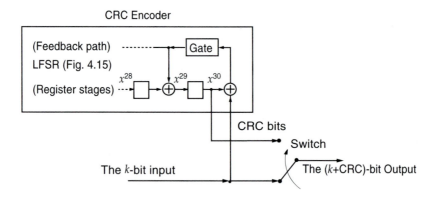

k = Message length field + Message body length
    = 8+(2 to 2002) in bits

Figure 4.16 CRC encoder for the sync channel

4.4 PAGING CHANNEL

The paging channel is an encoded, interleaved, spread, and modulated spread spectrum signal that is used by mobile stations operating within the coverage area of the base station. The base station uses the paging channel to transmit system overhead information and mobile station specific messages. The paging channel transmits information at a fixed data rate of 9600 or 4800 bps and its frame is 20 ms in duration. The paging channel structure for the data rate of either 9.6 kbps or 4.8 kbps is shown in Fig. 4.18.

The pilot PN sequences for the paging channel use the same pilot PN sequence offset as the pilot channel for a given base station.

The first paging channel frame will occur at the start of base station transmission time. The paging channel is divided into paging channel slots that are each 80 ms in duration.

For the nonslotted mode of operation, paging and control messages for a mobile station can be received in any of the paging channel slots. For the slotted mode of operation, a mobile station monitors the paging channel only during certain assigned slots. Figure 4.17 shows an example for a slot length of 1.28 seconds, in which the computed value of a paging slot is equal to 6. Consequently, the mobile station's slot cycle begins when the slot number equals 6. The mobile station will begin monitoring the paging channel at the start of the slot number 6. The minimum length slot cycle consists of 16 slots of 80 ms each as shown in Fig. 4.17, whence 1.28 (=16 x 80 ms) seconds.

The paging channel slot number is determined as

$$\text{Slot No.} = \lfloor t/4 \rfloor \bmod 2048$$

where t is the system time in frames and mod 2048 represents the maximum length slot cycle (2048 slots).

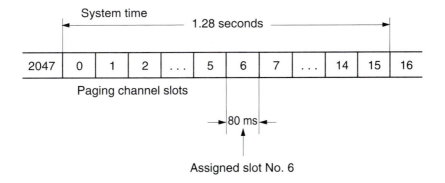

Figure 4.17 Mobile station's slotted mode operation

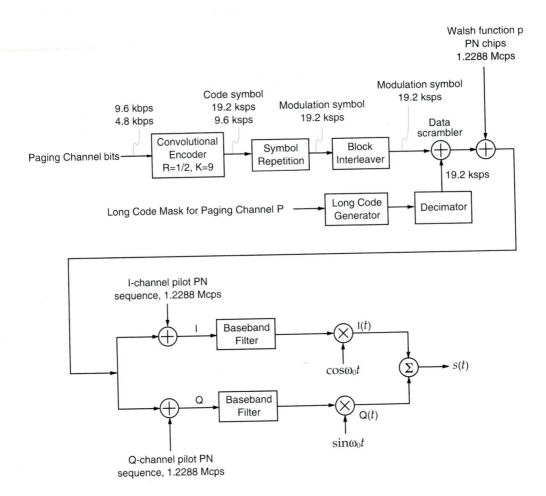

Figure 4.18 Paging channel structure for the data rate of either 9600 bps or 4800 bps

The modulation parameters of a paging channel are listed in Table 4.8.

Table 4.8   Paging Channel Modulation Parameters

Parameters	Data Rate (bps) 9600	4800	units
PN chip rate	1.2288	1.2288	Mcps
Code rate	1/2	1/2	bits/code symbols

Table 4.8   Paging Channel Modulation Parameters *(continued)*

Code repetition	1	2	mod symbol/code symbol
Modulation symbol rate	19,200	19,200	sps
PN chips/Modulation symbol	64	64	PN chips/mod symbol
PN chips/bit	128	256	PN chips/bit

4.4.1   Paging Channel Encoding

The paging channel data is convolutionally encoded prior to transmission as specified in Section 4.3.1. The (2, 1, 8) convolutional encoder is equally applicable to the paging, sync, and forward traffic channels. The same encoder structure as shown in Fig. 4.12 will be used for the paging channel. The generator sequences of convolutional encoder for the paging channel are given by $g_1^{(1)}$ =(111101011) and $g_1^{(2)}$ = (101110001). Since the convolutional encoding involves the modulo-2 addition of selected taps of a serially time-delayed data sequence, the generator polynomials also can be expressed as

$$g_1^{(1)}(x) = 1 + x + x^2 + x^3 + x^5 + x^7 + x^8$$

and $\qquad g_1^{(2)}(x) = 1 + x^2 + x^3 + x^4 + x^8$

Since the code rate is 1/2, two encoded symbols corresponding to each input bit to the encoder will be output of the code symbol c_0 encoded with $g_1^{(1)}$ and the code symbol c_1 encoded with $g_1^{(2)}$, respectively.

The first code symbol of the convolutional encoder output after initialization will be a code symbol encoded with generator sequence $g_1^{(1)}$.

Example 4.11 Consider the (2, 1, 8) convolutional encoder with generator functions $g_1^{(1)} =$ (111101011) and $g_1^{(2)} = $ (101110001) which was shown in Fig. 4.12. Choose the paging channel operating at a fixed rate of 4800 bps.

Given the 96-bit information sequence

d = (1100111010 0011011000 0011100011 0100101000 1010010111
    0101010011 1100101001 0110011101 0111000010 110111)

and the all-zero initial contents of the encoder, the code symbols c_0 and c_1 can be computed using the following encoding equations:

$$c_0 = d_0 + d_1 + d_2 + d_3 + d_5 + d_7 + d_8$$
$$c_1 = d_0 + d_2 + d_3 + d_4 + d_8$$

The 96-bit code symbol sequences c_0 and c_1 encoded with $g_1^{(1)}$ and $g_1^{(2)}$ are, respectively

c_0 = (1000010010 0111100110 1000010000 0110101101 1111011000
    1100001001 1100111010 1001000101 0111011011 000011)

c_1 = (1110100011 0110010000 0110101001 0010011000 0011101011
    1000100101 0010111011 1010110101 1100100011 110111)

Multiplexing and pairwise concatenating (c_0, c_1) results in a 192-bit encoder output symbol sequence c as shown below:

c = (1101010001100000
      1101001111101001
      0010100010010100
      0110010000010010
      1100100111100010
      1010111101101100
      0101111000000100
      1001001110100100
      1111110011011100
      0110010100110011
      0111101001101000
      1111010100011111)

c is arranged 16 bits in a row for up to total 12 rows in order to easily utilize Table 4.10 for the paging channel interleaver output array.

4.4.2   Symbol Repetition and Block Interleaving

For the paging channel, each convolutionally encoded symbol will be repeated prior to block interleaving whenever the information rate is lower than 9600 bps. Each code symbol at the 4800 bps rate is repeated 1 time, i.e., sent 2 consecutive times per symbol. For the data rate of 4800 bps, this repetition process results in a constant modulation symbol rate of 19.2 ksps.

All symbols after repetition on the paging channel are block interleaved. The paging and forward traffic channels will use the identical block interleaver spanning 20 ms equivalent to 384 modulation symbols at the modulation symbol rate of 19.2 ksps. The input and output symbol sequences for the two data rates at 4800 bps and 9600 bps are given in Tables 4.9, 4.10, 4.11, and 4.12. These tables are read down by columns from the left to the right as with the sync channel interleaver.

Using Table 4.12, the 384-bit interleaver output can be computed as follows:

1010101011010110 1010010001111100 1000000011101001 0100110110111010
0110111000010110 1100001000010011 1011111100111010 1110100010011001
1011010110001101 1001111001010100 1001001100100010 1111000000100011
1010101011010110 1010010001111100 1000000011101001 0100110110111010
0110111000010110 1100001000010011 1011111100111010 1110100010011001
1011010110001101 1001111001010100 1001001100100010 1111000000100011

In this symbol sequence, symbols with the same number denote repeated code symbols.

For the paging channel at the 9600 bps rate, each convolutionally encoded symbol must not be repeated before block interleaving. The input and output symbol sequences for the 9600 bps rate interleaving are given in Tables 4.11 and 4.12. The paging channel at the 9600 bps rate also will use the identical block interleaver spanning 20 ms equivalent to 384 modulation symbols at the modulation symbol rate of 19.2 ksps.

To exchange the order of the symbol repeater and block interleaver in the CDMA code channels is really immaterial. In other words, each convolutionally encoded symbol may be block interleaved prior to symbol repetition.

Table 4.9   Paging and Forward Traffic Channel Interleaver Input Operating at 4800 bps (TIA/EIA/IS-95)

1	13	25	37	49	61	73	85	97	109	121	133	145	157	169	181
1	13	25	37	49	61	73	85	97	109	121	133	145	157	169	181
2	14	26	38	50	62	74	86	98	110	122	134	146	158	170	182
2	14	26	38	50	62	74	86	98	110	122	134	146	158	170	182
3	15	27	39	51	63	75	87	99	111	123	135	147	159	171	183
3	15	27	39	51	63	75	87	99	111	123	135	147	159	171	183
4	16	28	40	52	64	76	88	100	112	124	136	148	160	172	184
4	16	28	40	52	64	76	88	100	112	124	136	148	160	172	184
5	17	29	41	53	65	77	89	101	113	125	137	149	161	173	185
5	17	29	41	53	65	77	89	101	113	125	137	149	161	173	185
6	18	30	42	54	66	78	90	102	114	126	138	150	162	174	186
6	18	30	42	54	66	78	90	102	114	126	138	150	162	174	186
7	19	31	43	55	67	79	91	103	115	127	139	151	163	175	187
7	19	31	43	55	67	79	91	103	115	127	139	151	163	175	187
8	20	32	44	56	68	80	92	104	116	128	140	152	164	176	188
8	20	32	44	56	68	80	92	104	116	128	140	152	164	176	188
9	21	33	45	57	69	80	93	105	117	129	141	153	165	177	189
9	21	33	45	57	69	81	93	105	117	129	141	153	165	177	189
10	22	34	46	58	70	82	94	106	118	130	142	154	166	178	190
10	22	34	46	58	70	82	94	106	118	130	142	154	166	178	190
11	23	35	47	59	71	83	95	107	119	131	143	155	167	179	191
11	23	35	47	59	71	83	95	107	119	131	143	155	167	179	191
12	24	36	48	60	72	84	96	108	120	132	144	156	168	180	192
12	24	36	48	60	72	84	96	108	120	132	144	156	168	180	192

Table 4.10   Paging and Forward Traffic Channel Interleaver Output Operating at 4800 bps (TIA/EIA/IS-95)

1	5	3	7	2	6	4	8	1	5	3	7	2	6	4	8
33	37	35	39	34	38	36	40	33	37	35	39	34	38	36	40
65	69	67	71	66	70	68	72	65	69	67	71	66	70	68	72
97	101	99	103	98	102	100	104	97	101	99	103	98	102	100	104
129	133	131	135	130	134	132	136	129	133	131	135	130	134	132	136
161	165	163	167	162	166	164	168	161	165	163	167	162	166	164	168
17	21	19	23	18	22	20	24	17	21	19	23	18	22	20	24
49	53	51	55	50	54	52	56	49	53	51	55	50	54	52	56
81	85	83	87	82	86	84	88	81	85	83	87	82	86	84	88
113	117	115	119	114	118	116	120	113	117	115	119	114	118	116	120
145	149	147	151	146	150	148	152	145	149	147	151	146	150	148	152
177	181	179	183	178	182	180	184	177	181	179	183	178	182	180	184
9	13	11	15	10	14	12	16	9	13	11	15	10	14	12	16
41	45	43	47	42	46	44	48	41	45	43	47	42	46	44	48
73	77	75	79	74	78	76	80	73	77	75	79	74	78	76	80
105	109	107	111	106	110	108	112	105	109	107	111	106	110	108	112
137	141	139	143	138	142	140	144	137	141	139	143	138	142	140	144
169	173	171	175	170	174	172	176	169	173	171	175	170	174	172	176
25	29	27	31	26	30	28	32	25	29	27	31	26	30	28	32
57	61	59	63	58	62	60	64	57	61	59	63	58	62	60	64
89	93	91	95	90	94	92	96	89	93	91	95	90	94	92	96
121	125	123	127	122	126	124	128	121	125	123	127	122	126	124	128
153	157	155	159	154	158	156	160	153	157	155	159	154	158	156	160
185	189	187	191	186	190	188	192	185	189	187	191	186	190	188	192

Table 4.11 Paging and Forward Traffic Channel Interleaver Input Operating at 9600 bps (TIA/EIA/IS-95)

1	25	49	73	97	121	145	169	193	217	241	265	289	313	337	361
2	26	50	74	98	122	146	170	194	218	242	266	290	314	338	362
3	27	51	75	99	123	147	171	195	219	243	267	291	315	339	363
4	28	52	76	100	124	148	172	196	220	244	268	292	316	340	364
5	29	53	77	101	125	149	173	197	221	245	269	293	317	341	365
6	30	54	78	102	126	150	174	198	222	246	270	294	318	342	366
7	31	55	79	103	127	151	175	199	223	247	271	295	319	343	367
8	32	56	80	104	128	152	176	200	224	248	272	296	320	344	368
9	33	57	81	105	129	153	177	201	225	249	273	297	321	345	369
10	34	58	82	106	130	154	178	202	226	250	274	298	322	346	370
11	35	59	83	107	131	155	179	203	227	251	275	299	323	347	371
12	36	60	84	108	132	156	180	204	228	252	276	300	324	348	372
13	37	61	85	109	133	157	181	205	229	253	277	301	325	349	373
14	38	62	86	110	134	158	182	206	230	254	278	302	326	350	374
15	39	63	87	111	135	159	183	207	231	255	279	303	327	351	375
16	40	64	88	112	136	160	184	208	232	256	280	304	328	352	376
17	41	65	89	113	137	161	185	209	233	257	281	305	329	355	379
18	42	66	90	114	138	162	186	210	234	258	282	306	330	354	378
19	43	67	91	115	139	163	187	211	235	259	283	307	331	355	379
20	44	68	92	116	140	164	188	212	236	260	284	308	332	356	380
21	45	69	93	117	141	165	189	213	237	261	285	309	333	357	381
22	46	70	94	118	142	166	190	214	238	262	286	310	334	358	382
23	47	71	95	119	143	167	191	215	239	263	287	311	335	359	383
24	48	72	96	120	144	168	192	216	240	264	288	312	336	360	384

Table 4.12   Paging and Forward Traffic Channel Interleaver Output Operating at 9600 bps (TIA/EIA/IS-95)

1	9	5	13	3	11	7	15	2	10	6	14	4	12	8	16
65	73	69	77	67	75	71	79	66	74	70	78	68	76	72	80
129	137	133	141	131	139	135	143	130	138	134	142	132	140	136	144
193	201	197	205	195	203	199	207	194	202	198	206	196	204	200	208
257	265	261	269	259	267	263	271	258	266	262	270	260	268	264	272
321	329	325	333	323	331	327	335	322	330	326	334	324	332	328	336
33	41	37	45	35	43	39	47	34	42	38	46	36	44	40	48
97	105	101	109	99	107	103	111	98	106	102	110	100	108	104	112
161	169	165	173	163	171	167	175	162	170	166	174	164	172	168	176
225	233	229	237	227	235	231	239	226	234	230	238	228	236	232	240
289	297	293	301	291	299	295	303	290	298	294	302	292	300	296	304
353	361	357	365	355	363	359	367	354	362	358	366	356	364	360	368
17	25	21	29	19	27	23	31	18	26	22	30	20	28	24	32
81	89	85	93	83	91	87	95	82	90	86	94	84	92	88	96
145	153	149	157	147	155	151	159	146	154	150	158	148	156	152	160
209	217	213	221	211	219	215	223	210	218	214	222	212	220	216	224
273	281	277	285	275	283	279	287	274	282	278	286	276	284	280	288
337	345	341	349	339	347	343	351	338	346	342	350	340	348	344	352
49	57	53	61	51	59	55	63	50	58	54	62	52	60	56	64
113	121	117	125	115	123	119	127	114	122	118	126	116	124	120	128
177	185	181	189	179	187	183	191	178	186	182	190	180	188	184	192
241	249	245	253	243	251	247	255	242	250	246	254	244	252	248	256
305	313	309	317	307	315	311	319	306	314	310	318	308	316	312	320
369	377	373	381	371	379	375	383	370	378	374	382	372	380	376	384

4.4.3 Paging Channel Data Scrambling

Data scrambling applies to the paging channel as well as the forward traffic channel. Data scrambling is performed on the block interleaver output at the modulation symbol rate of 19.2 ksps.

Referring to Fig. 4.18, data scrambling is accomplished by performing the modulo-2 sum of the interleaver output with the decimated binary value of the long code PN chips. The long code is periodic with period $2^{42}-1$ chips specified by the LFSR tap polynomial $p(x)$ of the code generator:

$$p(x) = 1 + x + x^2 + x^3 + x^5 + x^6 + x^7 + x^{10} + x^{16} + x^{17} + x^{19} + x^{21} + x^{22}$$
$$+ x^{25} + x^{26} + x^{27} + x^{31} + x^{33} + x^{35} + x^{42}$$

(4.14)

Each PN chip of the long code will be generated by the modulo-2 inner product of a 42-bit mask and the 42-bit state vector of the tap sequence generator as shown in Fig. 4.19.

The paging channel data will be scrambled utilizing the paging channel long code mask as shown in Fig. 4.20. The long code mask consists of a 42-bit binary sequence that creates the unique identity of the long code. The long code will provide limited privacy.

The data scrambling mechanism is shown in Fig. 4.21. The long code operating at a 1.2288 MHz clock rate is equivalent to the PN chip sequence which is the output of the long code generator. When the long code is divided into every 64 bits, the first bit of every 64 bits will be used for the data scrambling at a 19.2 ksps rate. The function of the decimator in Fig. 4.21 is to reduce the size of the long code by taking one out of every 64 (=1.2288 x 10^6/192 x 10^2) bits.

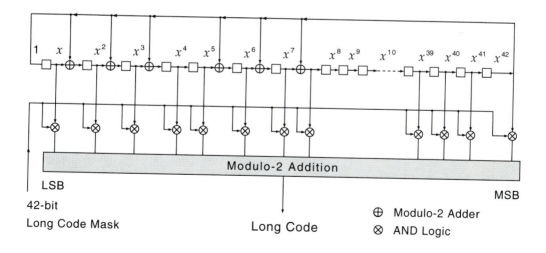

Figure 4.19 Long code generator

41	29	28	24	23	21	20	9	8	0
1100011001101		00000		PCN		000000000000		PILOT - PN	

PCN: Paging Channel Number

PILOT-PN: Pilot sequence offset index for the forward CDMA Channel

Figure 4.20 Paging channel long code mask

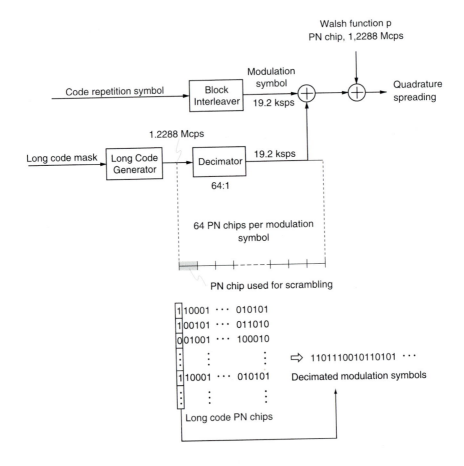

Figure 4.21 Data scrambling mechanism

Example 4.12 Consider an example of data scrambling by looking at Fig. 4.21. Data scrambling is accomplished by EX-ORing the interleaver output with the decimated value of the long code. Utilizing the paging channel long code mask, the long code can be computed as shown in the following. Suppose the pilot channel number (PCN) is 001 and the pilot PN sequence offset index is 100100011. Thus from Fig. 4.20, the long code mask sequence becomes
    110001100110100000001000000000000100100011.

Using Eq. 4.14, the code generator tap sequence can be written as
    11110111001000001111011001110001010101000000.

Under the initial contents 1000000000000000000000000000000000000000000, the output of the long code generator can be found using the long code mask and the generator tap sequence as shown in Fig. 4.19.

The generated long code sequence is computed as follows:

    1100010010000000 0000010000000101 0001100011011110 1111110100010101
    1001011101110001 1100001011100101 0000011111011010 0010101101011010
    0010011111111110 0000100110100101 0010010110111000 0001001000100010
    1011101001001100 1000111010011101 1001001110010000 0100111001010010
    1111001000110010 1111101000010000 0011110111110011 0001100010111011
    1010010101111010 0100000111110001 1101010110001110 1110100010010100
    0111100110100110 1011000010010100 0011011111000001 1000000011110001
    0000001101010000 1101111010111101 0001010101101111 1011110000010100
    1101100101101001 0000010010001010 0101100110000101 1001100010111111
    0001011100100101 0100010100100100 0001010001100100 0011000101110011
    1001111001001000 1001011110010001 0111000000000011 0000101101110010
    1101101111101110 0101100011101111 0001000101010001 0101010101111001
    0010001110100010 1100100000101011 1010101010011101 1111110100001101
    1110000101010011 1010100000001010 1010110010001011 1111010000011100
    0101011001100011 0111000111010101 1010011101100110 1000000100011110
    1100101001101010 1010101111100110 1110011000000010 1111010000000010

Since the first output (the leftmost column) of every 64 bits is used for data scrambling at a 19.2 ksps rate, the modulation symbols from the decimator for EX-ORing with the interleaver output symbols is 1101110010110101 · · · (leftmost column).

Interleaver Output:	1010101011010110 · · ·
Decimated code:	1101110010110101 · · ·
(a) EXOR output ($\oplus$):	0111011001100011 · · ·

This symbol sequence (a) will be the input being spread with a Walsh function at a fixed chip rate of 1.2288 Mcps to provide orthogonal channelization among all code channels on a given forward CDMA channel.

4.4.4   Paging Channel Orthogonal Spreading

For orthogonal spreading in the paging channel, one of 64 time-orthogonal Walsh functions, as defined in Table 4.7, is used. A paging channel that is actually spread using a Walsh function n $(1 \leq n \leq 7)$ is assigned to paging channel number n. The modulation symbol will be one of 7 mutually orthogonal waveforms generated using Walsh functions. The Walsh function spreading sequence repeats with a period of $52.083\,\mu$ s (=64/1.2288 Mcps) which is equal to the duration of one paging channel modulation symbol.

As seen in Fig. 4.1, paging channels are assigned to code channel numbers one through seven (inclusive) in sequence, i.e., W_1, W_2, $\cdots W_7$.

Example 4.13   Let us choose the paging channel 1 (W_1). The Walsh chip corresponding to the Walsh function index 1 will be the following 64-bit Walsh chip:

(b) 0101010101010101010101010101010101010101010101010101
01010101010101 (at 1.2288 Mcps rate).

Since PN chips/modulation symbol is 1.2288 Mcps/19.2 ksps=64, 1 modulation symbol corresponds to the 64-bit Walsh chip code. That is,

First bit of (a):	0000000000 $\cdots$ 00000000
(b):	0101010101 $\cdots$ 01010101
EX–OR (a $\oplus$ b):	0101010101 $\cdots$ 01010101
Second bit of (a):	1111111111 $\cdots$ 11111111
(b):	0101010101 $\cdots$ 01010101
EX–OR (a $\oplus$ b):	1010101010 $\cdots$ 10101010

We have only shown the EX-ORing operation for the first two symbols of the symbol sequence (a) at 19.2 ksps rate. Continuing similarly, the modulo-2 addition (by using the rest of input symbol bits) results in the following Walsh symbol output:

01010101 01010101 01010101 01010101 01010101 01010101 01010101 01010101
10101010 10101010 10101010 10101010 10101010 10101010 10101010 10101010
10101010 10101010 10101010 10101010 10101010 10101010 10101010 10101010
10101010 10101010 10101010 10101010 10101010 10101010 10101010 10101010
01010101 01010101 01010101 01010101 01010101 01010101 01010101 01010101
10101010 10101010 10101010 10101010 10101010 10101010 10101010 10101010

4.4.5   Paging Channel Quadrature Spreading

Following the orthogonal spreading, the paging channel is spread in quadrature as shown in Fig. 4.18. The 15-stage PN chip generators are used for quadrature spreading. The pilot PN sequences are modified by inserting an extra chip '0' in the 15th position, resulting in a period of $2^{15} = 32768$ chips.

The paging channel is quadrature spread by the I and Q pilot PN sequences as specified in Section 4.2.2.

Example 4.14   The in-phase pilot PN sequence P_I and the quadrature-phase PN sequence P_Q can be computed using the LFSR binary sequences $i(n)$ and $q(n)$ of length $2^{15}-1$.

(I) Using $i(n) = i(n-15) = 0$ for $1 \leq n \leq 14$, $i(15)=1$, and $i(n) = \displaystyle\sum_{K=15,10,8,7,6,2} i(n-k)$

for $16 \leq n \leq 32767$, the in-phase pilot PN sequence P_I is

found by computing $i(n)$ first and then inserting a 0 (zero) in $i(n)$ after 14 consecutive 0 outputs (see arrow ↓) as follows:

$$\downarrow$$
$P_I =$ 0000000000000000101010010011101000110111100110011001000
00111110000100011010010101101010101011101001001111

over $0 \leq n \leq 99$ only.

(II) Similarly, using $q(n) = q(n-15) = 0$ for $1 \leq n \leq 14$, $q(15)=1$, $q(n) = \displaystyle\sum_{k=15,12,11,10,9,5,4,3} q(n-k)$

for $16 \leq n \leq 32767$, we can compute $q(n)$ from which the quadrature-phase pilot PN sequence is found by inserting a 0 (zero) at the 15th place in $q(n)$ as shown below:

$$\downarrow$$
$P_Q =$ 0000000000000000100111101011101011010011100010100111
100111000110110001110100110001001001000110011000110

over $0 \leq n \leq 99$.

The in-phase channel sequence I and the quadrature-phase channel sequence Q can be computed from the following spreading operation:

$$I = \text{OSS} \oplus P_I \text{ and } Q = \text{OSS} \oplus P_Q \qquad (4.15)$$

where OSS designates the orthogonal symbol spreading.

Example 4.15   Using Eq 4.15, the I-channel sequence and Q-channel sequence can be computed as follows:

Modulo-2 addition with the first row of Walsh symbol output:

OSS: 01010101 01010101 01010101 01010101 01010101···

P_I :   00000000 00000001 01010010 01110100 01101111···

$I(\oplus)$:   01010101 01010100 00000111 00100001 00111010···

Modulo-2 addition with the second row of Walsh symbol output:

OSS: 10101010 10101010 10101010 10101010 10101010···

P_I :   11010010 10110101 01010111 01001001 1111···

$I(\oplus)$:   01111000 00011111 11111101 11100011 0101···

Whereupon, modulo-2 addition with the first row of Walsh symbol output:

OSS: 01010101 01010101 01010101 01010101 01010101 ···

P_Q :   00000000 00000001 00111101 01110101 10100111···

$Q(\oplus)$:   01010101 01010100 01101000 00100000 11110010···

Modulo-2 addition with the second row of Walsh symbol output:

OSS: 10101010 10101010 10101010 10101010 10101010···

P_Q :   00011101 00110001 00100100 01100110 0011 ···

$Q(\oplus)$:   10110111 10011011 10001110 11001100 1001 ···

We have thus computed I and Q sequences corresponding to the first two rows of the Walsh symbol output.

4.4.6   Paging Channel Baseband Filtering

Following the quadrature spreading, the I and Q channel sequences are applied to the respective baseband filters. The paging channel is similarly filtered as specified in Section 4.2.3. The baseband filters have a frequency response $S(f)$ that satisfies the limit given in Fig. 3.15.

Specifically, the normalized frequency response of the baseband filter is contained within $\pm\delta_1$ in the passband $0 \leq f \leq f_p$ and is less than or equal to $-\delta_2$ in the stopband $f \geq f_s$. The numerical values for these parameters are $\delta_1 = 1.5$ dB, $\delta_2 = 40$dB, $f_p = 590$ KHz, and $f_s = 740$ KHz. After baseband filtering, the binary I and Q data are mapped into phase according to Tables 4.2 and 4.3, as shown in Fig. 4.8.

4.4.7   QPSK for Paging Channel

The QPSK technique for the paging channel is similar to the pilot, sync, and forward traffic channel. The nonoffset QPSK problem was fully covered in Section 4.2.4. Let $I(t)$ and $Q(t)$ be two quadrature binary data coming out from the baseband filters. Then $I(t)$ and $Q(t)$ each can be determined according to the phase offset of $\theta(t)$, as shown in Table 4.2. Consequently, the QPSK waveform $s(t)$ can be depicted using Table 4.3.

Example 4.16 Consider the sample data, computed in Example 4.15, for I and Q quadrature streams as shown below:

I: 01111000   00011111   11111101 | 11100011 | 0101 $\cdots$

Q: 10110111   10011011   10001110 | 11001100 | 1001 $\cdots$

The QPSK waveform $s(t)$ based on two BPSK pulses, i.e., $I(t)$=11100011 and $Q(t)$=11001100, can be drawn as shown in Fig. 4.22.

Paging channel(4800 bps):
          I-channel output: 11100011 (-1, -1, -1, 1, 1, 1, -1, -1)
          Q-channel output: 11001100 (-1, -1, 1, 1, -1, -1, 1, 1)

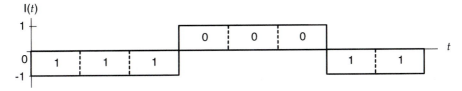

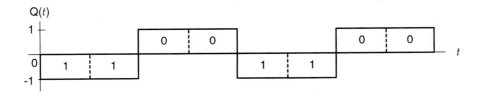

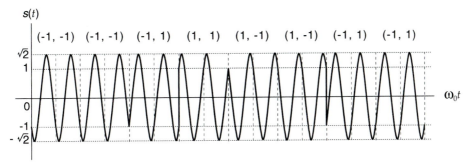

Figure 4.22 The paging channel QPSK waveform $s(t)$ plot

4.4.8   Paging Channel Slot and Message Capsule Structure

The paging channel is used to send information to mobile stations that have not been assigned to a traffic channel.

The paging channel is divided into 80 ms slots. Each 80 ms slot is composed of four paging channel frames, each 20 ms in length. The slots are grouped into cycles of 2040 slots (20 ms x 2040=163.84 seconds) which is referred to as the maximum slot cycles. The slots of each maximum slot cycle are numbered from 0 to 2047. Each maximum slot cycle begins at the start of the frame when the system time 80 ms per unit (modulo 2048) is zero.

A mobile station operating in the slotted mode monitors the paging channel using a slot cycle with a length that is a submultiple of the maximum slot cycle length. A mobile station operating in the nonslotted mode monitors the paging channel at all times. For each of its assigned slots, the mobile station begins monitoring the paging channel in time to receive the first bit of the assigned slot.

A 20 ms paging channel frame is divided into 10 ms long paging channel half frames, as shown in Fig. 4.23. The first bit in any paging channel half frame is called a synchronized capsule indicator (SCI) bit.

A paging channel message capsule is composed of a paging channel message and padding. A paging channel message consists of a message length field, a message body, and a CRC field. Padding consists of zero or more bits as required.

The base station may transmit synchronized or unsynchronized message capsules. A synchronized message capsule starts on the second bit of a padding channel half frame. An unsynchronized message capsule begins immediately after the previous message capsule.

After the end of a paging channel message, the base station acts as follows:

1. If 8 bits or more remain before the next SCI bit, the base station may transmit an unsynchronized message capsule immediately following that message. The base station does not include any padding bits in a paging channel message capsule that is followed by an unsynchronized paging channel message capsule.
2. If fewer than 8 bits remain before the next SCI bit, the base station includes sufficient padding bits in that message capsule to extend it through the bit preceding the next SCI bit, and transmits a synchronized message capsule immediately following that SCI bit. This implies that all bits transmitted on the paging channel are either SCI bits or are part of a message capsule. The base station sets all padding bits to '0'.

When a message capsule immediately follows an SCI bit, the base station sets that SCI bit to '1'. The base station sets all other SCI bits to '0'. The base station transmits the first message that begins in each slot in a synchronized message capsule. This permits mobile stations operating in the slotted mode to obtain synchronization immediately after becoming active. The overall structure of a paging channel message capsule is shown in Fig. 4.24.

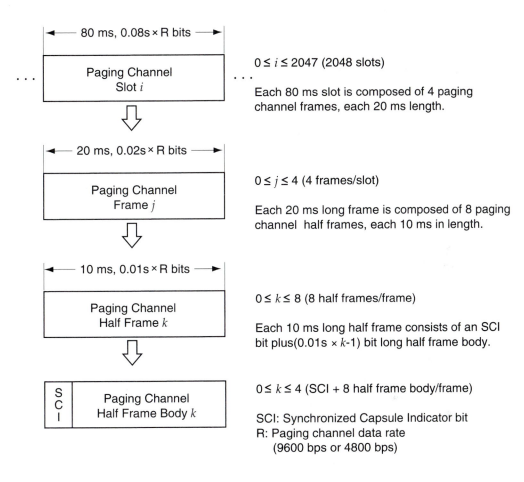

Figure 4.23 Paging channel slot and frame structure

4.4.9   CRC for Paging Channel Signaling Message

A 30-bit CRC is computed for each paging channel signaling message. A paging channel message consists of a message length field, a message body, and a CRC field. The message length field is 8 bits long. The base station limits the maximum paging channel message length to 148 octets, or 1184 bits.

The generator tap polynomial for the CRC is as follows:
$$g(x) = 1 + x + x^2 + x^6 + x^7 + x^8 + x^{11} + x^{12} + x^{13} + x^{15} + x^{20} + x^{21} + x^{29} + x^{30} \quad (4.16)$$
whose logic is shown in Fig. 4.15.

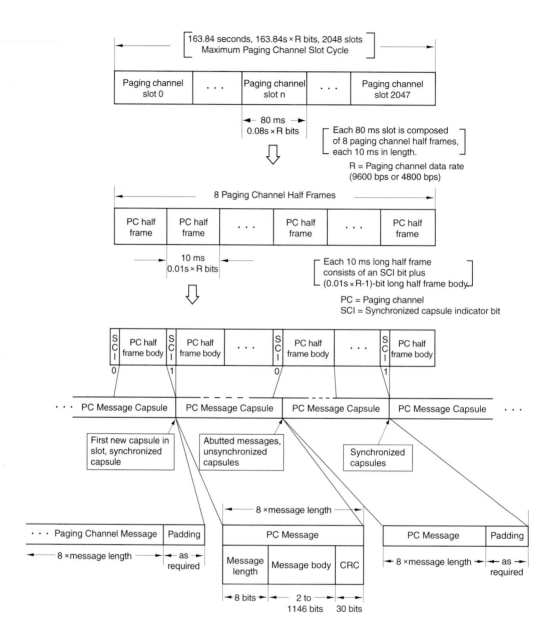

Figure 4.24 An example of paging channel structure

The CRC bits can be computed by the following procedure; the logic diagram is shown in Fig. 4.25.

1. Initial contents of the LFSR are set to binary one (1) in order to make the CRC field to nonzero values even for all-zero data.
2. Set the switch in the down position and close the gate in the feedback path of the CRC encoder.
3. Then the k input bits begin to transmit the encoder output as well as the feedback path of the shift register.
   The k-bit input includes the 8-bit message length + the message body length in bits.
4. The register is clocked k times with the k-bit input.
5. Set the switch to the up position and open the gate in the feedback path.
6. Clocking the register by an additional 30 times, the CRC field will be generated.
7. Thus, the paging channel message will be transmitted in the order of the 8-bit message length field, the message body length, and the 30-bit CRC field at the CRC encoder output.

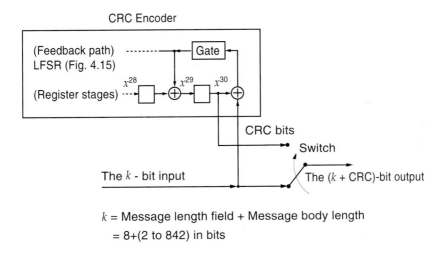

k = Message length field + Message body length
= 8+(2 to 842) in bits

Figure 4.25 CRC computation circuit for the paging channel

4.5 FORWARD TRAFFIC CHANNEL

The forward traffic channel is used for the transmission of user and signaling information to a specific mobile station during a call. The base station transmits information on the forward traffic channel at the variable data rates of 9600, 4800, 2400, and 1200 bps. The forward traffic channel frame is 20 ms in duration.

The data rate is selected on a frame-by-frame basis. Although the data rate may vary on a frame-by-frame basis, the modulation symbol rate is kept constant by code repetition at 19,200 symbols per second.

The pilot PN sequences for the forward traffic channel use the same pilot PN sequence offset as the pilot channel for a given base station.

The modulation symbols that are transmitted at the lower data rate are transmitted using lower energy. Specifically, the energy per modulation symbol, E_s, for the supported data rates is shown in Table 4.13, where E_b is the energy per information bit. Note that all symbols in an interleaver block are from the same frame. Thus they are all transmitted at the same energy.

Table 4.13   Transmitted Symbol Energy Versus Data Rate

Data Rate (bps)	Energy per Modulation Symbol
9600	$E_s = E_b/2$
4800	$E_s = E_b/4$
2400	$E_s = E_b/8$
1200	$E_s = E_b/16$

Forward traffic channel frames sent at the various transmission rates are shown as follows:

1. At the 9600 bps frame, the frame length is 192 bits (20 ms).
   Frame length (192 bits) = Information length (172 bits) + CRC (12 bits)
   + Encoder tail (8 bits)
2. At the 4800 bps frame, the frame length is 96 bits (20ms).
   Frame length (96 bits) = Information length (80 bits) + CRC (8 bits)
   + Encoder tail (8 bits)
3. At the 2400 bps frame, the frame length is 48 bits (20ms).
   Frame length (48 bits) = Information length (40 bits) + Encoder tail (8 bits)
4. At the 1200 bps frame, the frame length is 24 bits (20ms).
   Frame length (24 bits) = Information length (16 bits) + Encoder tail (8 bits)

Data scrambling applies to the forward traffic channel. The data scrambling is accomplished by performing the modulo-2 addition of the interleaver output symbol with the decimated binary value ('0' or '1') of the long code PN chips. A power control subchannel is continuously transmitted on the forward traffic channel. The subchannel transmits at a rate of one bit every 1.25 ms (i.e., 800 bps), as discussed later in this section.

The forward traffic channel has the overall structure shown in Fig. 4.26; the modulation parameters for the forward traffic channel are shown in Table 4.14.

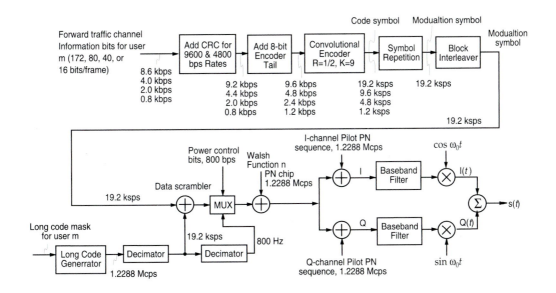

Figure 4.26 The overall structure of the forward traffic channel

Table 4.14   Forward Traffic Channel Modulation Parameters

Parameters	Data Rate (bps)				Units
	9600	**4800**	**2400**	**1200**	
PN chip rate	1.2288	1.2288	1.2288	1.2288	Mcps
Code rate	1/2	1/2	1/2	1/2	bits/code symbols
Code repetition	1	2	4	8	mod sym/code sym
Modulation symbol rate	19,200	19,200	19,200	19,200	sps
PN chips/Modulation symbol	64	64	64	64	PN/chips/mod sym
PN chips/bit	128	256	512	1024	PN chips/bit

4.5.1   CRC Computation for Forward Traffic Channel

Each 9600 bps and 4800 bps include a CRC encoder. This CRC encoder is also called the frame quality indicator. No CRC computation is required for the 2400 bps and 1200 bps transmission rates. For both the 9600 bps and 4800 bps rates, the CRC is calculated on all bits within the frame, except the CRC itself and the encoder tail bits. The 9600 bps transmission rate uses a 12-bit frame quality indicator (CRC).

The 4800 bps transmission rate uses an 8-bit frame quality indicator. The generator tap polynomials for frame quality indicators are, respectively

$$g(x) = 1 + x + x^4 + x^8 + x^9 + x^{10} + x^{11} + x^{12} \text{ for 9.6 kbps rate} \qquad (4.17)$$
and $\quad g(x) = 1 + x + x^3 + x^4 + x^7 + x^8 \text{ for 4.8 kbps rate} \qquad (4.18)$

The frame quality indicators (CRCs) are computed according to the following procedure using the logic diagram in Figs. 4.27 and 4.28.

1. All shift register contents are initially set to logical one and the switches are set in the up position.
2. The register is clocked 172 times (for the 192-bit frame) or 80 times (for the 96-bit frame) with the information bits (172 bits or 80 bits) as input.
3. After setting the switches in the down position, the register is clocked an additional 12 times (for the 192-bit frame) or 8 times (for the 96-bit frame). The 12 or 8 additional output bits will be the CRC bits.
4. The bits are transmitted in the order calculated.

The last eight bits of each forward traffic channel frame are called the encoder tail bits. These eight bits will be set to zero (0).

The transmission rate is increased by adding the CRC bits and the encoder tail bits.

1. For the 9600 bps frame, the 8600 bps data rate is raised to the 9600 bps transmission rate by adding the 12-bit CRC and the 8-bit encoder tail.
   Proof
   Since 192 bits/20 ms each frame, it follows that there are 9600 bits/1s.
   That is, since there are 172 information bits at the 8600 bps rate, we can see a 50 bps rate per bit. Then for the 12-bit CRC, we have 50 bps x 12 = 600 bps.
   Hence we have 8600 bps + 600 bps = 9200 bps. By adding eight encoder tail bits, it gives 50 bps x 8 = 400 bps. Finally we will have 9200 bps + 400 bps = 9600 bps.
2. For the 4800 bps frame, the 2000 bps original data rate is raised to the 4800 bps transmission rate by addition of the 8-bit CRC and the 8-bit encoder tail.
   Proof
   Since each frame contains 96 bits/20 ms, we know that there are 4800 bits/1s.
   It can be proved as follows: Since there are 80 information bits at the 4000 bps rate, we can see a 50 bps rate per bit. Whence for the 8-bit CRC, it gives 50 bps x 8 = 400 bps. At this point, the data rate can be raised to 4000 bps + 400 bps = 4400 bps. This data rate again will be raised by adding 8 encoder tail bits such that 4400 bps + (50 bps x 8) bps = 4800 bps as expected.

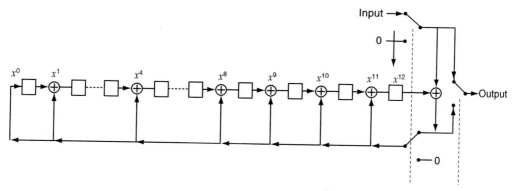

Figure 4.27 Forward traffic channel CRC calculation at the 9600 bps rate

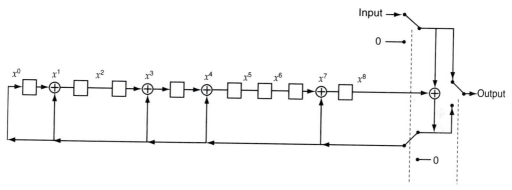

Figure 4.28 Forward traffic channel CRC calculation at the 4800 bps rate

3. For the 2400 bps frame, the 2000 bps original data rate is increased to the 2400 bps transmission rate by adding the 8-bit encoder tail.

   Proof

   Since this frame is shown by 48 bits/20 ms, it follows that there are 2400 (=48/20 x 10^{-3}) bits/s. In this frame, there are 40 information bits at the 2000 bps rate. Therefore, it gives 50 bps rate per bit. Whence for the 8-bit encoder tail, we obtain 50 bps x 8 = 400 bps. Thus, the transmission rate becomes 2000 bps + 400 bps = 2400 bps.

4. For the 1200 bps frame, the 800 bps original data rate is raised to the 1200 bps transmission rate by adding the 8-bit encoder tail.

   Proof

   In this frame, we have 16 information bits at the 800 bps original rate so that it will give 800/16 = 50 bps per bit. Therefore, the raised portion by the 8-bit encoder tail becomes 50 bps x 8 = 400 bps. The transmission rate then becomes 800 bps x 400 bps = 1200 bps.

Example 4.17 Consider the forward traffic channel (FTC) of the 4800 bps rate. The CRC computation is calculated on all bits within the frame, i.e., 80 information bits/frame at the 4800 bps rate. Let us show the CRC computation by using Fig. 4.28, as follows:

$$\text{Generator tap polynomial of Eq. 4.18:} \quad g = (11011001)$$

$$\begin{aligned}
\text{Assumed information sequence:} \quad d = (&1100011100\ 0010100100 \\
&1100110111\ 0101011001 \\
&1100110010\ 1001110101 \\
&1001111100\ 0110011110)
\end{aligned}$$

$$\text{Initial contents of register:} \quad \text{I.C.} = (11111111)$$

Note that all LFSR stages are initially set to logical ones because this initial setting causes the CRC for all-zero data to be nonzero.

By clocking the register 80 times (for the 96-bit frame) with the 80-bit information as input, the 8 additional output bits are generated as the CRC bits. Clocking the register 80 times, the CRC bits are obtained as 11001110. Thus, the generated bits sequence with the 8-bit CRC plus the 8-bit all-zero encoder tail is computed as shown below:

1100011100	0010100100	1100110111	0101011001	1100110010	
1001110101	1001111100	0110011110	01110011	00000000	(I)
			CRC	Encoder tail	

It is also recommended that the reader computes the CRC bits for the 9600 bps frame by using Eq. 4.17.

4.5.2   FTC Convolutional Encoding

The forward traffic channel (FTC) is convolutionally encoded prior to transmission. The code rate is 1/2 and the constraint length K is 9.

The generator sequence for this convolutional code is $g_1^{(1)} = 753$ (octal) = (111101011) (binary) and $g_1^{(2)} = 561$ (octal) = (101110001) (binary), respectively.

If the generator sequences are expressed in polynomial form, we have
$$g_1^{(1)}(x) = 1 + x + x^2 + x^3 + x^5 + x^7 + x^8 \qquad (4.19)$$
$$g_1^{(2)}(x) = 1 + x^2 + x^3 + x^4 + x^8 \qquad (4.20)$$

The (2,1,8) convolutional encoder with Eqs. 4.19 and 4.20 is illustrated as shown in either Fig. 4.12(a) or Fig. 4.12(b). Since the code rate is 1/2, two encoded symbols corresponding to each input bit to the encoder are output of the code symbol c_0 encoded with $g_1^{(1)}$ and the code symbol c_1 encoded with $g_1^{(2)}$, respectively.

Example 4.18 Compute the code symbols c_0 and c_1 encoded with $g_1^{(1)}(x)$ and $g_1^{(2)}(x)$ of Eqs 4.19 and 4.20. The encoder output equations are
$$c_0 = d_0 + d_1 + d_2 + d_3 + d_5 + d_7 + d_8$$
$$c_1 = d_0 + d_2 + d_3 + d_4 + d_8$$
Using Fig. 4.12 and the above equations along with the input sequence (I) and the all-zero initial contents of the convolutional encoder, each encoder output symbols can be found as

$c_0 = $ (1000101011 0111011111 1100101011 0010000101 0100010001
1110010001 1011011011 1010011011 0100100100 101101)

$c_1 = $ (1110001010 0111100011 1000111100 0010101100 0100001100
0111001011 0110011101 0100100010 0101100011 100011)

Pairwise multiplexing and concatenating (c_0, c_1) result in a single code symbol sequence as follows:

$c = $ (1101010010001100111000011 1111011010101111111100000
11011101101000001100100 0111001000110000000100101
00101011110100100100011 100111100011110110111001 (II)
1000011010001110001100011 1100001001011100010100111)

This represents the convolutional encoder output symbols.

Notice that the 192-symbol sequence c is divided into 8 blocks of 24 symbols each for easy application to the block interleaving table.

4.5.3 FTC Interleaving and Symbol Repetition

For the forward traffic channel, each convolutionally encoded symbol is either repeated prior to block interleaving, or can be block interleaved before symbol repetition whenever the information rate is lower than 9600 bps. Each code symbol at the 4800 bps rate is repeated 1 time (each symbol occurs 2 consecutive times). Each code symbol at the 2400 bps rate will be repeated 3 times (each symbol occurs 4 consecutive times). Each symbol at the 1200 bps data rate will be repeated 7 times (each symbol occurs 8 consecutive times). However, for all the data rates (9.6, 4.8, 2.4, and 1.2 kbps), this results in a constant modulation symbol rate of 19.2 ksps.

All symbols after repetition on the forward traffic channel are block interleaved and use the identical block interleaver spanning 20 ms equivalent to 384 modulation symbols at the 19200 sps symbol rate.

    The input and output symbol sequences to the forward traffic channel interleaver operating
at 4.8 kbps and 9.6 kbps are given in Tables 4.9 through 4.12. These tables are read down by columns from the left to the right as with the sync channel interleaver. In these tables, symbols with
the same number denote repeated code symbols.

Example 4.19 Compute the interleaver output at 4800 bps using Table 4.10. The code word
from the convolutional encoder is given by the code symbols (I) which are input to the 4800 bps
block interleaver. This code symbol sequence (I) is transformed into the modulation symbol
sequence based on Table 4.10. By reading down by columns from the left to the right, the interleaver output can be obtained as follows:

```
11100011001011011111110111
01011001000110001000011 0
010111101000011010111001
0101001001100011 01100111
1010001101010111 01100000   (III)
111010010111100000101111
100011011001001110100100
010111110100000010001001
```

    Tables 4.15 through 4.18 can be used for computation of the FTC interleaver input and
output operating at the data rate of 2400 bps or 1200 bps.

Table 4.15   Forward Traffic Channel Interleaver Input Operating at 2400 bps (After
TIA/EIA/IS-95)

1	7	13	19	25	31	37	43	49	55	61	67	73	79	85	91
1	7	13	19	25	31	37	43	49	55	61	67	73	79	85	91
1	7	13	19	25	31	37	43	49	55	61	67	73	79	85	91
1	7	13	19	25	31	37	43	49	55	61	67	73	79	85	91
2	8	14	20	26	32	38	44	50	56	62	68	74	80	86	92
2	8	14	20	26	32	38	44	50	56	62	68	74	80	86	92
2	8	14	20	26	32	38	44	50	56	62	68	74	80	86	92
2	8	14	20	26	32	38	44	50	56	62	68	74	80	86	92
3	9	15	21	27	33	39	45	51	57	63	69	75	81	87	93
3	9	15	21	27	33	39	45	51	57	63	69	75	81	87	93
3	9	15	21	27	33	39	45	51	57	63	69	75	81	87	93
3	9	15	21	27	33	39	45	51	57	63	69	75	81	87	93
4	10	16	22	28	34	40	46	52	58	64	70	76	82	88	94
4	10	16	22	28	34	40	46	52	58	64	70	76	82	88	94
4	10	16	22	28	34	40	46	52	58	64	70	76	82	88	94

Table 4.15 Forward Traffic Channel Interleaver Input Operating at 2400 bps (After TIA/EIA/IS-95) *(continued)*

4	10	16	22	28	34	40	46	52	58	64	70	76	82	88	94
5	11	17	23	29	35	41	47	53	59	65	71	77	83	89	95
5	11	17	23	29	35	41	47	53	59	65	71	77	83	89	95
5	11	17	23	29	35	41	47	53	59	65	71	77	83	89	95
5	11	17	23	29	35	41	47	53	59	65	71	77	83	89	95
6	12	18	24	30	36	42	48	54	60	66	72	78	84	90	96
6	12	18	24	30	36	42	48	54	60	66	72	78	84	90	96
6	12	18	24	30	36	42	48	54	60	66	72	78	84	90	96
6	12	18	24	30	36	42	48	54	60	66	72	78	84	90	96

Table 4.16 Forward Traffic Channel Interleaver Output Operating at 2400 bps (After TIA/EIA/IS-95)

1	3	2	4	1	3	2	4	1	3	2	4	1	3	2	4
17	19	18	20	17	19	18	20	17	19	18	20	17	19	18	20
33	35	34	36	33	35	34	36	33	35	34	36	33	35	34	36
49	51	50	52	49	51	50	52	49	51	50	52	49	51	50	52
65	67	66	68	65	67	66	68	65	67	66	68	65	67	66	68
81	83	82	84	81	83	82	84	81	83	82	84	81	83	82	84
9	11	10	12	9	11	10	12	9	11	10	12	9	11	10	12
25	27	26	28	25	27	26	28	25	27	26	28	25	27	26	28
41	43	42	44	41	43	42	44	41	43	42	44	41	43	42	44
57	59	58	60	57	59	58	60	57	59	58	60	57	59	58	60
73	75	74	76	73	75	74	76	73	75	74	76	73	75	74	76
89	91	90	92	89	91	90	92	89	91	90	92	89	91	90	92
5	7	6	8	5	7	6	8	5	7	6	8	5	7	6	8
21	23	22	24	21	23	22	24	21	23	22	24	21	23	22	24
37	39	38	40	37	39	38	40	37	39	38	40	37	39	38	40
53	55	54	56	53	55	54	56	53	55	54	56	53	55	54	56
69	71	70	72	69	71	70	72	69	71	70	72	69	71	70	72
85	87	86	88	85	87	86	88	85	87	86	88	85	87	86	88
13	15	14	16	13	15	14	16	13	15	14	16	13	15	14	16
29	31	30	32	29	31	30	32	29	31	30	32	29	31	30	32
45	47	46	48	45	47	46	48	45	47	46	48	45	47	46	48
61	63	62	64	61	63	62	64	61	63	62	64	61	63	62	64
77	79	78	80	77	79	78	80	77	79	78	80	77	79	78	80
93	95	94	96	93	95	94	96	93	95	94	96	93	95	94	96

Table 4.17   Forward Traffic Channel Interleaver Input Operating at 1200 bps (After TIA/EIA/IS-95)

1	4	7	10	13	16	19	22	25	28	31	34	37	40	43	46
1	4	7	10	13	16	19	22	25	28	31	34	37	40	43	46
1	4	7	10	13	16	19	22	25	28	31	34	37	40	43	46
1	4	7	10	13	16	19	22	25	28	31	34	37	40	43	46
1	4	7	10	13	16	19	22	25	28	31	34	37	40	43	46
1	4	7	10	13	16	19	22	25	28	31	34	37	40	43	46
1	4	7	10	13	16	19	22	25	28	31	34	37	40	43	46
1	4	7	10	13	16	19	22	25	28	31	34	37	40	43	46
2	5	8	11	14	17	20	23	26	29	32	35	38	41	44	47
2	5	8	11	14	17	20	23	26	29	32	35	38	41	44	47
2	5	8	11	14	17	20	23	26	29	32	35	38	41	44	47
2	5	8	11	14	17	20	23	26	29	32	35	38	41	44	47
2	5	8	11	14	17	20	23	26	29	32	35	38	41	44	47
2	5	8	11	14	17	20	23	26	29	32	35	38	41	44	47
2	5	8	11	14	17	20	23	26	29	32	35	38	41	44	47
3	6	9	12	15	18	21	24	27	30	33	36	39	42	45	48
3	6	9	12	15	18	21	24	27	30	33	36	39	42	45	48
3	6	9	12	15	18	21	24	27	30	33	36	39	42	45	48
3	6	9	12	15	18	21	24	27	30	33	36	39	42	45	48
3	6	9	12	15	18	21	24	27	30	33	36	39	42	45	48
3	6	9	12	15	18	21	24	27	30	33	36	39	42	45	48
3	6	9	12	15	18	21	24	27	30	33	36	39	42	45	48
3	6	9	12	15	18	21	24	27	30	33	36	39	42	45	48

Table 4.18   Forward Traffic Channel Interleaver Output Operating at 1200 bps (After TIA/EIA/IS-95)

1	2	1	2	1	2	1	2	1	2	1	2	1	2	1	2
9	10	9	10	9	10	9	10	9	10	9	10	9	10	9	10
17	18	17	18	17	18	17	18	17	18	17	18	17	18	17	18
25	26	25	26	25	26	25	26	25	26	25	26	25	26	25	26
33	34	33	34	33	34	33	34	33	34	33	34	33	34	33	34
41	42	41	42	41	42	41	42	41	42	41	42	41	42	41	42
5	6	5	6	5	6	5	6	5	6	5	6	5	6	5	6
13	14	13	14	13	14	13	14	13	14	13	14	13	14	13	14

Table 4.18   Forward Traffic Channel Interleaver Output Operating at 1200 bps (After TIA/EIA/IS-95) *(continued)*

21	22	21	22	21	22	21	22	21	22	21	22	21	22	21	22
29	30	29	30	29	30	29	30	29	30	29	30	29	30	29	30
37	38	37	38	37	38	37	38	37	38	37	38	37	38	37	38
45	46	45	46	45	46	45	46	45	46	45	46	45	46	45	46
3	4	3	4	3	4	3	4	3	4	3	4	3	4	3	4
11	12	11	12	11	12	11	12	11	12	11	12	11	12	11	12
19	20	19	20	19	20	19	20	19	20	19	20	19	20	19	20
27	28	27	28	27	28	27	28	27	28	27	28	27	28	27	28
35	36	35	36	35	36	35	36	35	36	35	36	35	36	35	36
43	44	43	44	43	44	43	44	43	44	43	44	43	44	43	44
7	8	7	8	7	8	7	8	7	8	7	8	7	8	7	8
15	16	15	16	15	16	15	16	15	16	15	16	15	16	15	16
23	24	23	24	23	24	23	24	23	24	23	24	23	24	23	24
31	32	31	32	31	32	31	32	31	32	31	32	31	32	31	32
39	40	39	40	39	40	39	40	39	40	39	40	39	40	39	40
47	48	47	48	47	48	47	48	47	48	47	48	47	48	47	48

4.5.4   FTC Data Scrambling

Data scrambling applies to the forward traffic channel and the paging channel as well. Data scrambling is performed on the block interleaver output at the modulation symbol rate of 19.2 ksps.

As shown in Fig. 4.26, data scrambling is accomplished by performing the modulo-2 addition of the block interleaver output with the decimated sampling value ('0' or '1') of the long code PN chips. This PN sequence is equivalent to the long code operating at 1.2288 MHz clock rate where only the first output bit of every 64 bits is used for the data scrambling at the 19.2 ksps. The long code may be generated as described in the following. The FTC public long code mask format is as shown in Fig. 4.29.

The long code mask consists of a 42-bit binary sequence that creates the unique identity of the long code.

The FTC public long code mask is explained as E_{41}–E_{32} = (1100011000) and E_{31}–E_0 = Permuted ESN bits.

Let the ESN bits be (E_{31}, E_{30}, E_{29}, $\cdots$, E_2, E_1, E_0). ESN denotes the 32-bit Electronic Serial Number assigned by the manufacturer, uniquely identifying the mobile station to any cellular system. The bit allocation of the ESN is sketched as shown in Fig. 4.30. At the time of initial issuance, the manufacturer will assign an MFR code within the eight most-significant bits (E_{31} through E_{24}) of the 32-bit ESN. E_{23} through E_{18} is allocated by the FCC, and E_{17} through E_0 is uniquely assigned by each manufacturer. Then the permuted ESN can be derived from the following format:

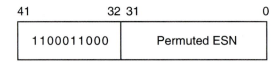

Figure 4.29 Forward traffic channel public long code mask

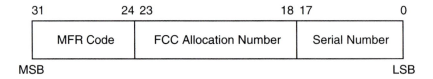

Figure 4.30 The bit allocation of ESN

Permuted ESN = $(E_0, E_{31}, E_{22}, E_{13}, E_4, E_{26}, E_{17}, E_8, E_{30}, E_{21}, E_{12}, E_3, E_{25},$
$E_{16}, E_7, E_{29}, E_{20}, E_{11}, E_2, E_{24}, E_{15}, E_6, E_{28}, E_{19}, E_{10}, E_1,$
$E_{23}, E_{14}, E_5, E_{27}, E_{18}, E_9)$

The permutation of ESN is designed to prevent high correlation between long codes.

The long code is periodic with period $2^{42}-1$ chips specifying by the register tap polynomial $p(x)$ of the long code generator:

$$p(x) = 1 + x + x^2 + x^3 + x^5 + x^6 + x^7 + x^{10} + x^{16} + x^{17} + x^{19} + x^{21} + x^{22} + x^{25} + x^{26}$$
$$+ x^{27} + x^{31} + x^{33} + x^{35} + x^{42}$$

or $P = (1 1 1 1 0 1 1 1 0 0 1 0 0 0 0 0 1 1 1 1 0 1 1 0 0 1 1 1 0 0 0 1 0 1 0 1 0 0 0 0 0 0)$ (4.21)

Each PN chip of the long code is generated by the modulo-2 inner product of a 42-bit mask and the 42-bit state vector of the tap sequence generator as shown in Fig. 4.19.

The forward traffic channel data is scrambled utilizing the FTC long code generator as shown in Fig. 4.19. The data scrambling mechanism is depicted in Fig. 4.31. The long code operation at 1.2288 MHz clock rate is equivalent to the PN chip sequence which is the output of the long code generator. When the long code is divided into the 64-bit length each, every first-bit per 64 bits is the decimated hash value and these decimated values are used for data scrambling at a 19.2 ksps rate. The function of the decimator is to reduce the size of long code by taking one out of 64 bits.

Voice privacy is provided in the CDMA system by means of the private long code mask used for PN spreading. Voice privacy control is provided on the traffic channels only. All calls are initiated using the public long code mask for PN spreading. The transition to private long code mask is not performed if authentication is not performed.

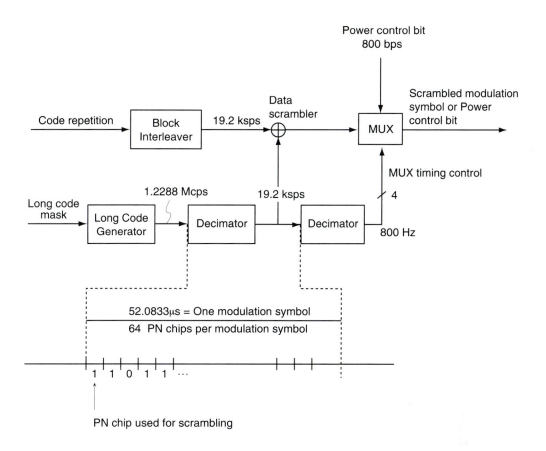

Figure 4.31 Data scrambling mechanism and timing

To initiate a transition to the private or public long code mask, either the base station or the mobile station sends a long code transition request order on the traffic channel. The mobile station or the base station takes action in response to receipt of this order. The base station also can cause a transition to the public long code mask by sending the handoff direction message with the PRIVATE-LCN bit set appropriately.

Example 4.20   Consider again the case of the 4800 bps transmission rate. Referring to Fig. 4.31, data scrambling is accomplished by EX-ORing the interleaver output sequence with the decimated sampling values of the long code. The FTC public long code mask consists of the mask pattern $E_{41}-E_{32} = (1100011000)$ and $E_{31}-E_0 = $ 32-bit permuted ESN. Using the FTC public long code mask, the long code is computed as explained below:

The bit allocation of ESN consists of MFR code, FCC allocation number, and MFR serial number. Assume that

MRF Code:	01001100 (8 bits)
FCC Number:	101100 (6 bits)
MRF Serial Number:	111010011101100111 (18 bits)

Then the 32-bit ESN is $(E_{31}, E_{30}, E_{29}, \cdots, E_2, E_1, E_0)$ = (01001100101100111 010011101100111).

Applying the permuted ESN format, we have

Permuted ESN bits = (100101111100010010101100111011101101)

Thus, the FTC public long code mask can be written as follows:

Public long code mask = (1100011000100101111100010010101100111011101101)

or Long code mask (LSB-MSB) = (1011011100110101001000111110100100011100011)

The register tap sequence of Eq. 4.21 is

P = (111101110010000011110110011100010101000000)

Initial contents of register =(100000000000000000000000000000000000000000)

Now, using Fig. 4.19, the long code sequence operating at 1.2288 MHz clock rate can be generated as follows:

```
1    1011011100110101 0010001111101001 0001100011111011 1010100000100011
2    1000000110001100 0000000101000100 0011100010010010 0110011100101001
3    0011010111100110 1111110001111101 0010010000000100 0011001100110011
4    1010010010011011 1010100000010100 1101101111000010 1011100110011010
            .                                .
            .                                .
            .                                .
23   1010101011110111 0000010000101000 1110110001100011 0110010011011110
24   1010100001000011 1000111101100101 0100101000110110 1101111000100010
            .                                .
            .                                .
            .                                .
```

Observing the previous code sequence, the long code of period $2^{42}-1$ bits is divided into 64 bits length each and is arranged in the row-by-row array. The leftmost column, which is composed of all the first bit of each row, must denote the decimated binary values. If this decimated binary sequence is divided into 24 bits per row, it will be given as follows:

```
1101111001101111101101111
1010111110011101110001010
001010100000011100100110   (IV)
            .              .
            .              .
            .              .
```

Data scrambling is then performed on the modulation symbols from the block interleaver at the 19.2 ksps rate and the decimated binary sequence from the long code PN chips operating at 1.2288 MHz clock rate.

Thus data scrambling is accomplished by performing the modulo-2 addition of the block interleaver output sequence (III) with this decimated binary sequence (IV). That is,

Data scrambling = Block interleaver output sequence (III) $\oplus$ Decimated binary sequence (IV)

Interleaver Output(III):	11100011001011011111101111 0101100100011000100000110 $\cdots$
Decimated sequence(IV):	1101111001101111101101111 101011111001110111000101 $\cdots$
Scrambled data(III $\oplus$ IV):	0011110101000010010000000 111101101000010101000011 $\cdots$

The first decimated sequence is input to another decimator in order to set the starting position for the power control bit. A discussion on the power control bits appears in the following section.

4.5.5   Power Control Subchannel

The power control subchannel on the forward traffic channel continuously transmits the power control bits at a rate of 800 bps, i.e., one bit ('0' or '1') every 1.25 ms (=1/800).

A '0' power control bit implies to the mobile station an increase in the mean output power level and a '1' power control bit indicates to the mobile station a decrease in the mean output power level. The mobile station will adjust its mean output power level in response to each valid power control bit received on the forward traffic channel. A power control bit is considered valid if it is received in a 1.25 ms time slot that is the second time slot following a time slot in which the mobile station transmitted. For example, if the signal is received on the reverse traffic channel (RTC) in power control group number 5, then the corresponding power control bit is transmitted on the forward traffic channel during power control group number 7 (=5+2). In other words, the transmission of the power control bit occurs on the forward traffic channel in the second power control group following the corresponding reverse traffic channel power control group in which the signal strength was estimated.

The base station RTC receiver estimates the received signal strength of the particular mobile station it is assigned to over a 1.25 ms period, equivalent to 6 (=96 x 1.25/20) modulation symbols. The base station receiver uses the estimate to determine the value of the control bit ('0' or '1'). The base station transmits the power control bit on the corresponding forward traffic channel using the puncturing technique described below.

The length of one power control bit corresponds exactly to two modulation symbols of the forward traffic channel (i.e., 104.166μ s = $(1.25 \times 10^{-3}/24) \times 2$). Each power control bit replaces two consecutive FTC modulation symbols by the technique of symbol puncturing and is transmitted with energy not less than E_b (the energy per information bit), as shown in Fig. 4.32. The power control bits are inserted into the FTC data stream after data scrambling (see Fig. 4.31).

There are 16 possible starting positions for the power control bit as shown in Fig. 4.33. Each position corresponds to one of the first 16 modulation symbols (numbered 0 through 15) out of the 24 modulation symbols over a 1.25 ms period. In each 1.25 ms period, a total of 24 bits from the long code are used for scrambling. These 24 bits are numbered 0 through 23, where bit 0 is the first to be used and bit 23 the last in each 1.25 ms period. The 4-bit binary number (expressed by decimal) with values 0 through 15 formed by scrambling bits 23, 22, 21, and 20 will be used to determine the position of the power control bit as shown in Fig. 4.33. Bit 20 is the least significant bit, and bit 23 is the most significant bit. In the example shown in Fig. 4.33, the values of bits 23, 22, 21, and 20 are 0111 (14 decimal), and the power control bit starting position is the 14th. Figure 4.31 shows the relationship between the scrambled modulation symbols at the 19.2 ksps rate and the punctured power control subchannel at the 800 bps rate.

Example 4.21 Consider MUX operation following data scrambling on the forward traffic channel (see Fig. 4.34). The power control bits are inserted into the FTC data stream after data scrambling. The scrambled data sequence from the previous example is

```
0011110101000010010000000
1111011010000101010000011
0111010010000001100111111
        .                .
        .                .
        .                .
```

which is input to MUX. The base station transmits the power control bit in accordance with the puncturing technique. The length of one power control bit corresponds to exactly two modulation symbols of the forward traffic channel.

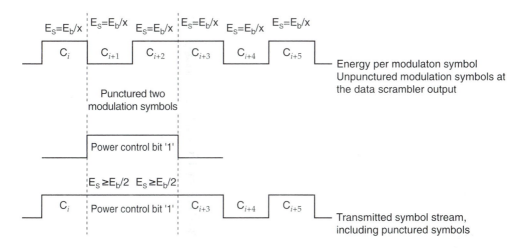

(a) Power control bit '1' (A decrease in the mean output power level)

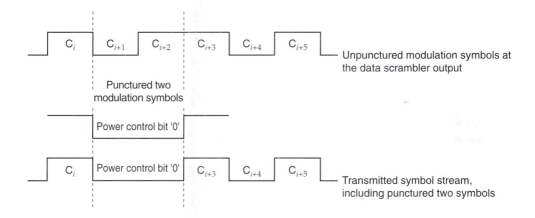

(b) Power control bit '0' (A increase in the mean output power level)

Transmission rate	Values of x
9.6 kbps	2
4.8 kbps	4
2.4 kbps	8
1.2 kbps	16

Figure 4.32 Power control bit and punctured modulation symbols

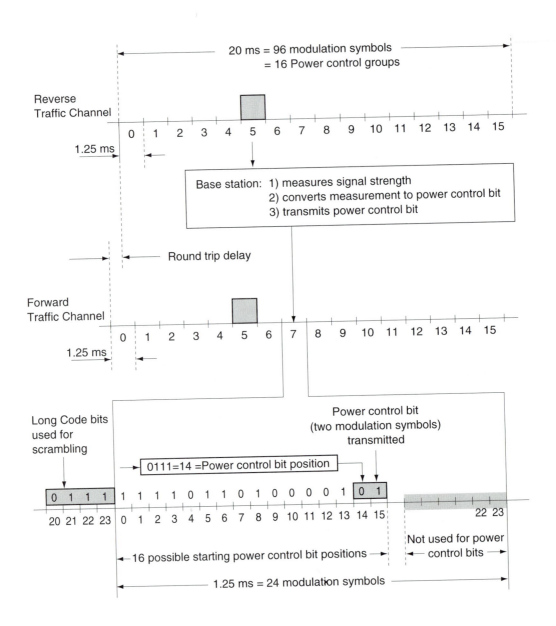

Figure 4.33 Randomization of power control bit position

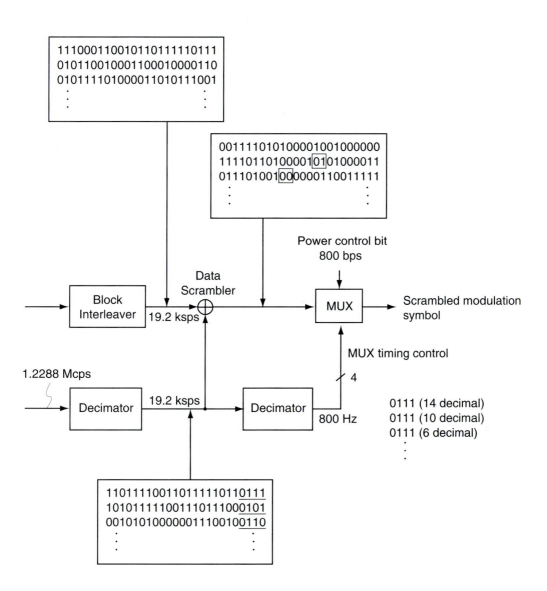

Figure 4.34 MUX operation following the data scrambling for Example 4.21

Since the first 24-bit decimated long code is 110111100110111110110111, the last 4 bits, i.e., 0111 (14 decimal) determine the power control bit position for the second scrambled data sequence 111101101000010101000011 as follows:

MUX input: 111101101000001 $\boxed{01}$ 01000011

MUX timing control bits: 0111 (14 decimal)

Power control bit: 1 (assumed)

MUX output: 111101101000001 $\boxed{11}$ 01000011

This MUX output represents the scrambled modulation symbol sequence.

Code channel number zero (W_0) is always assigned to the pilot channel. The sync channel is assigned to code channel number (W_{32}). Paging channels are assigned to code channel numbers one through seven ($W_1, W_2, \cdots, W_7$) in sequence. The remaining 55 traffic channels are assigned as $W_8, W_9, \cdots, W_{31}, W_{33}, \cdots, W_{63}$.

4.5.6   FTC Orthogonal Spreading

The forward traffic channel transmitted on the forward CDMA link is spread with a Walsh function at a fixed chip rate of 1.2288 Mcps to provide orthogonal channelization among all code channels on a given forward CDMA channel. The forward traffic channel that is spread using Walsh code n is assigned to code index number n ($n = 0$ to 63). One of the time-orthogonal Walsh codes defined in Table 4.7 is used for 9.6 kbps, 4.8 kbps, 2.4 kbps and 1.2 kbps, respectively. The Walsh code spreading sequence is repeated with a period of $52.083\,\mu$s (=64/1.2288 Mcps), which is equal to the duration of one FTC modulation symbol.

Example 4.22   Consider the orthogonal spreading of the multiplexed output from MUX with the Walsh code number W_8 at a fixed ship rate of 1.2288 Mcps. The 64-bit Walsh code corresponding to Walsh function index number W_8 is
    00000000 11111111 00000000 11111111 00000000 11111111 00000000 11111111 (a)
    MUX output at the rate of 19.2 ksps is
                        111101101000011101000011 (b)
Since 1.2288 Mcps/19.2 ksps = 64 bits, one symbol of the multiplexed output corresponds to 64 bits of Walsh code output. Thus, EX-ORing the first symbol '1' of the MUX output with the 64-bit Walsh code is:

First symbol of (b)   11111111   . . .   11111111   . . .   11111111

Walsh code (a)   00000000 11111111 00000000 11111111 00000000 11111111 00000000 11111111

(a) $\oplus$ (b)   11111111 00000000 11111111 00000000 11111111 00000000 11111111 00000000

Therefore, as an example, the first 8 symbols (i.e., 11110110) of the multiplexed output sequence (b) can be spread into the following Walsh code:

11111111 00000000 11111111 00000000 11111111 00000000 11111111 00000000
11111111 00000000 11111111 00000000 11111111 00000000 11111111 00000000
11111111 00000000 11111111 00000000 11111111 00000000 11111111 00000000
11111111 00000000 11111111 00000000 11111111 00000000 11111111 00000000
00000000 11111111 00000000 11111111 00000000 11111111 00000000 11111111
11111111 00000000 11111111 00000000 11111111 00000000 11111111 00000000
11111111 00000000 11111111 00000000 11111111 00000000 11111111 00000000
00000000 11111111 00000000 11111111 00000000 11111111 00000000 11111111

4.5.7   FTC Quadrature Spreading and Filtering

The forward traffic channel after orthogonal spreading is quadrature spread by the pilot PN sequence as specified in 4.1.2.

The I and Q pilot PN sequences, P_I and P_Q, are generated from the maximum length LFSR sequences $i(n)$ and $q(n)$, $1 \leq n \leq 32767$, under the initial contents (100000000000000) of the LFSR.

In order to obtain the I and Q pilot PN sequences of period 2^{15}, a '0' is inserted in $i(n)$ and $q(n)$ after 14 consecutive '0' outputs. This occurs only once in each period. Therefore, the pilot PN sequences have one run of 15 consecutive '0' outputs instead of 14.

Since the chip rate for the pilot PN sequence is 1.2288 Mcps, the pilot PN sequence period is 26.666 (=32768/1228800) ms, and exactly 75 pilot PN sequence repetitions occur every 2 seconds.

Example 4.23 Compute the quadrature pilot PN sequence which is composed of the in-phase component P_I and the quadrature component P_Q.

For the generation of pilot PN chip code, we set up the following primary:

Initial condition:	100000000000000
In-phase tap vector:	100001011100010
Quadrature-phase tap vector:	100111100011100

Then P_I and P_Q can be easily computed using the above primary.

On the other hand, P_I and P_Q are also computed using the reciprocal polynomials $i(n)$ and $q(n)$ of Eqs 4.5 and 4.6 as follows:

(I) Using $i(n) = i(n-15) = 0$ for $1 \leq n \leq 14$, $i(15) = 1$, and $i(n) = \sum_{K=15,10,8,7,6,2} i(n-k)$

for $16 \leq n \leq 32767$, the in-phase pilot PN sequence P_I is found by computing $i(n)$ first and then inserting a 0 (zero) in $i(n)$ after 14 consecutive 0 outputs (see arrow $\downarrow$) as follows:

$$\downarrow$$
$\mathcal{P}_I =$ 00000000000000010101001001110100011011110011001000
                001111000010001101001010110101010101110100100011111
over $0 \le n \le 99$ only.

(II) Similarly, using $q(n)=q(n-15)=0$ for $1 \le n \le 14$, $q(15) = 1$, and $q(n) = \sum_{K=15,12,11,10,9,5,4,3} q(n-k)$ for $16 \le n \le 32767$, we can compute $q(n)$ from which the quadrature-phase pilot PN sequence $\mathcal{P}_Q$ is found by inserting a 0 (zero) at the 15th place in $q(n)$ as shown below:

$$\downarrow$$
$\mathcal{P}_Q =$ 00000000000000010011111010111010110100111000101001 1
                1001110001101100011101001100010010010001100110011
over $0 \le n \le 99$.

Example 4.24 Consider quadrature spreading by means of modulo-2 addition of the Walsh code output with the quardature pilot PN sequence computed in the previous example.

$\qquad$ I = WCO $\oplus$ $\mathcal{P}_I$ and $Q = $ WCO $\oplus$ $\mathcal{P}_Q$ $\qquad\qquad\qquad\qquad$ (4.22)

where WCO denotes Walsh code output.

$\qquad$ Using Eq. 4.22, the output of quadrature spreading (i.e., I-channel sequence and Q-channel sequence) can be computed as follows:

$\qquad$ (I) EX-ORing the Walsh code output with the in-phase PN sequence $\mathcal{P}_I$:

WCO: 11111111 00000000 11111111 00000000 11111111 00000000 11111111 00000000

$\mathcal{P}_I$:   00000000 00000001 01010010 01110100 01101111 00110010 00001111 00001000

$I(\oplus)$:   11111111 00000001 10101101 01110100 10010000 00110010 11110000 00001000

$\qquad$ (II) EX-ORing the Walsh code output with the quadrature-phase PN sequence $\mathcal{P}_Q$:

WCO :   11111111 00000000 11111111 00000000 11111111 00000000 11111111 00000000

$\mathcal{P}_Q$:   00000000 00000001 00111101 01110101 10100111 00010100 11100111 00011011

$Q(\oplus)$:   11111111 00000001 11000010 01110101 01011000 00010100 00011000 00011011

$\qquad$ Our computations shown above are for I and Q corresponding to the first row of the Walsh code output.

$\qquad$ Following quadrature spreading, the I and Q streams are subject to apply to the I and Q baseband filters. The FTC filtering is exactly identical as specified in Section 4.2.3. The baseband filters have a frequency response $S(f)$ that satisfies the limit given in Fig. 3.15.

4.5.8   QPSK for Forward Traffic Channel

The QPSK technique for the forward traffic channel is exactly identical to that specified in Section 4.2.4. Let $I(t)$ and $Q(t)$ be two quadrature binary data coming out from the baseband filters. In nonoffset QPSK, the two baseband streams coincide in time so that the carrier phase can be changed once every T_b seconds.

A QPSK waveform $s(t)$ is obtained by amplitude modulating $I(t)$ and $Q(t)$ each onto the cosine and sine functions of a carrier wave.

The sum of these two orthogonal BPSK waveforms yields the QPSK waveform $s(t)$.

Let $s(t)$ denote the QPSK waveform as represented by

$$s(t) \quad = I(t) \cos \omega_0 t + Q(t) \sin \omega_0 t$$
$$= \sqrt{2} \cos (\omega_0 t - \theta (t))$$

where $I(t) = \sqrt{2} \cos \theta (t)$ and $Q(t) = \sqrt{2} \sin \theta (t)$.

$I(t)$ and $Q(t)$ each can be determined according to the phase offset of $\theta (t)$ as shown in Table 4.2. Consequently, the QPSK waveform $s(t)$ can be depicted using Table 4.3.

Example 4.25   Consider a sample data stream of I and Q quadrature BPSK:

I: 11111111 00000001 10101101 01110100 10010000 00110010 11110000 00001000

Q: 11111111 00000001 11000010 01110101 01011000 00010100 00011000 00011011

The QPSK waveform $s(t)$ based on these two sample BPSK pulses, i.e., $I(t) = 10010000$ and $Q(t) = 01011000$, can be illustrated as shown in Fig. 4.35.

4.5.9   FTC Channel Structure and Message Structure

The base station sends signaling messages to the mobile station during the traffic channel operation. When sending a forward traffic channel message, the base station sends it as signaling traffic using some specified signaling traffic formats (See Table 4.19). The base station may use one or more forward traffic channel frames to send the message.

The first signaling traffic bit in an FTC frame is a Start of Message (SOM) bit. If an FTC message begins in the frame, the base station will set this bit to '1'. If the frame contains bits of an FTC message that began in the previous frame, the base station sets this bit to '0'. The base station uses the remaining signaling traffic bits of the frame to send FTC message bits. If the frame used to send the last bits of a message contains any unused signaling traffic bits, the base station sets each of these bits to '0' as padding bits.

An FTC message consists of a message length field, a message body, and a CRC field as shown in Fig. 4.36. The base station sets an FTC message structure to the length in octets, including the message length field, the message body, and the CRC field. The message length field is 8 bits in length. The base station limits the maximum FTC message length to 148 octets or 1184 bits.

Forward Traffic Channel (4800 bps):
                    I-channel output: 10010000 (-1, 1, 1, -1, 1, 1, 1, 1)
                    Q-channel output: 01011000 (1, -1, 1, -1, -1, 1, 1, 1)

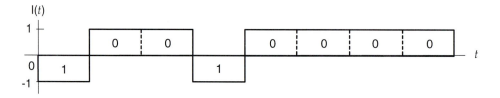

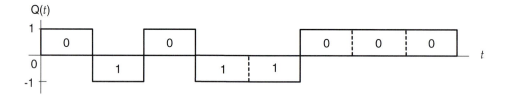

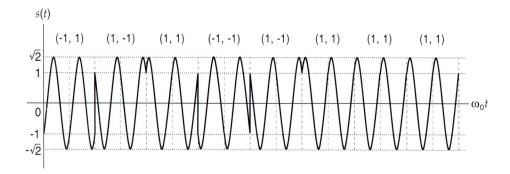

Figure 4.35 The forward traffic channel QPSK waveform $s(t)$ plot

Message length	Message body	CRC
← 8 bits →	← 16~1160 bits →	← 16 bits →

Figure 4.36 Forward traffic channel message structure

4.5.10 FTC Message CRC Field

The base station will set the CRC field (16 bits in length) computed for the FTC message. The CRC computation includes the message length field and the message body.

The generator polynomial for the CRC is the standard CRC-CCITT format as shown below.

$$g(x) = 1 + x^2 + x^{12} + x^{16} \qquad (4.23)$$

The CRC value can be computed by the following procedure and its logic diagram is shown in Fig. 4.37.

1. All initial contents of the LFSR are set to logical one ('1') in order to make the CRC field to nonzero values even for all-zero data.
2. Set the switch in the down position and close the gate in the feedback path of the CRC encoder.
3. Then the k input bits begin to flow into the encoder output as well as the feedback path of the shift register.
   The k-bit information input consists of the 8-bit message length + the message body length in bits.
4. The register will be clocked k time with the k-bit input.
5. Set the switch to the up position and open the gate in the feedback path.
6. Clocking the register by an additional 16 times, the CRC field will be generated as the 16 additional output bits.
7. Thus, the FTC message is transmitted in the order of the 8-bit message length field, the message body length, and 16-bit CRC field at the CRC encoder output.

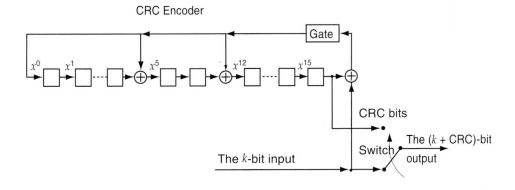

k = Message length field + Message body length
= 8+(16 to 1160) in bits

Figure 4.37 FTC signaling CRC computation

Lastly, the types of signaling messages sent over the FTC are summarized in Table 4.19.

Table 4.19   Forward Traffic Channel Messages

Message Name	Message type (binary)
Order message	00000001
Authentication challenge message	00000010
Alert with information message	00000011
Data burst message	00000100
Handoff direction message	00000101
Analog handoff direction message	00000110
In-traffic system parameters message	00000111
Neighbor list update message	00001000
Send burst DTMF message	00001001
Power control parameters message	00001010
Retrieve parameters message	00001011
Set parameters message	00001100
SSD update message	00001101
Flash with information message	00001110
Mobile station registered message	00001111

CHAPTER 5

Mobile Station Call Processing

Call processing can be classified into two parts, mobile station call processing and base station call processing. Call processing refers to the techniques of message flow protocols between the mobile station and the base station.

This chapter describes mobile station call processing based on TIA/EIA/IS-95. After power is applied to the mobile station, it enters the system determination substate of the mobile station initialization state with a power-up indication. As illustrated in Fig. 5.1, call processing consists of the following states:

- Mobile Station Initialization State—In this state, the mobile station selects and acquires a system.
- Mobile Station Idle State—In this state, the mobile station monitors messages on the paging channel.
- System Access State—In this state, the mobile station sends messages to the base station on the access channel.
- Mobile Station Control on the Traffic Channel State—In this state, the mobile station communicates with the base station using the forward and reverse traffic channels.

5.1 MOBILE STATION INITIALIZATION STATE

In the mobile station initialization state, the mobile station first selects a system to use, either for analog mode operation or for CDMA mode operation. If the selected system is a CDMA system, the mobile station proceeds to acquire and then synchronize to the CDMA system. If the selected system is an analog system, the mobile station begins analog mode operation. Upon entering the system determination substate with a power-up indication, the mobile station will set the first-idle ID status to enabled, and set the authentification random challenge variable to 0.

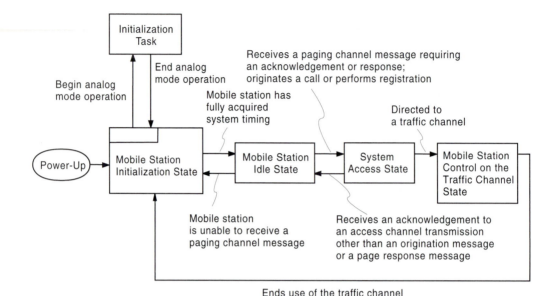

Figure 5.1 Mobile station call processing states

As illustrated in Fig. 5.2, the mobile station initialization state consists of the following substates, i.e. (1) system determination substate, (2) pilot channel acquisition substate, (3) sync channel acquisition substate, and (4) timing change substate. The function of each substate is listed below.

Substate	Function
System Determination	The mobile station selects which system to use.
Pilot Channel Acquisition	The mobile station acquires the pilot channel of a CDMA system.
Sync Channel Acquisition	The mobile station obtains system configuration and timing information for a CDMA system.
Timing Change	The mobile station synchronizes its timing to that of a CDMA system.

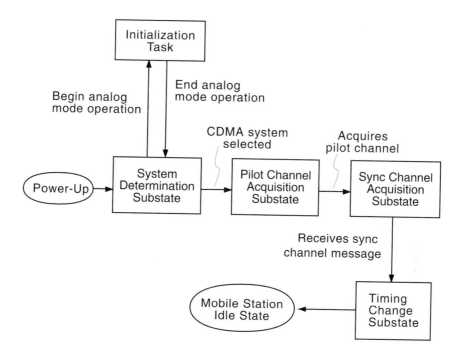

Figure 5.2 Mobile station initialization states

5.1.1 System Determination Substate

In this substate, the mobile station will perform the following:

- The mobile station determines which system to use.
- If the mobile station is to use System A (or B), it sets the selected serving system to SYS-A (or SYS-B).
- If the mobile station is to use an analog system, it enters the initialization task.
- If the mobile station is to use a CDMA system, it sets the CDMA channel number either to the primary or secondary CDMA channel number for the selected serving system. Then the mobile station enters the pilot channel acquisition substate.
- If the mobile station fails to acquire a CDMA system on the first CDMA channel it tries, the mobile station attempts to acquire on the alternate CDMA channel (primary or secondary) before performing the system selection process again.

5.1.2   Pilot Channel Acquisition Substate

In this substate, the mobile station acquires the pilot channel of the selected CDMA system. Upon entering the pilot channel acquisition substate, the mobile station tunes to the CDMA channel number equal to CDMACHs, sets its code channel for the pilot channel, and searches for the pilot channel. If the mobile station acquires the pilot channel within $T_{20m} = 15$ seconds, the mobile station will enter the sync channel acquisition substate. If the mobile station does not acquire the pilot channel within T_{20m} seconds, the mobile station enters the system determination substate. T_{20m} denotes the maximum time to remain in the pilot channel acquisition substate.

5.1.3   Sync Channel Acquisition Substate

In this substate, the mobile station receives and processes the sync channel message to obtain system configuration and timing information. Upon entering the sync channel acquisition substate, the mobile station sets its code channel number equal to 32 for the sync channel.

If the mobile station does not receive a valid sync channel message within $T_{21m} = 1$ second, the mobile station enters the system determination substate. T_{21m} denotes the maximum time to receive a valid sync channel message. If the mobile station receives a valid sync channel message within T_{21m} seconds but the protocol revision level supported by the mobile station is less than the minimum protocol revision level supported by the base station, the mobile station will enter the system determination substate. If the mobile station receives a valid sync channel message within T_{21m} seconds and the protocol revision level supported by the mobile station is greater than or equal to the minimum protocol revision level supported by the base station, the mobile station will store the following information from the message:

- Protocol revision level
- Minimum protocol revision level
- Network identification
- Pilot PN sequence offset
- Long code state
- System time
- Paging channel data rate

The mobile station may also store the following information from the message:

- Number of leap seconds that have occurred since the start of system time
- Offset of local time from system time
- Daylight saving time indicator

The mobile station then enters the timing change substate.

5.1.4   Timing Change Substate

In this substate, the mobile station synchronizes its long code timing to the CDMA system long code timing derived from the long code state obtained from the sync channel message, and synchronizes its system timing to the CDMA system timing derived from the current value of CDMA system time as received in the sync channel message.

The stored system time is equal to the system time corresponding to 4 sync channel superframes 320 ms past the end to the last 80 ms superframe of the received sync channel message minus the pilot PN sequence offset. In other words, the sync channel message contains the long code state valid at a time equal to 320 ms–pilot PN sequence offset after the end of the message. The stored long code state is equal to the system long code state corresponding to the internal value stored by the mobile station in temporary memory which is not sent over the air.

5.2 MOBILE STATION IDLE STATE

In this state, the mobile station monitors the paging channel. The mobile station can receive messages, receive an incoming call (i.e., terminated call), initiate a call (i.e., originated call), initiate a registration, or initiate a message transmission.

Upon entering the mobile station idle state, the mobile station sets its code channel, sets the paging channel data rate, and performs paging channel supervision.

The paging channel supervision shall be specified as follows:

The mobile station checks the 30-bit CRC of all received paging channel messages. If the mobile station is operating in the nonslotted mode in the mobile station idle state, it monitors the paging channel at all times. The mobile station will reset a timer for $T_{30m} = 3$ seconds whenever a valid message is received on the paging channel, whether addressed to the mobile station or not. If the timer expires, the mobile station declares a loss of the paging channel. T_{30m} denotes the maximum time to receive a valid paging channel message.

If the mobile station is operating in the slotted mode in the mobile station idle state, it will set a timer for T_{30m} seconds at the start of the first slot in which the paging channel is monitored, and will reset the timer for T_{30m} seconds whenever a valid message is received, whether addressed to the mobile station or not. If the timer expires while the mobile station is monitoring the paging channel, the mobile station will declare a loss of the paging channel. The mobile station will disable the timer when not monitoring the paging channel.

If at any time, the mobile station exits the mobile station idle state, it will enter the system determination substate of the mobile station initialization state.

While in the mobile station idle state, the mobile station performs the following procedures:

- Paging channel monitoring procedures (as specified in Section 5.2.1)
- Message acknowledgment procedures (Section 5.2.2)
- Registration procedures (Section 5.2.3)
- Idle handoff procedures (Section 5.2.4)
- Response to overhead information operation (Section 5.2.5)
- Page match operation (Section 5.2.6)
- Order and message processing operation (Section 5.2.7)
- Origination operation (Section 5.2.8)
- Message transmission operation (Section 5.2.9)
- Power-down operation (Section 5.2.10)

5.2.1 Paging Channel Monitoring Procedures

The paging channel is divided into 80 ms slots called paging channel slots. Paging and control messages for a mobile station operating in the nonslotted mode can be received in any of the paging channel slots. Therefore, the nonslotted mode of operation requires the mobile station to monitor all slots.

The paging channel protocol also provides for scheduling the transmission of messages for a specific mobile station in certain assigned slots.

A mobile station that monitors the paging channel only during certain assigned slots is referred to as operating in the slotted mode. In the slots during which the paging channel is not being monitored, the mobile station can stop or reduce its processing for power conservation. A mobile station may not operate in the slotted mode unless it is in the mobile station idle state. In other words, when a mobile station monitors the paging channel in any state other than the mobile station idle state, it operates in the nonslotted mode. A mobile station operating in the slotted mode generally monitors the paging channel for one or two slots per slot cycle. The mobile station can specify its preferred slot cycle using the slot-cycle-index field in the registration message, origination message, or page response message. The mobile station also can specify its preferred slot cycle using the slot-cycle-index field of the terminal information record of the status message when in the mobile station control on the traffic channel state. The length of the slot cycle T is given by

$$T = 2^i$$

where i is the selected slot cycle index. There are 16 x T slots in a slot cycle. As shown in Fig. 5.3, the minimum length of the slot cycle consists of 16 slots of 80 ms each, which equal 1.28 seconds. The paging channel slot number, modulo the maximum length slot cycle (2048 slots) is given by

$$\text{Slot No.} = \lfloor t/4 \rfloor \mod 2048$$

where t denotes the system time in frames. If the mobile station is directed by the user to modify the preferred slot cycle index, the mobile station performs parameter-change registration.

For each of its assigned slots, the mobile station begins monitoring the paging channel in time to receive the first bit of the assigned slot. To determine its assigned slots, the mobile station uses the hash function to select a number in the range 0 to 2047. The mobile station's assigned slots are those slots in which ($[t/4]$ – PGSLOT) mod (16 x T)=0, where t is the system time in frames and T is the slot cycle length in units of 1.28 seconds given by $T=2^i$. PGSLOT denotes the value obtained from the hash function, used to determine the mobile station's assigned paging channel slots. Figure 5.3 shows an example for a slot cycle length of 1.28 seconds, in which the computed value of the paging channel slot number is equal to 6, so that one of the mobile station's slot cycles begins when the slot number equals 6. The mobile station begins monitoring the paging channel at the start of the slot in which the slot no=6. The next slot in which the mobile station must begin monitoring the paging channel is 16 slots later, i.e., the slot in which the slot number is 22 (=6+16).

During operation in the slotted mode, the mobile station insures that its stored configuration parameter values are current. Configuration parameters are received in the configuration messages, i.e., system parameters message, neighbor list message and CDMA channel list message.

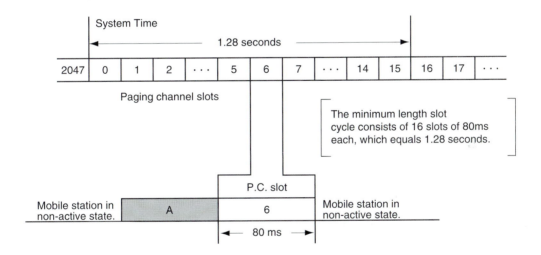

A: Reacquistion of CDMA system
6: Mobile station's assigned paging channel slot
Systam Time: System time is synchronous to UTC time and same time origin
　　　　　　as GPS time. All base stations and mobile stations use the same
　　　　　　system time, but offset by the propagation delay from the base
　　　　　　station to the mobile station must be taken into account.
UTC: Universal Coordinated Time
GPS: Global Positioning Time

Figure 5.3 Mobile station idle slotted mode structure

Associated with the set of configuration messages sent on each paging channel is a configuration message sequence number. When the contents of one or more of the configuration messages change, the configuration message sequence number is incremented. For each of the configuration messages received, the mobile station stores the configuration message sequence number contained in the configuration message. The mobile station examines the stored values of the configuration message sequence numbers to determine whether the configuration parameters stored by the mobile station are current.

5.2.2   Message Acknowledgment Procedures

Acknowledgment procedures referred to as layer 2 procedures facilitate the reliable exchange of messages between the base station and the mobile station. Acknowledgments of messages received on the paging channel are sent on the access channel. The mobile station uses the following layer 2 fields to support the acknowledgment mechanism:

- Acknowledgments address type field
- Acknowledgment sequence number field
- Acknowledgment required field
- Valid acknowledgment field

All other message fields and the processing thereof are referred to as layer 3.

When sending a message that includes an acknowledgment, the mobile station sets the valid acknowledgment field to '1'. For acknowledgment of a page message or slotted page message, the mobile station sets the acknowledgment sequence number field equal to the message sequence number field of the record containing the mobile station's identification number (MIN), and sets the acknowledgment address type field to '000'.

When sending a message that does not include an acknowledgment, the mobile station sets the valid acknowledgment field to '0' and sets the acknowledgment address type field and acknowledgment sequence number field equal to the address type and message sequence number fields, respectively, of the last message received that required acknowledgment. If no such message has been received, the mobile station sets the acknowledgment address type field to '000' and sets the acknowledgment sequence number field to '111'.

Unless otherwise specified in the requirements for processing a specific message, the mobile station transmits an acknowledgment in response to any message received that is addressed to the mobile station and that has the acknowledgment required field set to '1'. The mobile station transmits a page response message including an acknowledgment in response to each record of a page message or slotted page message addressed to the mobile station's MIN. If a specific message is required in response to any other message requiring acknowledgment, the acknowledgment will be included with the response. If no specific message is required to be transmitted in response to a received message requiring acknowledgment, the mobile station includes the acknowledgment in a mobile station acknowledgment order.

If no message requiring acknowledgment has been received, the mobile station does not include an acknowledgment in any transmitted message until a message is received that requires acknowledgment. After a message including an acknowledgment has been sent, the mobile station does not include an acknowledgment in any subsequent transmitted message until another message is received that requires acknowledgment.

The mobile station detects duplicate received messages using the following rules. The mobile station considers two messages (or order records) containing the mobile station's address field to be duplicates if all of the following are true:

- The messages (or records) were received (1) on the same paging channel; and (2) within $T_{4m} = 2.2$ seconds of each other. T_{4m} denotes a period in which two messages received by the mobile station on the same paging channel and carrying the same sequence numbers are considered duplicates.
- The messages (or records) contain (1) the same value in both the message sequence number field and the acknowledgment required field; and (2) identical address type and address fields.

The mobile station considers two page records to be duplicates if all of the following are true:

- Both records were received (1) on the same paging channel; and (2) were in messages received within T_{4m} seconds of each other, or in the same message.
- Both records contain (1) the same values in the message sequence number field; and (2) are addressed to the same MIN.

The mobile station discards, without further processing, any message or page record that is a duplicate of one previously received.

Paging channels are considered different (1) if they are from different base stations, or (2) on different code channels, or (3) on different CDMA channels.

The mobile station considers messages to be different if they are not duplicates according to the rules given above.

5.2.3   Registration and its Procedure

Registration is the process by which the mobile station notifies the base station of its location, status, identification, slot cycle, station class, and other characteristics. The mobile station informs the base station of its location and status so that the base station can efficiently page the mobile station when establishing a mobile terminated call. The mobile station supplies the base station with the set of MINs that are active. This allows the base station to know which mobile station to contact when handling a mobile terminated call. If operating in the slotted mode, the mobile station supplies the slot-cycle-index parameter so that the base station knows which slots the mobile station is monitoring. The mobile station supplies the station class mark and common air interface revision (CAI-REV) number so that the base station knows the capabilities of the mobile station.

The CDMA system supports the following different forms of registration:

1. Power-up registration

   Power-up registration is performed when the mobile station is turned on. The mobile station registers when it powers on, switches from using the alternate serving system, or switches from using the analog system. To prevent multiple registrations when power is quickly turned on and off, the mobile station delays $T_{57m} = 20$ seconds before registering after entering the mobile station idle state. T_{57m} denotes the limit of the power-up registration timer.

2. Power-down registration

   Power-down registration is performed when the user directs the mobile station to power off. If power-down registration is performed, the mobile station does not power down until after completing the registration attempt. The mobile station does not perform power down registration if it has not previously registered in the system that corresponds to the current system identifier and network identification. A base station is a member of a cellular system and a network. A network is a subset of a system. Systems are labeled with an identifier called the system identification (SID). Networks within a system are given a network identification (NID). Thus, a network is uniquely identified by the pair (SID, NID).

   The NID number 0 is a reserved value indicating all base stations that are not included in a specific network. The NID number 65535 ($2^{16}-1$) is a reserved value the mobile station may use for roaming status determination to indicate that the mobile station considers all NIDs within an SID to be home (nonroaming).

   The mobile station has a list of one or more home (nonroaming) (SID, NID) pairs. A mobile station is roaming if the stored (SID, NID) pair does not match one of the mobile station's home (SID, NID) pairs. Two types of roaming are defined: A mobile station is a foreign NID roamer if the mobile station is roaming and there is some (SID, NID) pair in the mobile station's (SID, NID) list for which SID is equal to the stored SID. A mobile station is a foreign SID roamer if there is no (SID, NID) pair in the mobile station's (SID, NID) list for which SID is equal to the stored SID.

3. Timer-based registration

   The mobile station registers when a timer expires. Timer-based registration causes the mobile station to register at regular intervals. Its use also allows the system to automatically deregister mobile stations that did not perform a successful power-down registration. Timer-based registration uses a paging channel slot counter. Timer-based registration is performed when the counter reaches a maximum value that is controlled by the base station via the registration period field of the system parameters message. The base station disables time-based registration by setting the registration period to zero.

   The mobile station computes and stores the timer expiration count as

   $$E = \lfloor 2^{r/4} \rfloor^{260}$$

where E denotes the timer expiration count and r represents the registration period (the time interval between timer-based registrations).

Whenever the mobile station changes the timer-based registration indicator from NO to YES, it sets the timer-based registration count to a pseudorandom value between 0 and E–1, using a linear congruential generator defined by

$$Z_n = a\, Z_{n-1} \bmod m$$

where $a = 7^5 = 16807$ and $m = 2^{31}-1 = 2147483647$.

If the mobile station is operating in the nonslotted mode, it increments the time-based registration counter once per 80 ms whenever the time-based registration indicator equals YES. If the mobile station is operating in slotted mode, it increments the timer-based registration counter when it begins to monitor the paging channel.

4. Distance-based registration

The mobile station registers when the distance between the current base station and the base station in which it last registered exceeds a threshold. The mobile station determines that it has moved a certain distance by computing a distance measure based on the difference in latitude and longitude between the current base station and the base station where the mobile station last registered. If this distance measure exceeds the threshold value, the mobile station registers.

The mobile station stores the base station latitude, the base station longitude, and the registration distance of the base station where the access channel was used for the mobile station's last registration. The mobile station computes the current base station's distance from the last registration point as:

$$\text{Distance} = \left[\frac{(\Delta\; latitude)^2 + (\Delta\; longitude)^2}{16} \right]$$

where

Δ latitude = A – B

Δ longitude = (C – D) cos (π /180 x B/14400)

A: Latitude value of current-base-station stored in a mobile station's temporary memory

B: Base station's latitude registration value stored in the mobile station's semi-permanent memory

C: Longitude value of current-base-station stored in the mobile station's temporary memory

D: Base station's longitude registration value stored in mobile station's semi-permanent memory.

5. Zone-based registration

Zones are groups of base stations within a given system and network. A base station's zone assignment is identified by the registration zone number field of the system parameters message. The mobile station registers when it enters a new zone. A zone is added to the list whenever a registration occurs, and is deleted upon expiration of a timer. After a system access, timers are enabled for every zone except the one that was successfully registered by the access. Timers also are enabled at the start of a call.

A mobile station can be registered in more than one zone. Zones are uniquely identified by a zone number plus the system identification (SID) and the network identification (NID) of the zone. The mobile station stores a list of the zones in which the mobile station has registered. Each entry in the zone list includes the zone number and the (SID, NID) pair for the zone. The mobile station is capable of storing at least $N_{9m} = 7$ entries in the zone list. N_{9m} denotes the maximum supported zone list whose value is 7. A base station is considered to be in the zone list only if the base station's registration zone, SID and NID are found in an entry in the zone list. The mobile station provides storage for one entry of the zone list in semi-permanent memory.

The base station controls the maximum number of zones in which a mobile station may be considered registered, by means of the total-zones field of the system parameters message. When an entry is added to the zone list, or the total zones is decreased, the mobile station removes entries from the zone list if there are more entries than allowed by the setting of total zones. Whenever the zone list contains more than the total zone entries, the mobile station deletes the excess entries.

6. Parameter-change registration

Parameter-change registration is performed when a mobile station modifies any of the following stored parameters:

- The preferred slot cycle index
- The station class mark
- The call termination enabled indicator

The mobile station has a list of one or more nonroaming (SID, NID) pairs. A mobile station is roaming if the stored (SID, NID) pair does not match one of the mobile station's nonroaming (SID, NID) pairs.

Parameter-change registration is independent of the roaming status of the mobile station because the registration-enabled indicator does not govern the parameter-change registration.

7. Ordered registration

The mobile station registers when the base station requests it. The base station can command the mobile station to register by sending a registration request order. Ordered registration is performed in the mobile station order and message processing operation. During this processing operation, the mobile station processes all messages except overhead messages and page messages.

Ordered registration is performed after receiving a registration request order while in the *mobile station order and message processing operation.* The mobile station enters the update overhead information substate of the system access state with a registration indication within $T_{33m} = 0.3$ seconds after the registration request order is received. T_{33m} is the maximum time to enter the update overhead information substate of the system access state to respond to messages received while in the mobile station idle state (except authentication messages).

8. Implicit registration

When a mobile station successfully sends an origination message or page response message, the base station infers the mobile station's location. This is considered an implicit registration. Successful registration or implicit registration is performed after the mobile station receives an acknowledgment for a registration message, origination message, or page response message sent on the access channel.

9. Traffic channel registration

Whenever the base station has registration information for a mobile station that has been assigned to a traffic channel, the base station can notify the mobile station that it is registered.

The first five forms of registration, as a group, are called autonomous registration and are enabled by roaming status. Parameter-change registration is independent of roaming status. Ordered registration is initiated by the base station through an order message. Implicit registration does not involve the exchange of any registration messages between the base station and the mobile station. While a mobile station is assigned a traffic channel, the base station can obtain registration information by sending the status request order to obtain status messages from the mobile station. The mobile station can be notified that it is registered through the mobile station registered message.

Any of the various forms of autonomous registration and parameter-change registration can be enabled and disabled. The forms of registration that are enabled and the corresponding registration parameters are communicated in the system parameters message.

The mobile station provides a means of enabling and disabling each timer. A timer that has been enabled is referred to as active. While in the mobile station idle state, the mobile station maintains or updates all active registration timers. When a timer is disabled, it is not considered expired. If any timer expires while in this idle state, the mobile station preserves the expiration status so that further action can be taken in the mobile station idle state.

The mobile station provides the following registration timers:

• Power-up/initialization timer

The mobile station maintains a power-up/initialization timer. While the power-up/initialization timer is active, the mobile station does not make registration access attempts.

- Time-based registration timer

  The mobile station registers when a timer expires. This is a registration method in which the mobile station registers whenever a counter reaches a predetermined value. The counter is incremented an average of once per 80 ms period. Time-based registration uses a paging channel slot counter equivalent to a timer with time increments of 80 ms.

- Zone list entry timers

  An autonomous registration method in which the mobile station registers whenever it enters a zone that is not in the mobile station's zone list. The mobile station maintains a zone list entry timer for each entry in the registration zone list. When an entry is removed from the zone list, the corresponding zone list entry timer is disabled. The timer duration is determined from the stored value of the zone timer length. The mobile station provides a means to examine each timer's value while the timer is active, so that the age of list entries can be computed.

- SID/NID list entry timers

  The mobile station maintains a SID/NID list entry timer for each entry in the SID/NID list. When an entry is removed from the registration SID/NID list, the corresponding SID/NID list entry timer is disabled. The time duration is determined from the stored value of the zone timer length. The mobile station provides a means to examine each timer's value while the timer is active, so that the age of list entries can be computed.

5.2.4   Idle Handoff Procedures

An idle handoff occurs when a mobile station has moved from the coverage area of one base station into the coverage area of another base station during the mobile station idle state. The mobile station determines that an idle handoff should occur when it denotes a sufficiently strong pilot channel signal other than that of the current base station's pilot channel signal.

Pilot channels are identified by their offsets relative to the zero offset PN sequence. Pilot offsets are grouped into sets describing their status with regard to pilot searching.

The following sets of pilot offsets are defined for a mobile station in the mobile station idle state.

- Active set
  The pilot offset of the forward CDMA channel whose paging channel is being monitored.

- Neighbor set
  The offsets of the pilot channels that are likely candidates for idle handoff. The members of the neighbor set are specified in the neighbor list message.

- Remaining set
  The set of all possible offsets in the current system on the current CDMA frequency assignment, excluding the pilots in the neighbor set and the active set.

The mobile station supports a neighbor set size of at least 20 pilots.

In the mobile station idle state, the mobile station continuously searches for the strongest pilot channel signal on the current CDMA frequency assignment whenever it monitors the paging channel.

The pilot search should be governed by the following:

• Active set

The search window size for the pilot in the active set should be the number of PN chips specified in Table 5.1 corresponding to the search window size for the active set (SRCH-WIN-A). The mobile station should center the search window for the pilot of the active set around the earliest arriving usable multipath component of the pilot. If the mobile station receives a value greater than or equal to 13 for SRCH-WIN-A, it may store and use the value 13 in SRCH-WIN-A.

• Neighbor set

The search window size for each pilot in the neighbor set should be the number of PN chips specified in Table 5.1 corresponding to the search window size for the neighbor set (SRCH-WIN-N). The mobile station should center the search window for each pilot in the neighbor set around the pilot's PN sequence offset using timing defined by the mobile station's time reference.

Table 5.1 Search Window Sizes

SRCH-WIN-A SRCH-WIN-N SRCH-WIN-R	Window Size (PN ships)	SRCH-WIN-A SRCH-WIN-N SRCH-WIN-R	Window Size (PN ships)
0	4	8	60
1	6	9	80
2	8	10	100
3	10	11	130
4	14	12	160
5	20	13	226
6	28	14	320
7	40	15	452

- Remaining set

  The search window size for each pilot in the remaining set should be the number of PN chips specified in Table 5.1 corresponding to the search window size for the remaining set (SRCH-WIN-R). The mobile station should center the search window for each pilot in the remaining set around the pilot's PN sequence offset using timing defined by the mobile station's time reference. The mobile station should only search for remaining set pilots whose pilot PN sequence offset indices are equal to integer multiples of the pilot offset index increment.

If the mobile station determines that one of the neighbor set or remaining set pilot channel signals is sufficiently stronger than the active set pilot channel, the mobile station should perform an idle handoff as specified in the following.

While performing an idle handoff, the mobile station operates in the nonslotted mode until the mobile station has received at least one valid message on the new paging channel. Following the reception of this message the mobile station resumes slotted mode operation as discussed in Section 5.2.1. After performing an idle handoff, the mobile station discards all unprocessed messages received on the old paging channel.

If the new base station is listed in the neighbor list message from the old base station, the mobile station uses the 3-bit neighbor configuration field to determine the actions required to transition to the new base station. If the new base station is not listed in the neighbor list message, the mobile station performs the handoff operation using the same procedure as for a pilot in the list with the neighbor configuration field set to '011'.

If the neighbor configuration field is '000', the mobile station sets the access parameters message sequence number to NULL and sets the pilot PN sequence offset to the pilot index of the base station transmitting the new paging channel. If the mobile station has not stored configuration parameters for the new paging channel, or if the stored information is not current, the mobile station sets the configuration message sequence number, the system parameters message sequence number, the neighbor list message sequence number, and the CDMA channel list message sequence number to NULL. The mobile station then begins monitoring the paging channel of the new base station, using the same code channel and CDMA channel.

If the neighbor configuration field is '010', the mobile station sets the access parameters message sequence number to NULL and sets the pilot PN sequence offset to the pilot offset index of the base station transmitting the new paging channel. If the mobile station has not stored configuration parameters for the primary paging channel of the new base station, or if the stored information is not current, the mobile station sets the current message sequence number, the system parameters message sequence number, the neighbor list message sequence number, and CDMA channel list message sequence number to NULL.

Set the number of paging channels supported on the current CDMA channel to '1'. The mobile station tunes to the first CDMA channel given in the CDMA channel list message for the old base station and begins monitoring the primary paging channel of the new base station. If the neighbor configuration field is '011', the mobile station enters the system determination substate of the mobile station initialization state.

5.2.5 Response to Overhead Information Operation

The paging channel messages are summarized into the message groups as shown in Table 5.2. Messages of each group are sent either periodically or on an as-needed basis.

Table 5.2 Paging Channel Message

Message Name	Message Type (binary)
System parameter message	00000001
Access parameter message	00000010
Neighbor list message	00000011
CDMA channel list message	00000100
Slotted page message	00000101
Page message	00000110
Order message	00000111
Channel assignment message	00001000
Data burst message	00001001
Authentication challenge message	00001010
SSD update message	00001011
Feature notification message	00001100
Null message	- - -

The paging channel is used to send control information to mobile stations that have not been assigned to a traffic channel. The paging channel is divided into 80 ms time slots. The slots are grouped into maximum slot cycles of 2048 slots (163.84 seconds). Each maximum slot cycle begins at the start of the frame when a system time, in units of 80 ms, modulo 2048 is zero. A mobile station operating in the slotted mode monitors the paging channel using a slot cycle with a length that is a submultiple of the maximum slot cycle length.

The overhead messages on the paging channel are

1. System parameter message
2. Access parameter message
3. Neighbor list message
4. CDMA channel list message

The response to an overhead information operation is performed whenever the mobile station receives an overhead message. The mobile station updates internally stored information from the received message's data fields.

Configuration parameters and access parameters are received in the configuration messages and the access parameters message. The configuration messages are

1. System parameter message
2. Neighbor list message
3. CDMA channel list message

Associated with the set of configuration messages sent on each paging channel is a configuration message sequence number. When the contents of one or more of the configuration messages change, the configuration message sequence number is incremented. For each of the configuration messages received, the mobile station stores the configuration message sequence number contained in the configuration message. The mobile station also stores the most recently received configuration message sequence number contained in any message. The mobile station examines the stored values of the configuration message sequence numbers to determine whether the configuration parameters stored by the mobile station are current.

The configuration message sequence number is also included in the page message and the slotted page message. This allows the mobile station to determine whether the stored configuration parameters are current without waiting for a configuration message. In the page message and the slotted page message, the access parameters message sequence number equals the value of that sequence number being transmitted in the current access parameters message. If a mobile station receives a page message and the slotted page message with an access parameters message sequence number that does not match the sequence number of the last access parameters message it received and stored, then the mobile station receives the current message and updates to the new parameters before accessing the system.

Access parameter messages are independently sequence-numbered by the access parameters message sequence number. The access parameters message sequence number tells mobile stations if the access parameters message sent by a particular base station has changed. In the access parameters message, the value of the access parameters message sequence number increments by 1 whenever any field in the message changes. The mobile station stores the most recently received access parameters message sequence number.

Paging channels are considered different if they are transmitted by different base stations, if they are transmitted on different code channels, or if they are transmitted on different CDMA channels. Configuration and access parameters from one paging channel are not used while monitoring a different paging channel. The mobile station ignores any overhead message whose pilot PN sequence offset index field is not equal to the pilot offset index of the base station whose paging channel is being monitored.

The mobile station may store the configuration parameters from paging channels it has recently monitored. When a mobile station starts monitoring a paging channel that it has recently monitored, the mobile station can determine whether the stored parameters are current by examining the stored configuration message sequence number in a configuration message, a slotted page message, or a page message.

The mobile station considers the stored configuration parameters to be current only if all the following conditions are true:

• All three stored configuration message sequence numbers are equal to the configuration message sequence number; and
• The stored configuration message sequence number is not equal to NULL; and

• No more than $T_{31m} = 600$ seconds have elapsed since the mobile station last received a valid message on the paging channel for which the parameters were stored. T_{31m} indicates the maximum time for which configuration parameters are considered valid.

The mobile station defines a special value NULL to be stored in place of sequence numbers for messages that have not been received or are marked as not current. The special value NULL is not equal to any valid message sequence number.

If the stored parameters are current, the mobile station processes the parameters as described with regard to the system parameters message, neighbor list message, and CDMA channel list message.

1. System parameter message

   Whenever a system parameters message is received on the paging channel, the received configuration message sequence number is compared to that stored in the system parameters message sequence number of the mobile station. If the comparison results in a match, the mobile station ignores the message. If the comparison results in a mismatch, then the mobile station must process the remaining fields in the message as described in IS-95.

2. Access parameters message

   Whenever an access parameters message is received on the paging channel, the received access parameters message sequence number is compared to the stored access parameters message sequence number. If the comparison results in a match, the mobile station ignores the message. If the comparison results in a mismatch, then the mobile station processes the remaining fields in the message as described in IS-95.

3. Neighbor list message

   Whenever a valid neighbor list message is received on the current paging channel, the received configuration message sequence number is compared to that stored in the neighbor list message sequence number of the mobile station. If the comparison results in a match, the mobile station ignores the message. If the comparison results in a mismatch, then the mobile station must process the remaining fields in the message as discussed in IS-95.

4. CDMA channel list message

   Whenever a CDMA channel list message is received on the paging channel, the received configuration message sequence number is compared to that stored in the CDMA channel list message sequence number of the mobile station. If the comparison results in a match, the mobile station ignores the message. If the comparison results in a mismatch, then the mobile station must process the remaining fields in the message as described in IS-95.

5.2.6 Mobile Station Page Match Operation

The page messages on the paging channel are the page message and the slotted page message. The mobile station page match operation is performed whenever the mobile station receives a page message. The mobile station searches each message to determine whether it contains the mobile station's identification number (MIN). If so, the mobile station transmits a page response message to the page message on the access channel.

The mobile station compares the received value to the stored value of the configuration message sequence number. If the comparison results in a mismatch, then the mobile station sets the stored configuration message sequence number to the received configuration message sequence number. The mobile station also compares the received access parameters message sequence number with that stored in the mobile station. If the comparison results in a mismatch, then the mobile station sets the stored access parameters message sequence number to NULL.

The mobile station compares its MIN with the MIN in each record of the page message. If both MIN1 and MIN2 are present in a record and both MIN1 and MIN2 match MIN1 and MIN2 for the mobile station, then a page match is declared. Note that MIN1 denotes the 24-bit identification number which corresponds to the 7-digit directory telephone number assigned to a mobile station, while MIN2 is the 10-bit identification number which corresponds to the 3-digit area code assigned to a mobile station. If MIN1 but MIN2 is present in a record, MIN1 matches MIN1 for the mobile station, and a nonroaming (SID, NID) pair matches the SID and NID of the base station, then a page match is declared. Any other combination is considered a mismatch.

If a page match is declared, and the mobile station is configured to receive mobile station terminated calls in its present roaming status, the mobile station enters the update overhead information substate of the system access state with a page response indication within $T_{33m} = 0.3$ seconds after the page message is received. If the mobile station is not configured to receive mobile station terminated calls in its present roaming status, the mobile station ignores the record.

5.2.7 Mobile Station Order and Message Processing Operation

During this operation, the mobile station processes all messages except overhead messages and the page messages.

The mobile station compares the address field of the message to the corresponding mobile station identification data (i.e., MIN or ESN). If the identification data matches the address field, the mobile station processes the message; otherwise the mobile station ignores the message.

The following cases occur for a message received on the paging channel whose address field matches the mobile station's identification data:

1. If the message requires acknowledgment, and is not the lock until power-cycled order or the unlock order, the mobile station acknowledges the message. The mobile station enters the update overhead information substate of the system access state with an order/message response indication within T_{33m} seconds, unless otherwise specified for a particular message.

2. If the message does not require acknowledgment, the mobile station transmits a response only if it is required by the message or order. If a response is required, the mobile station enters the update overhead information substate of the system access state with an order/message response indication within T_{33m} seconds, unless otherwise specified for a particular message.

3. If the message is a message that cannot be processed by the mobile station, the mobile station responds with a mobile station reject order by indicating the reason for rejection. The mobile station enters the update overhead information substate of the system access state with an order/message response indication within T_{33m} seconds, unless otherwise specified for a particular message.

Sixteen (16) different messages and orders can be received. If any field value of the message is outside its permissible range, the mobile station sends a mobile station reject order with the order qualifier field equal to 00000100 (message field not in valid range) or 00000010 (message not acceptable in this state), as appropriate.

5.2.8 Mobile Station Origination Operation

The mobile station origination operation is performed when the mobile station is directed by the user to initiate a call. The mobile station enters the update overhead information substate of the system access state with a call origination indication within T_{33m} seconds. In the system access state, the mobile station sends messages to the base station on the access channel(s) and receives messages from the base station on the paging channel.

5.2.9 Mobile Station Message Transmission Operation

If the mobile station supports the mobile station message transmission operation, the operation is performed when the user directs the mobile station to transmit a message. The mobile station then enters the update overhead information substate of the system access state with a message transmission indication within T_{33m} seconds. Support of this operation is optional.

5.2.10 Mobile Station Power-Down Operation

This operation is performed when the user directs the mobile station to power down. The mobile station updates stored parameters and performs other registration procedures. If no power-down registration is performed, the mobile station may power down.

5.3 SYSTEM ACCESS STATE

In the system access state, the mobile station sends messages to the base station on the access channel(s) and receives messages from the base station on the paging channel.

As illustrated in Fig. 5.4, the system access state consists of the following six substates: Update Overhead Information substate, Mobile Station Origination Attempt substate, Page Response substate, Mobile Station Order/Message Response substate, Registration Access substate, and Mobile Station Message Transmission substate.

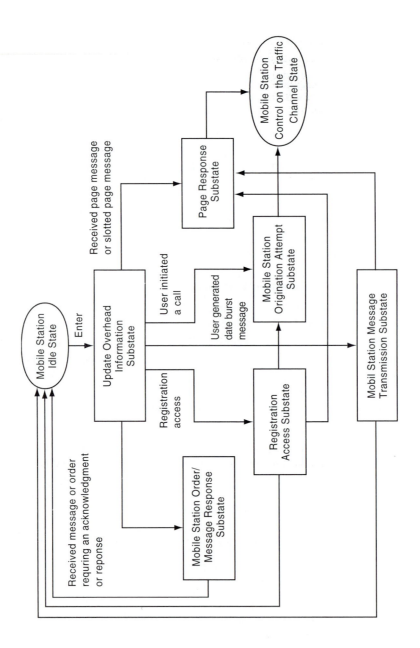

Figure 5.4 System access state

5.3.1   Access Procedures

The mobile station transmits on the access channel using a random access procedure. Many parameters of the random access procedure are supplied by the base station in the access parameters message. The entire process of sending one message and receiving (or failing to receive) an acknowledgment for that message is called an access attempt, as shown in Fig. 5.5. Each transmission in the access attempt is called an access probe. The mobile station transmits the same message in each access probe of an access attempt. Each access probe consists of an access channel preamble and an access channel message capsule.

Within an access attempt, access probes are grouped in access probe sequences. Each access probe sequence consists of up to 1 + Number of Access Probes, all transmitted on the same access channel. The maximum number of probes in an access probe sequence is 1 + Number of Access Probes, ranging from 1 to 16. The mobile station transmits all 1 + Number of Access Probes in an access probe sequence unless an acknowledgment is received from the base station. The access channel used for each access probe sequence is chosen pseudorandomly from among all the access channels associated with the current paging channel. The first access probe of each access probe sequence is transmitted at a specified power level relative to the nominal open loop power level. Each subsequent access probe is transmitted at a power level a specified amount higher than the previous access probe.

The timing of access probes and access probe sequences is expressed in terms of access channel slots. The transmission of an access probe begins at the start of an access channel slot.

There are two types of messages sent on the access channel, i.e., a response message which is a response to a base station message or request message sent autonomously by the mobile station. Different procedures are used for sending a response message and for sending a request message. The timing of the start of each access probe sequence is determined pseudorandomly. For each access probe sequence, a backoff delay (the inter-probe sequence backoff (RS), from 0 to 1 + Access Channel Backoff slots, is generated pseudorandomly. For request access probe sequences only, an additional delay is imposed by the use of a persistence test. A persistence test is not needed for response access attempts because the base station controls the arrival rate of response messages directly by controlling the rate at which it transmits messages requiring responses. For each slot after the backoff delay (RS) the mobile station performs a pseudorandom test, with parameters that depend on the reason for the access attempt and the access overload class of the mobile station. If the test passes, the first access probe of the sequence begins in that slot. If the test fails, the access probe sequence is deferred until at least the next slot.

Timing between access probes of an access probe sequence is also generated pseudorandomly. After transmitting each access probe, the mobile station waits a specified period, TA = (2 + Acknowledgment Timeout) x 80 ms, from the end of the slot to receive an acknowledgment from the base station. If an acknowledgment is received, the next access attempt ends. If no acknowledgment is received, the next access probe is transmitted after an additional backoff delay (RT) from 0 to 1 + Probe Backoff slots.

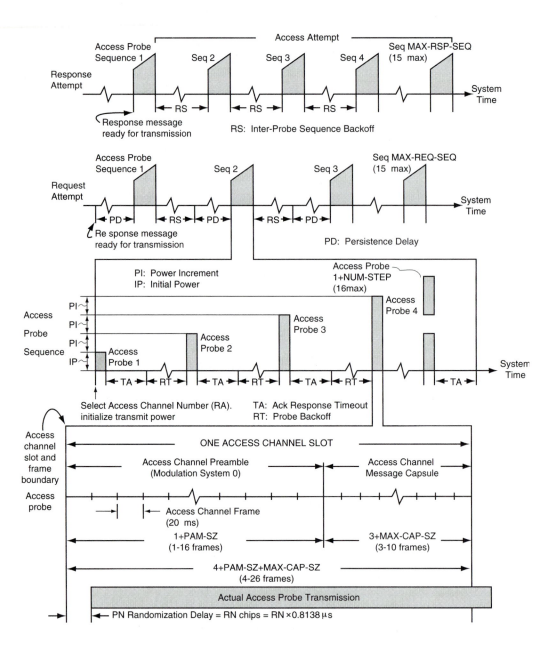

Figure 5.5   Access channel request and response attempts

The precise timing of the access channel transmission in an access attempt is determined by a procedure called PN randomization. For each access attempt, the mobile station computes a delay (RN) from 0 to $2^{TR}-1$ PN chips using a hash function that depends on its ESN. TR denotes the time randomization for access channel probes, used by the mobile station to calculate the number PN before each access channel probe. The mobile station delays its transmit timing by RN PN chips. This transmit timing adjustment includes delay of the direct sequence spreading long code and of the quadrature spreading I and Q pilot PN sequences, so it effectively increases the apparent range from the mobile station to the base station. This increases the probability that the base station will be able to separately demodulate transmissions from multiple mobile stations in the same access channel slot, especially when many mobile stations are at a similar range from the base station. Use of a nonrandom algorithm for PN randomization permits the base station to separate the PN randomization from the actual propagation delay from the mobile station, so it can accurately estimate the timing of reverse traffic channel transmissions from the mobile station.

Each time the mobile station performs an access attempt, it computes a number RN from 0 to $2^{TR}-1$ using the hashing technique mentioned above. For the duration of this access attempt, the mobile station delays its transmit timing, including long code direct sequence spreading and I and Q pilot PN sequence quadrature spreading by RN PN chips.

When the mobile station performs an access attempt, it transmits one or more access probe sequences. If the access attempt is an access channel request, the mobile station transmits no more than the maximum number of access probe sequences; if the access attempt is an access channel response, the mobile station transmits no more than the maximum number of access probe sequences.

Before transmitting each access probe sequence, the mobile station generates a random number, RA, from 0 to the number of access channels supported by the current paging channel. The mobile station uses this random number, RA, as the access number for all access probes in that access probe sequence.

Before transmitting each access probe sequence other than the first access probe sequence, the mobile station generates a random number RS from 0 to (1 + Probe Sequence Backoff Range). The mobile station then delays the transmission of the access probe sequence for RS slots.

If the access attempt is an access channel request, then before transmitting the first access probe in each access sequence, and after a delay of RS if applicable, the mobile station transmits the first access probe of a probe sequence in a slot only if the test passes for that slot. To perform the persistence test, the mobile station generates a random number RP, $0<RP<1$, using the technique of the linear congruential generator. The persistence test is said to pass when RP is less than the current value of the probability P for the type of this access attempt. If P equals 0, the access attempt fails, and the mobile station ends the access attempt, updates its registration variables, and enters the system determination substate of the mobile station initialization state.

If the access channel request is a registration, the probability P is computed by

$$P = \left[\begin{array}{ll} 2^{-PV(n)/4} \cdot 2^{-PM} & \text{if } PV(n) \neq 63, n = 0, 1, \cdots, 9 \\ 0 & \text{otherwise} \end{array} \right.$$

$$P = \left[\begin{array}{ll} 2^{-PV(n)} \cdot 2^{-PM} & \text{if } PV(n) \neq 7, n = 10, 11, \cdots, 15 \\ 0 & \text{otherwise} \end{array} \right.$$

where n is the overload class assigned to the mobile station. $PV(n)$ denotes the persistence value for the mobile station's overload class. PM designates the persistence modifier for registration accesses.

If the access channel request is a message transmission, the probability P is computed by

$$P = \left[\begin{array}{ll} 2^{-PV(n)/4} \cdot 2^{-PMT} & \text{if } PV(n) \neq 63, n = 0, 1, \cdots, 9 \\ 0 & \text{otherwise} \end{array} \right.$$

$$P = \left[\begin{array}{ll} 2^{-PV(n)} \cdot 2^{-PMT} & \text{if } PV(n) \neq 7, n = 10, 11, \cdots, 15 \\ 0 & \text{otherwise} \end{array} \right.$$

where n is the overload class assigned to the mobile station. $PV(0-9)$ represents the persistence value for access overload classes 0 through 9. PMT denotes the persistence modifier for access channel attempts for message transmissions (3 bits).

If the access channel request is other than a registration or a message transmission, the probability P will be computed by

$$P = \left[\begin{array}{ll} 2^{-PV(n)/4} & \text{if } PV(n) \neq 63, n = 0, 1, \cdots, 9 \\ 0 & \text{otherwise} \end{array} \right.$$

$$P = \left[\begin{array}{ll} 2^{-PV(n)} & \text{if } PV(n) \neq 7, n = 10, 11, \cdots, 15 \\ 0 & \text{otherwise} \end{array} \right.$$

where n is the overload class assigned to the mobile station. $PV(n)$ represents the persistence value for access overload classes.

The persistence parameters limit the rate of access channel request transmissions during overload conditions. The single parameter $PV(0-9)$ corresponds to mobile station overload classes 0-9 (normal subscribers), $PV(10)$ corresponds to the mobile station overload class 10 (test mobile stations), and $PV(11)$ to overload class 11 (emergency mobile stations). $PV(12)$, $PV(13)$, $PV(14)$, and $PV(15)$ are reserved. Each mobile station calculates the probability of transmitting in a slot by using the appropriate PV parameter. Registration request transmissions are further limited by including the registration persistence parameter PM in the transmission probability calculation.

The mobile station transmits the first probe and successive access probes in the access probe sequence at a power level greater than that of the previous probe. Between access probes, the mobile station disables its transmitter.

After transmitting each probe, the mobile station waits for the acknowledgment response timeout, TA = (2 + Acknowledgment Timeout) x 80 ms from the end of the access channel slot. If no acknowledgment is received within TA seconds, the mobile station performs the following:

1. If the number of access probes stored or fewer access probes have been transmitted in this access probe sequence, the mobile station generates a random number RT (inter-probe

backoff) from 0 to 1 + Access Channel Probe Backoff. The mobile station then delays RT additional access channel slots, and then transmits the next access probe.

2. Otherwise, if fewer than the maximum number of access probe sequences for an access channel request or the maximum number of access probe sequences for an access channel response have been transmitted in this access attempt, the mobile station begins the randomization procedures for another access probe sequence.

3. Otherwise, the mobile station updates its registration variables and enters the system determination substate of the mobile station initialization state.

5.3.2   Acknowledgment Procedures

The acknowledgment procedures facilitate the reliable exchange of messages between the base station and the mobile station. The mobile station uses the following five fields to support this mechanism. These fields are the acknowledgment address type, the acknowledgment sequence number, the message sequence number, the acknowledgment required indicator, and the valid acknowledgment indicator. These fields are referred to as layer 2 fields, and the acknowledgment procedures are referred to as layer 2 procedures. All other message fields and the processing thereof are referred to as pertaining to layer 3.

All access channel messages share the following four fields:

1. Acknowledgment sequence number

   The mobile station sets this field to the value of the message sequence number field from the most recently received paging channel message requiring acknowledgment. If no such message has been received, the mobile station sets this field to '111'.

2. Message sequence number

   Different sequence numbers are used on each channel, and different sequence numbers are used for messages requiring acknowledgment and messages not requiring acknowledgment. The message sequence number is 2 bits in the paging channel and access channel messages; 3 bits in the forward and reverse traffic channel messages.

3. Acknowledgment required indicator (1 bit)

   This is one of the layer 2 acknowledgment fields for signaling messages. If the sender of the message requires an acknowledgment, it is set to '1'; if not, it is set to '0'.

4. Valid acknowledgment indicator

   This is also one of the layer 2 acknowledgment fields. A value of '1' specifies that the acknowledgment sequence number field has a sequence number that acknowledges a message. A value of '0' indicates that the acknowledgment sequence number field is ignored.

The mobile station generates a single set of the message sequence numbers for messages sent on the access channel. The mobile station sets the message sequence number field of the first message sent on the access channel after powering on to '000'. The mobile station then increments the message sequence number, modulo 8, for each new access attempt; even if the contents of the new message are identical to those of the previous message.

The mobile station monitors the paging channel while in the system access state. When a mobile station receives a message with the valid acknowledgment indicator field set to '1' and the acknowledgment sequence number field set to the message sequence number of the message currently being sent, the mobile station considers the message to have been acknowledged and ends the access attempt. The mobile station does not begin a new access attempt until the previous access attempt has ended.

Upon exiting the system access state, the mobile station aborts any access attempt in progress and discards the associated message. While in the system access state, the message station should continue its pilot search, but will not perform idle handoffs.

5.3.3   Update Overhead Information Substate

In this substate, the mobile station monitors the paging channel until it has received the current configuration messages. The mobile station compares sequence numbers to determine whether all the configuration messages are up to date. To make sure it has the latest access parameters, the mobile station receives at least one message containing the access parameters message sequence number field and waits, if necessary, for an access parameters message.

Upon entering the update overhead information substate, the mobile station sets the system access state timer to a value of $T_{41m} = 4$ seconds. T_{41m} is the maximum time to obtain updated overhead messages arriving on the paging channel. The mobile station sets the PAGES to NO. If this substate was entered with a page response indication, the mobile station sets the current access parameters message sequence number to the stored access parameters message sequence number; otherwise, it sets the current access parameters message sequence number to NULL.

If the state timer expires while in this substate, the mobile station enters the system determination substate of the mobile station initialization state. While in this substate, the mobile station monitors the paging channel. If the mobile station declares a loss of the paging channel, it then enters the mobile station idle state. If the mobile station receives the local control order or the lock until power-cycled order or the maintenance required order containing an address field matching the corresponding mobile station identification data, the mobile station records the reason for any of these orders in the mobile station's semi-permanent memory. For the lock until power-cycled order, the mobile station should notify the user of the locked condition. The mobile station then enters the system determination substate of the mobile station initialization state, and will not enter the system access state again until after the next mobile station powerup or until it has received an unlock order. For the maintenance required order, the mobile station remains in the unlocked condition. The mobile station should notify the user of the maintenance required condition.

If the mobile station receives the system parameters message or the access parameters message or neighbor list message, the mobile station processes the parameters from the message.

If the mobile station receives the slotted page message or page message, the mobile station sets the current access parameters message sequence number to the received access parameters message sequence number. If this substate was not entered with an origination or page response indication, the mobile station compares its MIN with the MIN in each record of the message. If a match is declared, the mobile station sets PAGES to YES. If the mobile station receives any other message with a message type specified in Table 5.2, it will process all layer 2 fields of the message and ignore all other fields. The mobile station ignores all other messages.

When the stored configuration parameters are current and the current access parameters message sequence number and the stored access parameters message sequence number are equal and are not NULL, the mobile station should disable the system access state timer and do one of the following:

- If PAGED is equal to YES or if this substate was entered with a page response indication, the mobile station determines whether the message resulting in the page match was received on the current paging channel. If the message was received on the current paging channel, the mobile station enters the page response substate; otherwise, the mobile station enters the mobile station idle state. Note that PAGED designates the indicator for a page match detected while the mobile station is in the system access state.
- If this substate was entered with an origination indication, the mobile station enters the mobile station origination attempt substate.
- If this substate was entered with an order/message response indication, the mobile station determines if the message resulting in the response was received on the current paging channel. If the message was received on the current paging channel, the mobile station enters the mobile station order/ message response substate; otherwise, the mobile station discards the response and enters the mobile station idle state.
- If this substate was entered with a registration indication, the mobile station enters the registration access substate.
- If this substate was entered with a message transmission indication, the mobile station enters the mobile station message transmission substate.

5.3.4   Page Response Substate

In this substate, the mobile station sends a page response message in response to a page message or slotted page message from the base station. If the base station responds to the page response message with an authentication request, the mobile station responds in this substate.

Upon entering the page response substate, the mobile station sends a page response message, using the access procedures specified in Section 5.3.1. If message authentication is enabled, the mobile station calculates the values of the authentication mode (2 bits) and random challenge value (8 bits) using the current stored random challenge value.

While in this substate, the mobile station monitors the paging channel. If the mobile station declares a loss of the paging channel, it disables its transmitter and enters the mobile station idle state. If the mobile station receives an acknowledgment to any message sent by the mobile station in this substate, the mobile station ends the access attempt. If the acknowledgment was not included in a channel assignment message, the authentication challenge message, the base station challenge confirmation order, or the SSD update message, the mobile station sets the system access state timer to T_{42m} = 12 seconds. T_{42m} denotes the maximum time to receive a delayed layer 3 response following the receipt of an acknowledgment for an access probe.

If the access attempt for the page response message ends by the receipt of an acknowledgment from the base station, the mobile station updates its registration variables. If the system access state timer expires while in this substate, the mobile station enters the mobile station idle state.

If the mobile station receives the authentication challenge message or the base station challenge confirmation order, then the mobile station processes the message as follows: If the mobile station receives this message while an access attempt is in progress, the mobile station ignores the message. If the mobile station receives this message after the acknowledgment to any message sent by the mobile station in this substate, or if the acknowledgment is included in this message, the mobile station will disable the system access state timer and respond to the message as specified in the unique challenge-response procedure, using the access procedures specified in Section 5.3.1.

If the mobile station receives the channel assignment message, then the mobile station terminates any access attempt in progress. It then processes the message as follows:

If the received assignment mode equals '000', the mobile station stores the forward traffic channel code channel (stored code channel = received code channel, 8 bits), the frame offset (4 bits) gives the amount that forward traffic channel and reverse traffic channel frames are delayed relative to normal system timing of 1.25 ms (stored frame offset = received frame offset), and the message encryption mode indicator (stored message encryption mode = received message encryption mode)—(1) when set to '0000', encryption is disabled, and all messages sent on the forward and reverse traffic channels are unencrypted; (2) when set to '0001', standard encryption is enabled and messages are encrypted. If the received frequency included indicator (1 bit) equals '1', the frequency assignment (11 bits) (i.e., stored CDMA channel frequency assignment = received CDMA channel frequency assignment), then the mobile station enters the traffic channel initialization substate of the mobile station control on the traffic channel state.

If the assignment mode (3 bits) equals '001', the mobile station performs the following actions: if a CDMA channel (CDMA channel frequency assignment (11 bits)) is listed in the assignment, the mobile station sets the stored CDMA channel assignment equal to the received CDMA channel frequency assignment and tunes to the new frequency assignment. The mobile station then sets the stored access parameters message sequence number (6 bits) to NULL and sets the stored pilot PN sequence offset index (9 bits) to the pilot PN sequence offset of the strongest pilot in the received pilot PN sequence offset index. If the mobile station has not stored configuration parameters for the primary paging channel of the new base station, or if the stored information is not current, the mobile station sets the stored configuration message sequence number (6 bits), the stored system parameters message sequence number, the stored neighbor list message number, and the stored CDMA channel list message sequence number to NULL. The mobile station then begins monitoring the primary paging channel of the selected base station. If the received response on analog control channel indicator (1 bit) is equal to '1', the mobile station enters the update overhead information substate with a page response retransmission indication. If the received response is equal to '0', the mobile station enters the mobile station idle state. If the received assignment mode equals '010' and the received response equals '1', the mobile station enters the initialization task with a page response indication. If the received assignment mode equals '010' and the received response equals '0', the mobile station enters the initialization task with a wait for page indication.

For the lock until power-cycled order, the mobile station disables its transmitter and records the reason for this order in the mobile station's semi-permanent memory. The mobile station should notify the user of the locked condition and then enters the system determination substate of the mobile station initialization state. It will not enter the system access state again until after the next mobile station power-up or until it has received an unlock order. This requirement takes precedence over any other mobile station requirement specifying entry to the system access state.

The mobile station records the reason for the maintenance required order in the mobile station's semi-permanent memory. The mobile station remains in the unlocked condition and should notify the user of the maintenance required condition.

For the release order, the mobile station enters the mobile station idle state or the system determination substate of the mobile station initialization state.

If the mobile station receives the SSD update message while an access attempt is in progress, the mobile station ignores the message. If the mobile station receives this message after the acknowledgment to any message sent by the mobile station in this substate, or if the acknowledgment is included in this message, the mobile station disables the system access state timer and responds to the message using the access procedures specified in Section 5.3.1.

5.3.5   Mobile Station Order/Message Response Substate

In this substate, the mobile station sends a message that is a response to a message received from the base station. If the base station responds to the mobile station's message with an authentication request, the mobile station responds in this substate. Upon entering this substate, the mobile station sends the response message using the access procedures specified in Section 5.3.1.

While in this substate, the mobile station monitors the paging channel. If the mobile station declares a loss of the paging channel, it disables its transmitter and enters the mobile station idle state. If the mobile station receives an acknowledgment to any message sent by the mobile station in this substate, it ends the access attempt. If the acknowledgment was not included in an authentication challenge message, the base station challenge confirmation order, or the SSD update message, the mobile station enters the mobile station idle state.

If the mobile station receives any of the following messages addressed to the mobile station, then the mobile station processes the message as follows:

1. Authentication challenge message

   If the mobile station receives the authentication challenge message while an access attempt is in progress, the mobile station ignores the message. If the mobile station receives this message after the acknowledgment to any message sent by the mobile station in this substate, or if the acknowledgment is included in this message, the mobile station responds to the message as specified in the unique challenge-response procedure, using the access procedures specified in Section 5.3.1.

2. Base station challenge confirmation order

   If the mobile station receives this message while an access attempt is in progress, the mobile station ignores the message. If the mobile station receives this message after the acknowledgment to any message sent by the mobile station in this substate, or if the acknowledgment is included in this message, the mobile station responds to the message as specified for updating the shared secret data (SSD) using the access procedures.

3. Feature notification message and local control order

   No requirements.

4. Lock until power-cycled order

   The mobile station disables its transmitter and records the reason for the lock until power-cycled order in the mobile station's semi-permanent memory. The mobile station should notify the user of the locked condition. The mobile station then enters the system determination substate of the mobile station initialization state, and will not enter the system access state again until after the next mobile station power-up or until it has received an unlock order. This requirement takes precedence over any other mobile station requirement specifying entry to the system access state.

5. Maintenance required order

   The mobile station records the reason for the maintenance required order in the mobile station's semi-permanent memory. The mobile station remains in the unlocked condition. The mobile station should notify the user of the maintenance required condition.

6. SSD update message

   If the mobile station receives this message while an access attempt is in progress, the mobile station ignores the message. If the mobile station received this message after the acknowledgment to any message sent by the mobile station in this substate, or if the acknowledgment is included in this message, the mobile station responds to the message as specified for updating the shared secret data (SSD), according to the access procedures specified in Section 5.3.1.

7. Any other message

If the mobile station receives any other message with a message type specified in Table 5.1, it will process all layer 2 fields of the message and ignore all other fields. The mobile station ignores all other messages.

5.3.6 Mobile Station Origination Attempt Substate

In this substate, the mobile station sends an origination message. If the base station responds to the origination message with an authentication request, the mobile station responds in this substate.

Upon entering this substate, the mobile station sends the origination message using the access procedures specified in Section 5.3.1. The mobile station includes in the origination message as many of the dialed digits as possible without exceeding the message capsule size. If message authentication is enabled, the mobile station calculates the values of the authentication mode (2 bits) and random challenge value (8 bits) fields using the current value of stored random challenge value. If the authentication mode is equal to '00', authentication is disabled. If the authentication mode is equal to '01', authentication data are included in access channel messages where appropriate. The eight most significant bits of the random challenge value (0 or 32 bits) are used for the computation of the authentication mode (2 bits).

While in this substate, the mobile station monitors the paging channel. If the mobile station declares a loss of the paging channel, it disables its transmitter and enters the mobile station idle state. If the mobile station receives an acknowledgment to any message sent by the mobile station in this substate, it ends the access attempt. If the acknowledgment was not included in a channel assignment message, the authentication challenge message, the base station challenge confirmation order, or the SSD update message, the mobile station sets the system access state timer to $T_{42m} = 12$ seconds.

If the access attempt for the origination message ends by the receipt of an acknowledgment from the base station, the mobile station should update its registration variables.

If the system access state timer expires while in this substate, the mobile station enters the mobile station idle state. If the mobile station is directed by the user to disconnect the call, the mobile station aborts any access attempt in progress and enters the system determination substate of the mobile station initialization state.

If the mobile station receives any of the following messages addressed to the mobile station, then the mobile station processes the message as follows:

1. Authentication challenge message

If the mobile station receives this message while an access attempt is in progress, the mobile station ignores the message. If the mobile station receives this message after the acknowledgment to any message sent by the mobile station in this substate, or if the acknowledgment is included in this message, the mobile station disables the system access state timer and responds to the message as specified in the unique-response procedure according to the access procedures specified in Section 5.3.1.

2. Base station challenge confirmation order

If the mobile station receives this message while an access attempt is in progress, the mobile station ignores the message. If the mobile station receives this message after the acknowledgment to any message sent by the mobile station in this substate, or if the acknowledgment is included in this message, the mobile station disables the system access state timer and responds to the message as specified for updating the shared secret data (SSD) according to the access procedures specified in Section 5.3.1.

3. Channel assignment message

The mobile station terminates any access attempt in progress. It then processes the message as follows:

If the received assignment mode (3 bits) equals '000', the mobile station stores the forward traffic code channel (stored code channel = received code channel), the frame offset (stored frame offset = received frame offset, 4 bits), the message encryption mode indicator (stored message encryption mode = received message encryption mode), and, if the frequency included indicator (1 bit) equals '1', the frequency assignment (stored CDMA channel = received CDMA channel frequency assignment), and then enters the traffic channel initialization substate of the mobile station control on the traffic channel state.

If the received assignment mode equals '001', the mobile station performs the following actions: If a CDMA channel (CDMA channel frequency assignment) is specified in the assignment, the mobile station sets the stored CDMA channel = received CDMA frequency assignment and tunes to the new frequency assignment. The mobile station sets the stored access parameters message sequence number (6 bits) to NULL and sets the stored pilot PN sequence offset index (9 bits) to the pilot PN sequence offset of the strongest pilot in the list (received pilot PN sequence offset index). If the mobile station has not stored configuration parameters for the primary paging channel of the new base station, or if the stored information is not current, the mobile station sets the configuration message sequence number (6 bits), the system parameters message sequence number, the neighbor list message sequence number, and the channel list message sequence number to NULL. The mobile station then begins monitoring the primary paging channel of the selected base station. If the received response on the analog control channel indicator (1 bit) is equal to '1', the mobile station enters the update overhead information substate with an origination indication.

If the received assignment mode (3 bits) equals '010' and the received response on the analog control channel indicator (1 bit) equals '1', the mobile station enters the initialization task with an origination indication.

If the received assignment mode (3 bits) equals '011', the mobile station stores the system identification, voice mobile station attenuation code, voice channel number, supervisory audio tone color code, and message encryption mode indicator and enters the confirmation initial voice channel task with an origination indication.

4. Feature notification message

If the received release order is equal to '1', the mobile station terminates any access attempt in progress and enters the mobile station idle state or the system determination substate of the mobile station initialization state. If the received release order is equal to '0', the mobile station resets the system access state timer to $T_{42m} = 12$ seconds.

5. Intercept order

The mobile station terminates any access attempt in progress and enters the mobile station idle state.

6. Local control order

No requirements.

7. Lock until power-cycled order

The mobile station disables its transmitter and records the reason for the lock until power-cycled order in the mobile station's semi-permanent memory. The mobile station should notify the user of the locked condition. The mobile station then enters the system determination substate of the mobile station initialization state, and will not enter the system access state again until after the next mobile station power-up or until it has received an unlock order. This requirement takes precedence over any other mobile station requirement specifying entry to the system access state.

8. Maintenance required order

The mobile station records the reason for the maintenance required order in the mobile station's semi-permanent memory. The mobile station remains in the unlocked condition. The mobile station should notify the user of the maintenance required condition.

9. Release order

The mobile station enters the mobile station idle state or the system determination substate of the mobile station initialization state.

10. Reorder order

The mobile station terminates any access attempt in progress and enters the mobile station idle state.

11. SSD update message

If the mobile station receives this message while an access attempt is in progress, the mobile station ignores the message. If the mobile station receives this message after the acknowledgment to any message sent by the mobile station in this substate, or if the acknowledgment is included in this message, the mobile station disables the system access state timer and responds to the message as specified for updating the shared secret data (SSD), using the access procedures specified in Section 5.3.1.

12. Any other message

> If the mobile station receives any other message with a message type specified in Table 5.2, it processes all layer 2 fields of the message and ignores all other fields. The mobile station ignores all other messages.

5.3.7 Registration Access Substate

In this substate, the mobile station sends a registration message to the base station. If the base station responds with an authentication request, the mobile station responds in this substate. Upon entering the registration access substate, the mobile station sends the registration message, using the access procedures specified in Section 5.3.1. If message authentication is enabled, the mobile station calculates the values of the authentication data (18 bits) and random challenge value (8 bits) fields using the current value of random challenge (0 or 32 bits).

While in this substate, the mobile station monitors the paging channel. If the mobile station declares a loss of the paging channel, it disables its transmitter and enters the mobile station idle state. If the mobile station receives an acknowledgment to any message sent by the mobile station in this substate, it ends the access attempt. If the acknowledgment was not included in an authentication challenge message, the base station challenge confirmation order, the SSD update message, or the release order, or if the registration access was initiated due to a user direction to power down, the mobile station will do one of the following:

- The mobile station will update registration variables and may power down.
- Otherwise, the mobile station enters the mobile station idle state.

If the access attempt for a registration message ends by the receipt of an acknowledgment from the base station, the mobile station updates its registration variables.

If the mobile station is directed by the user to originate a call, the mobile station aborts any access attempt in progress and enters the mobile station origination attempt substate.

If the mobile station receives a page message or a slotted page message, the mobile station compares its MIN with the MIN in each record of the message. If a match is declared, the mobile station aborts any access attempt in progress and enters the page response substate.

If the mobile station receives any of the following messages addressed to the mobile station, then the mobile station will process the message as described.

1. Authentication challenge message

2. Base station challenge confirmation order

> If the mobile station receives either of above two messages 1 and 2 while an access attempt is in progress, or if the registration access was initiated due to a user direction to power down, the mobile station ignores the message. If the mobile station receives either message after the acknowledgment to any message sent by the mobile station in this substate, or if the acknowledgment is included in either message, the mobile station responds to the message as specified in the unique challenge-response procedure for the authentication challenge

message and in the procedure of updating the shared secret data (SSD) for the base station challenge confirmation order, using the access procedures specified in Section 5.3.1.

3. Feature notification message

   No requirements.

4. Local control order

   No requirements.

5. Lock until power-cycled order

   The mobile station disables its transmitter and records the reason for the lock until power-cycled order in the mobile station's semi-permanent memory. The mobile station should notify the user of the locked condition. The mobile station then enters the system determination substate of the mobile station initialization state, and does not enter the system access state again until after the next mobile station power-up or until it has received an unlock order. This requirement takes precedence over any other mobile station requirement specifying entry to the system access state.

6. Maintenance required order

   The mobile station records the reason for the maintenance required order in the mobile station's semi-permanent memory. The mobile station remains in the unlocked condition. The mobile station should notify the user of the maintenance required condition.

7. Registration accepted order

   No requirements.

8. Registration rejected order

   No requirements.

9. Release order

   The mobile station enters the mobile station idle state or the system determination substate of the mobile station initialization state.

10. SSD update message

    If the mobile station receives this message while an access attempt is in progress, or if the registration access was initiated due to a user direction to power down, the mobile station ignores the message. If the mobile station receives this message after the acknowledgment to any message sent by the mobile station in this substate, or if the acknowledgment is included in this message, the mobile station responds to the message as specified for updating the shared secret data using the access procedures specified in Section 5.3.1.

11. Any other message

    If the mobile station receives any other message with a message type specified in Table 5.2, it will process all layer 2 fields of the message and ignore all other fields. The mobile station ignores all other messages.

5.3.8   Mobile Station Message Transmission Substate

In this substate, the mobile station sends a data burst message to the base station. If the base station responds with an authentication request, the mobile station responds in this substate, although support of this substate is optional.

Upon entering the mobile station message transmission substate, the mobile station transmits the data burst message using the access procedures specified in Section 5.3.1.

While in this substate, the mobile station monitors the paging channel. If the mobile station declares a loss of the paging channel, it disables its transmitter and enters the mobile station idle state. If the mobile station receives an acknowledgment to any message sent by the mobile station in this substate, it ends the access attempt. If the acknowledgment was not included in an authentication challenge message, the base station challenge confirmation order, or the SSD update message, the mobile station enters the mobile station idle state.

If the mobile station receives a page message or a slotted page message, the mobile station compares its MIN with the MIN in each record of the message. If a match is declared, the mobile station aborts any access attempt in progress and enters the page response substate. The mobile station may store the message for later transmission.

If the mobile station receives any of the following messages addressed to the mobile station, then the mobile station will process the messages as described.

1. Authentication challenge message

   If the mobile station receives this message while an access attempt is in progress, the mobile station ignores the message. If the mobile station receives this message after the acknowledgment to any message sent by the mobile station in this substate, or if the acknowledgment is included in this message, the mobile station responds to the message as specified in the unique challenge-response procedure, using the access procedures specified in Section 5.3.1.

2. Base station challenge confirmation order

   If the mobile station receives this message while an access attempt is in progress, the mobile station ignores the message. If the mobile station receives this message after the acknowledgment to any message sent by the mobile station in this substate, or if the acknowledgment is included in this message, the mobile station responds to the message as specified by updating the shared secret data using the access procedures discussed in Section 5.3.1.

3. Data burst message

   No requirements.

4. Local control order

   No requirements.

5. Lock until power-cycled order

   The mobile station disables its transmitter and records the reason for the lock until power-cycled order in the mobile station's semi-permanent memory. The mobile station should notify the user of the locked condition. The mobile station enters the system determination substate of the mobile station initialization state, and will not enter the system access state again until after the next mobile station power-up or until it has received an unlock order. This requirement takes precedence over any other mobile station requirement specifying entry to the system access state.

6. Maintenance required order

The mobile station records the reason for the maintenance required order in the mobile station's semi-permanent memory. The mobile station remains in the unlocked condition. The mobile station should notify the user of the maintenance required condition.

7. SSD update message

If the mobile station receives this message while an access attempt is in progress, the mobile station ignores the message. If the mobile station receives this message after the acknowledgment to any message sent by the mobile station in this substate, or if the acknowledgment is included in this message, the mobile station responds to the message as specified by updating the shared secret data using the access procedures specified in Section 5.3.1.

8. Any other messages

If the mobile station receives any other message with a message type specified in Table 5.2, it will process all layer 2 fields of the message and ignore all other fields. The mobile station ignores all other messages.

5.4 MOBILE STATION CONTROL ON THE TRAFFIC CHANNEL STATE

In this state, the mobile station communicates with the base station using the forward and reverse traffic channels.

As illustrated in Fig. 5.6, the mobile station control on the traffic channel state consists of the following substates:

• Traffic Channel Initialization Substate

In this substate, the mobile station verifies that it can receive the forward traffic channel and begins transmitting on the reverse traffic channel.

• Waiting for Order Substate

In this substate, the mobile station waits for an alert with information message.

• Waiting for Mobile Station Answer Substate

In this substate, the mobile station waits for the user to answer the call.

• Conversation Substate

In this substate, the mobile station's primary service option application exchanges primary traffic packets with the base station.

• Release Substate

In this substate, the mobile station disconnects the call.

Prior to describing the mobile station control on the traffic channel state, we begin with the forward traffic channel power control, the service option, and acknowledgment procedures.

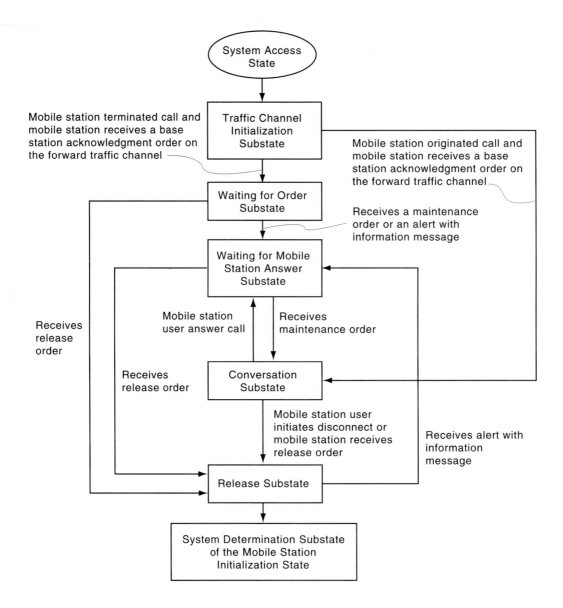

Figure 5.6 Mobile station control on the traffic channel state

5.4.1   Forward Traffic Channel Power Control

To support the forward traffic channel power control, the mobile station reports frame error rate statistics to the base station which enables either periodic reporting or threshold reporting.

The mobile station maintains a counter for the total number of received frames and a counter for the number of received bad frames. The mobile station performs the following actions for each received frame:

- The mobile station increments the total number of stored frames by 1.
- If the received frame is bad, the mobile station increments the number of stored bad frames by 1.
- If either (1) the threshold report mode indicator is equal to '1' and the number of stored bad frames is equal to the stored power control reporting threshold, or (2) the periodic report mode indicator is equal to '1' and the total number of stored frames is equal to $\lfloor 2^{a/2} \times 5 \rfloor$, the mobile station will send a power measurement report message to the base station, a denotes the power control reporting frame count.
- If the total number of stored frames is equal to $\lfloor 2^{a/2} \times 5 \rfloor$, the mobile station sets the total number of stored frames and the number of stored bad frames to zero.

To initialize the forward traffic channel power control, the mobile station sets a counter for the total number of received frames and a counter for the number of received bad frames to zero. The mobile station stores the following parameters from the power control parameters message:

- Power control reporting threshold
- Power control reporting frame count
- Threshold report mode indicator
- Periodic report mode indicator
- Power report delay

5.4.2   Service Option

During the traffic channel operation, the mobile station and base station support primary traffic services. Such a service option has a set of requirements that govern the way in which the primary traffic bits from forward and reverse traffic channel frames are processed by the mobile station and base station.

Either the mobile station or the base station can request a service option. The mobile station can request a particular service option at the time of call origination, when responding to a page, or during the traffic channel operation. If the service option request is acceptable to the base station, the mobile station and base station begin using the new service option. If the mobile station requests a service option that is not acceptable to the base station, the base station can reject the requested service option or request an alternative service option. If the base station requests an alternative service option, the mobile station can accept or reject the base station's alternative service option or request another service option. This service option negotiation ends when the mobile station and the base station find a mutually acceptable service option, or when the mobile station rejects a service option request from the base station or the base station rejects a service option request from the mobile station.

The mobile station and base station use the service option request order either to request a service option or suggest an alternative service option, and the service option response order to accept or reject a service option request.

In addition, the mobile station can request a service option in the origination message or the page response message, and the base station can request a service option in the page message or the slotted page message. The mobile station and base station use the service option control order to invoke service option specific functions.

The mobile station uses a variable (service option request number) to record the number of the service option for which the mobile station has sent an outstanding request, either in an origination message, a page response message, or a service option request order. The service option request number is set to NULL when the mobile station does not have an outstanding service option request. The mobile station uses another variable to record the number of the service option which is currently active. The active service option number is set to NULL when there is no active service option.

5.4.3   Processing Service Option Orders

When the mobile station receives a service option request order, it performs the following:

- If the mobile station accepts the requested service option, the mobile station sets the service option request number to NULL and sends a service option response order accepting the requested service option within $T_{58m} = 5$ seconds. T_{58m} denotes the maximum time for the mobile station to respond to a service option request. The mobile station interprets the message action time of the service option request order in accordance with the requirements for the requested service option and begins using the requested service option in accordance with those requirements. The mobile station sets the active service option number to the requested service option number when the service option becomes active.

- If the mobile station does not accept the requested service option and has an alternative service option to request, the mobile station sets the service option request number to the alternative service option number and then sends a service option request order requesting the alternative service option within T_{58m} seconds.

- If the mobile station does not accept the requested service option and does not have an alternative service option to request, the mobile station sets the service option request number to NULL and sends a service option response order to reject the request within T_{58m} seconds. The mobile station continues to process primary traffic as it did prior to receiving the service option request order and remains in the current state.

When the mobile station receives a service option response order, it performs the following processes:

- If the service option number specified in the order is equal to the service option request number, the mobile station sets that request number to NULL. The mobile station interprets the message action time of the service option response order in accordance with the requirements for the specified service option, and begins using the specified service option in accordance with those requirements. The mobile station sets the stored service option current number to the specified service option number when the service option becomes active.

- If the order indicates a service option rejection, the mobile station sets the service option request number to NULL. The mobile station continues to process primary traffic as it did prior to receiving the service option response order and remains in the current state.

- If the order does not indicate a service option rejection and the service option specified in the order is not equal to the service option request number, then the mobile station sets the service option request number to NULL and sends a mobile station reject order (order qualification code (8 bits), i.e., 00000100) within T_{58m} seconds. The mobile station continues to process primary traffic as it did prior to receiving the service option response order and remains in the current state.

If there is an active service option which is not equal to NULL, the mobile station interprets the message action time of the service option control order in accordance with the requirements for the active service option and processes the service option control order in accordance with these requirements; otherwise, the mobile station sends a mobile station reject order (order qualification code (8 bits), i.e., 00000001) within $T_{56m} = 0.2$ seconds.

To perform service option request initialization, the mobile station sets the service option request number to the specified service option number.

5.4.4   Acknowledgment Procedures

The acknowledgment procedures facilitate the reliable exchange of messages between the base station and the mobile station. The mobile station uses the fields such as the acknowledgment sequence number, the message sequence number, and the acknowledgment required indicator to detect duplicate messages and to provide a reference for acknowledgments. These message fields are referred to as layer 2 fields, and the acknowledgment procedures are referred to as layer 2 procedures. All other message fields are referred to as layer 3 fields, and the processing of layer 3 is referred to as layer 3 processing.

On both the forward traffic channel and the reverse traffic channel, the procedure for messages requiring acknowledgment is a selective repeat scheme in which a message is retransmitted only if an acknowledgment for it is not received.

A traffic channel message requires acknowledgment when the field of acknowledgment required indicator is set to '1'.

The layer 2 protocol does not guarantee delivery of messages in any order. If the mobile station requires that the base station receive a set of messages in a certain order, the mobile station must wait for an acknowledgment to each message before transmitting the next message in the set. For messages requiring acknowledgment whose relative ordering is not important, the mobile station may transmit up to four such messages before receiving an acknowledgment for the first message.

The mobile station stores a message sequence number for messages requiring acknowledgment. The mobile station stores an acknowledgment status indicator for each possible value of the reverse traffic channel message sequence number field. The mobile station will not send a new message requiring acknowledgment when an acknowledgment status indicator [(message sequence acknowledgment number + 4) mod 8] is equal to YES.

When the mobile station receives any message on the forward traffic channel, it sets the acknowledgment status indicator to NO. When the mobile station sends a new message requiring acknowledgment on the reverse traffic channel, it sets the acknowledgment status indicator to YES and sets the message sequence number field of the message to the message sequence acknowledgment number. The mobile station then increments the stored message sequence acknowledgment number, modulo 8. The mobile station will not retransmit a message for which it has received an acknowledgment.

If the mobile station has not received an acknowledgment within $T_{1m} = 0.4$ seconds after transmitting the message, the mobile station retransmits the message (see Fig. 5.7). If the mobile station retransmits a message, the mobile station uses the same message sequence number for retransmission. The mobile station will not retransmit a message sooner than $T_{1m} = 0.4$ seconds after the previous transmission of the same message. T_{1m} denotes the maximum time that the mobile station waits for an acknowledgment.

The mobile station stores a retransmission counter for each transmitted message requiring acknowledgment. The mobile station sets the retransmission counter to zero prior to the first transmission of the message. After each transmission of the message, the mobile station increments the retransmission counter if no acknowledgment has been received. When the retransmission counter is equal to $N_{1m} = 3$, the mobile station declares an acknowledgment failure. N_{1m} denotes the maximum number of times that a mobile station transmits a message requiring an acknowledgment on the reverse traffic channel.

Messages received on the forward traffic channel contain the message sequence number fields that are incremented by the same rules as messages transmitted on the reverse traffic channel. Separate sequence numbers are maintained for forward traffic channel messages that require acknowledgment and for messages that do not require acknowledgment.

The mobile station acknowledges a received message by transmitting a message with the acknowledgment sequence number field set equal to the message sequence number field of the received message. A message transmitted with the acknowledgment sequence number field set in this manner is referred to as including an acknowledgment of the received message.

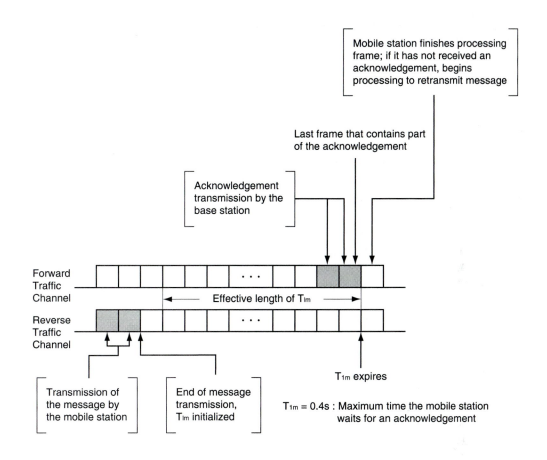

Figure 5.7 Time limit for acknowledgment of reverse traffic channel messages

Whenever a message requiring acknowledgment is received, the mobile station sets the acknowledgment sequence number field of subsequent reverse traffic channel messages to the received message sequence number. If no message has been received, the mobile station sets this field to '111'. After receiving a message requiring acknowledgment, the mobile station transmits a message including an acknowledgment within $T_{2m} = 0.2$ seconds as shown in Fig. 5.8. When a received message requires acknowledgment and no message is available within $T_{2m} = 0.2$ seconds after the message is received, the mobile station transmits a mobile station acknowledgment order including the acknowledgment. The mobile station acknowledgment order will be sent as a message not requiring acknowledgment. T_{2m} denotes the maximum time allowed for the mobile station to send an acknowledgment.

A traffic channel message does not require acknowledgment when the acknowledgment required indicator field is set to '0'. The mobile station will store a message sequence number for messages not requiring acknowledgment. For each new message sent that does not require acknowledgment, the mobile station sets the message sequence number field of the message to a number for no-required acknowledgment and then increments it (no-required acknowledgment number), modulo 8.

The mobile station considers all messages received within $T_{3m} = 0.32$ seconds that do not require acknowledgment and have the same message sequence number to be duplicates, as shown in Fig. 5.9. If the mobile station receives multiple copies of a message as determined by the message sequence number, it discards the duplicate copies. T_{3m} denotes the period in which two messages arriving on the forward traffic channel, not requiring an acknowledgment, and carrying the same sequence number, are considered duplicates.

The mobile station resets the acknowledgment procedures as follows: If the acknowledgment status indicator for message sequence number n, $0 \le n \le 7$, is equal to YES for any n, the mobile station saves the corresponding messages and retransmits them after completing the reset of the acknowledgment procedures. For each such message the mobile station sets the retransmission counter to zero. The mobile station sets the next message sequence number for requiring acknowledgment to 0, the next message sequence number for not requiring acknowledgment to 0, and sets the acknowledgment status indicator [n] for $0 \le n \le 7$ to NO.

For acknowledgment sequence number reset, the mobile station sets the acknowledgment sequence number field of all reverse traffic channel messages to '111' until the first message requiring acknowledgment is received.

For duplicate detection reset, the mobile station sets the received message indicator for message sequence number n, $0 \le n \le 7$ to NO.

The mobile station stores the following parameters from the in-traffic system parameters message:

- System identification
- Network identification
- Search window size for the active set, candidate set, neighbor set and remaining set
- Pilot detection threshold
- Pilot drop threshold
- Active set versus candidate set comparison threshold
- Drop timer value
- Maximum age for retention of neighbor set members

The mobile station determines its roaming status. The mobile station should indicate to the user whether the mobile station is roaming.

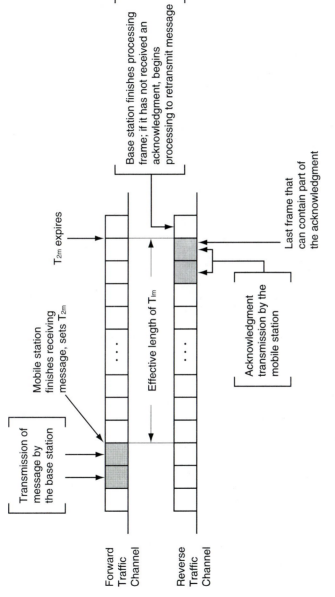

Figure 5.8 Time limit for acknowledgment of reverse traffic channel messages

$T_{2m} = 0.2s$: Maximum time allowed for the mobile stations to send an acknowledgment

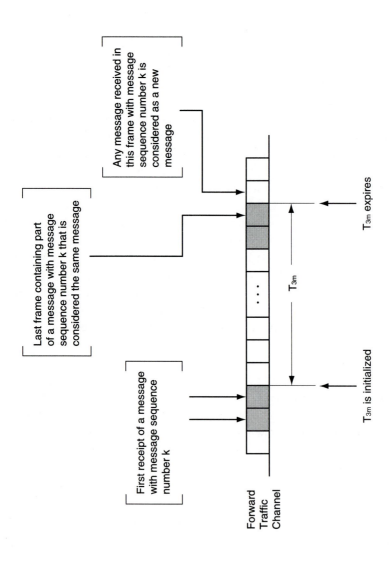

Last frame containing part of a message with message sequence number k that is considered the same message

Any message received in this frame with message sequence number k is considered as a new message

First receipt of a message with message sequence number k

Forward Traffic Channel

T_{3m} is initialized

T_{3m}

T_{3m} expires

$T_{3m} = 0.32s$: Period in which two messages received by the mobile station on the forward traffic channel, not requiring an acknowledgment, and carrying the same sequence numbers are considered duplicates.

Figure 5.9 Time window for detecting duplicate messages not requiring acknowledgment

Consider message action times. A message without a use-time field or with a use-time field set to '0' has an implicit action time. A message whose use-time field is set to '1' has an explicit action time which is specified in the action-time field of the message. A message with a future action time is called a pending message. Unless otherwise specified, a message having an implicit action time takes effect no later than the first 80 ms boundary (relative to the system time) occurring at least 80 ms after the end of the frame containing the last bit of the message. A message with an explicit action time takes effect when the system time (in 80 ms units) modulo 64 becomes equal to the message's action-time field. The difference in time between the action time and the end of the frame containing the last bit of the message is at least 80 ms. The mobile station supports one pending message at any given time, not including pending service option control orders. The number of pending service option control orders that the mobile station is required to support is specific to the service option.

Upon receiving a long code transition request order, the mobile station performs the following procedures. If the long code transition request order requests a transition to the private long code, and the mobile station is able to generate the private long code, and the mobile station accepts the request, the mobile station sends a long code transition response order 00000011 within $T_{56m} = 0.2$ seconds, where T_{56m} denotes the default maximum time to respond to a received message or order on the forward traffic channel. The mobile station uses the private long code on both the forward traffic channel and the reverse traffic channel. If the use-time equals '0', the mobile station begins using the private long code at the first 80 ms boundary (relative to the start of system time) after $N_{4m} = 20$ frames from the end of the response transmission. N_{4m} denotes the number of frames after the transmission of a response to a long code transition request order determining the 80 ms boundary for the private long code transition. The mobile station should indicate to the user that the voice privacy mode is active. If the long code transition request order requests a private long code transition, and the mobile station is not able to generate the private long code or the mobile station does not accept the request, the mobile station sends a long code transition response order 00000010 within $T_{56m} = 0.2$ seconds.

If the long code transition request order requests a transition to the public long code and the mobile station accepts the request, the mobile station sends a long code transition response order 00000010 within T_{56m} seconds. The mobile station should indicate to the user that the voice privacy mode is inactive. If the long code transition request order requests a public long code transition, and the mobile station code does not accept the request, the mobile station sends a long code transition response order 00000011 within T_{56m} seconds.

5.4.5 Traffic Channel Initialization Substate

In this substate, the mobile station verifies that it can receive the forward traffic channel and begins transmitting on the reverse traffic channel.

1. Upon entering the traffic channel initialization substate, the mobile station performs the following actions:
   - The mobile station performs the registration initialization, resets the acknowledgment procedures, and initializes the forward traffic power control.
   - The mobile station sets the active option number to NULL to indicate that there is no active service option.
   - If the call is mobile-station originated and the origination message requests a special service option, the mobile station performs service option request initialization specifying the special service option number.
   - If the call is mobile-station originated and the origination message does not request a special service option, the mobile station performs service option request initialization specifying 1 (the default service option number).
   - If the call is mobile-station terminated, the mobile station performs service option request initialization specifying the service option number requested in the page response message.

2. While in the traffic channel initialization substate, the mobile station performs the following actions:
   - The mobile station performs pilot strength measurements, but will not send pilot strength measurement messages.
   - The mobile station performs registration timer maintenance.

   If the mobile station does not support the assigned CDMA channel or the assigned forward traffic code channel, the mobile station enters the system determination substate of the mobile station initialization state.

3. If the mobile station supports the assigned CDMA and the assigned forward traffic code channel, the mobile station performs the following actions:
   - The mobile station tunes to the assigned CDMA channel.
   - The mobile station sets its code channel for the assigned forward traffic code channel.
   - The mobile station sets its forward and reverse traffic channel offsets to the assigned frame offset as determined by the current traffic channel frame offset, in units of 1.25 msec.
   - The mobile station sets its forward and reverse traffic channel long code masks to the public long code mask.

   If the mobile station does not receive $N_{5m} = 2$ consecutive good frames within $T_{50m} = 0.2$ seconds after entering this substate, the mobile station enters the system determination substate of the mobile station initialization state. N_{5m} denotes the number of received consecutive good forward traffic channel frames before a mobile station is allowed to enable its transmitter after entering the traffic channel initialization substate of the mobile station control on the traffic channel state. T_{50m} denotes the maximum time to obtain N_{5m} consecutive good forward traffic channel frames when in the traffic channel initialization substate.

4. If the mobile station receives N_{5m} consecutive good frames within T_{50m} seconds entering this substate, the mobile station performs the following additional functions while it remains in the traffic channel initialization substate:

   - The mobile station performs forward traffic channel supervision. If a loss of the forward traffic channel is declared, the mobile station enters the system determination substate of the mobile station initialization state.
   - The mobile station adjusts its transmit power.
   - The mobile station transmits the traffic channel preamble.
   - The mobile station performs the acknowledgment procedures. If an acknowledgment failure is declared, the mobile station disables its transmitter and enters the system determination substate of the mobile station initialization state.

   If the mobile station does not receive a base station acknowledgment order within T_{51m} = 2 seconds after entering this substate, the mobile station disables its transmitter and enters the system determination substate of the mobile station initialization state. T_{51m} denotes the maximum time for the mobile station to receive a base station acknowledgment order when in the traffic channel initialization substate.

5. If the mobile station receives a base station acknowledgment order within T_{51m} seconds after entering this substate, the mobile station performs the following actions:

   - If the call is mobile-station terminated, the mobile station enters the waiting for order substate.
   - If the call is mobile-station originated, the mobile station enters the conversation substate.

5.4.6   Waiting for Order Substate

In this substate, the mobile station waits for an alert with information message. Upon entering the waiting for order substate, the mobile station sets the substate timer to T_{52m} = 5 seconds. T_{52m} denotes the maximum time to receive a message in the waiting for order substate of the mobile station control on the traffic channel state that transits the mobile station to a different substate or state.

1. While in the waiting for order substate, the mobile station performs the following actions:

   - If the substate timer expires, the mobile station disables its transmitter and enters the system determination substate of the mobile station initialization state.
   - The mobile station performs forward traffic channel supervision. If a loss of the forward traffic channel is declared, the mobile station enters the system determination substate of the mobile station initialization state.
   - The mobile station adjusts its transmit power, performs the forward traffic channel power control, and performs handoff processing.

- If there is an active service option (the active service option number is not equal to NULL), the mobile station processes the received primary traffic bits in accordance with the requirements for the active service option; otherwise, the mobile station discards the received primary traffic bits.
- If there is an active service option (i.e., the active service option number is not equal to NULL), the mobile station will transmit primary traffic bits in accordance with the requirements for the active service option; otherwise, the mobile station transmits null traffic channel data.
- The mobile station performs registration timer maintenance.
- If the mobile station is directed by the user to transmit a message, the mobile station sends a data burst message.
- If the mobile station is directed by the user to request a service option, the mobile station performs the service option request initialization specifying the requested service option number and sends a service option request order.
- If there is an active service option (i.e., the active service option number is not equal to NULL), the mobile station sends a service option control order to invoke a service option specific function in accordance with the requirements for the active service option.
- If the mobile station is directed by the user to request a private long code transition and has the long code mask, the mobile station sends a long code transition request order 00000001 as a message requiring acknowledgment.
- If the mobile station is directed by the user to request a public long code transition, the mobile station sends a long code transition request order 00000000 as a message requiring acknowledgment.
- If the mobile station is directed by the user to operate in analog mode, the mobile station sends the request analog service order as a message requiring acknowledgment.
- If the mobile station is directed by the user to power down, the mobile station enters the release substate with a power-down indication.
- The mobile station performs the acknowledgment procedures. If an acknowledgment failure is declared, the mobile station disables its transmitter and enters the system determination substate of the mobile station initialization state.
- If the mobile station receives a message which is included in the following list and every message field value is within its permissible range, the mobile station processes the message as described below and in accordance with the message's action time.

1. Alert with information message
   If the message contains a signal information record, the mobile station should alert the user in accordance with the signal information record; otherwise, the mobile station should use the standard alert. The mobile station will enter the waiting for mobile station answer substate.

2. Analog handoff direction message

   The base station directs the mobile station to perform a CDMA to analog handoff by sending an analog handoff direction message. The mobile station then processes the message as specified previously and enters the waiting for order task with a handoff from CDMA indication.

3. Audit order

4. Authentication challenge message

   The mobile station resets the substate timer for T_{52m}=5 seconds. The mobile station will then process the message and respond within T_{32m}=5 seconds. T_{32m} denotes the maximum time to respond to an SSD update message, the base station challenge confirmation order, and the authentication challenge message.

5. Base station acknowledgment order

6. Base station challenge confirmation order

   The mobile station will reset the substate timer for T_{52m} seconds. The mobile station then processes the message and responds with an SSD update confirmation order or SSD update rejection order within T_{32m} seconds.

7. Data burst message

8. Handoff direction message

   The mobile station processes the message and resets the substate timer for T_{52m} seconds.

9. In-Traffic system parameters message

   The mobile station processes the message as already specified in "Processing the In-Traffic System Parameters Message."

10. Local control order

11. Lock until power-cycled order

    The mobile station disables its transmitter and records the reason for the lock until power-cycled order in the mobile station's semi-permanent memory. The mobile station should notify the user of the locked condition. The mobile station then enters the system determination substate of the mobile station initialization state, and will not enter the system access state again until after the next mobile station power-up or until it has received an unlock order. This requirement takes precedence over any other mobile station requirement specifying entry to the system access state.

12. Long code transition request order

    If the long code transition request order requests a transition to the private long code, and the mobile station is able to generate the private long code, and the mobile station accepts the request, the mobile station sends a long code transition response order 00000011 within T_{56m} = 0.2 seconds.

13. Maintenance order

    The mobile station enters the waiting for mobile station answer substate.

14. Maintenance required order

   The mobile station records the reason for the maintenance required order in the mobile station's semi-permanent memory. The mobile station remains in the unlocked condition. The mobile station should notify the user of the maintenance required condition.

15. Message encryption mode order

   In an effort to enhance the authentication process and to protect sensitive subscriber information, a method is provided to encrypt certain fields of selected traffic channel signaling messages. Signaling message encryption is controlled for each call individually. The initial encryption mode for the call is established by the value of the encrypt-mode field in the channel assignment message. If the encrypt mode is set to '0000', message encryption is off. To turn encryption on after channel assignment, the base station sends the message encryption mode or with the encrypt-mode field set to '0001'.

16. Mobile station registered message

17. Neighbor list update message

   The mobile station supports a neighbor set size of at least $N_{8m} = 20$ pilots. N_{8m} denotes the minimum supported neighbor set size. When the mobile station is first assigned a forward traffic channel, the mobile station initializes the neighbor set to contain the pilots specified in the most recently received neighbor list message. The mobile station maintains a counter (neighbor list age) for each pilot in the neighbor set. The mobile station initializes the counter to zero when it moves the pilot from the active set or the candidate set to the neighbor set. The mobile station initializes this counter to the neighbor set maximum age for retention in the set when it moves the pilot from the remaining set to the neighbor set. The mobile station increments the counter for each pilot in the neighbor set upon receipt of a neighbor list update message and deletes from the neighbor set all pilots whose counters exceed the neighbor set maximum age.

18. Parameter update order

   The mobile station resets the substate timer for T_{52m} seconds. The mobile station increments a modulo-64 count held in the mobile station. The mobile station sends a parameter update confirmation order within T_{56m} seconds. The mobile station sets the order qualification code field of the parameter update confirmation order to the same value as the order qualification code field of the parameters update order.

19. Pilot measurement request order

   If the mobile station receives the pilot measurement request order, then the mobile station sends, within T_{56m} seconds, a pilot strength measurement message.

20. Power control parameters message

   The mobile station stores the following parameters (stored value = received value) from the power control parameters message:

   • Power control reporting threshold
   • Power control reporting count

- Threshold report mode indicator
- Periodic report mode indicator
- Power report delay

The mobile station sets the total frames received and the bad frames received.

21. Release order

The mobile station enters the release substate with a base station release indication.

22. Retrieve parameters message

The mobile station sends, within T_{56m} seconds, a parameters response message.

23. Service option control order

If there is an active service option, the mobile station interprets the message action time of the service option control order in accordance with the requirements for the active service option and processes the service option control order in accordance with those requirements; otherwise, the mobile station sends a mobile station reject order 00000001 within T_{56m} seconds.

24. Service option request order

If the mobile station accepts the requested service option, the mobile station sets the service option request number to NULL and sends a service option response order accepting the requested service option within T_{58m} seconds. If the mobile station does not accept the requested service option and has an alternative service option to request, the mobile station sets the service option request number to the alternative service option number and sends a service option request order requesting the alternative service option within T_{58m} seconds. If the mobile station does not accept the requested service option and does not have an alternative service option to request, the mobile station sets the service option request number to NULL and sends a service option response order to reject the request within $T_{55m} = 2$ seconds.

25. Service option response order

If the service option number specified in the order is equal to the service option request number, the mobile station sets the service option request number to NULL.

If the order indicates a service option reject on, the mobile station sets the service option request number to NULL. If the order does not indicate a service option rejection and the service option specified in the order is not equal to the service option request number, the mobile station sets the service option request number to NULL and sends a mobile station reject order 00000100 within $T_{58m} = 5$ seconds.

26. Set parameter message

If the mobile station can set all of the parameters specified by the parameter-ID fields in the message, the mobile station will set them; otherwise, the mobile station sends, within T_{56m} seconds, a mobile station reject order.

27. SSD update message

    The mobile station resets the substate timer for T_{52m} seconds. The mobile station then processes the message and response with a base station challenge order within T_{32m} seconds.

28. Status request order

    The mobile station sends, within T_{56m} seconds, a status message.

    - If the mobile station receives any other message, it processes all layer 2 fields of the message. If the mobile station receives a message that cannot be processed, the mobile station discards the message and sends a mobile station rejection order within T_{56m} seconds.

5.4.7   Waiting for Mobile Station Answer Substate

In this substate, the mobile station waits for the user to answer the mobile station terminated call.

Upon entering the waiting for mobile station substate, the mobile station sets the substate timer for $T_{53m} = 65$ seconds. T_{53m} denotes the maximum time to receive a message in the mobile station answer substate that transits the mobile station to a different substate. While in the waiting for mobile station answer substate, the mobile station performs the following actions:

- If the substate timer expires, the mobile station disables its transmitter and enters the system determination substate of the mobile station initialization state.
- The mobile station performs forward traffic channel supervision. If a loss of the forward traffic channel is declared, the mobile station enters the system determination substate of the mobile station initialization state.
- The mobile station adjusts its transmit power. When operating in the CDMA transmission mode, mobile stations provide two independent means for output power adjustment, i.e., open loop estimation (solely a mobile station operation) and closed loop correction (involving both the mobile station and the base station).
- The mobile station performs handoff processing while in the mobile station control on the traffic channel state.

    1. Soft handoff: A handoff in which the mobile station commences communications with a new base station without interrupting communications with the old base station. Soft handoff can only be used between CDMA channels having identical frequency assignments.
    2. CDMA to CDMA hard handoff: A handoff in which the mobile station is transitioned between disjoint sets of base stations, different frequency assignment, or different frame offsets.
    3. CDMA to analog handoff: A handoff in which the mobile station is directed from a forward traffic channel to an analog voice channel.

- If there is an active service option (i.e., the active service option number is not equal to NULL), the mobile station processes the received primary traffic bits in accordance with the requirements for the active service option; otherwise, the mobile station discards the received primary traffic bits.

- If there is an active service option (i.e., the active service option number is not equal to NULL), the mobile station transmits primary traffic bits in accordance with the requirements for the active service option; otherwise, the mobile station transmits null traffic channel data.

- The mobile station performs registration timer maintenance. The mobile station provides a means of enabling and disabling each timer. When a timer is disabled, it will not be considered expired. A timer that has been enabled is referred to as active. The mobile station provides the following registration timers:

  1. Power-up/initialization timer

  2. Timer-based registration timer

  3. Zone list entry timers

  4. SID/NID list entry timers

- If the mobile station is directed by the user to answer the call, the mobile station sends a connect order to the base station as a message requiring acknowledgment. The mobile station enters the conversation substate.

- If the mobile station is directed by the user to transmit a message, the mobile station sends a data burst message.

- If the mobile station is directed by the user to request a service option, the mobile station performs the service option request initialization specifying the requested service option number and sends a service option request order.

- If there is an active option, the mobile station may send a service option control order to invoke a service option specific function in accordance with the requirements for the active service option.

- If the mobile station is directed by the user to request a private long code transition and has the long code mask, the mobile station sends a long code transition request order 00000001 as a message requiring acknowledgment.

- If the mobile station is directed by the user to request a public long code transition, the mobile station sends a long code transition request order 0 0 0 0 0 0 0 0 as a message requiring acknowledgment.

- If the mobile station is directed by the user to operate in analog mode, the mobile station sends the request analog service order as a message requiring acknowledgment.

- If the mobile station is directed by the user to power down, the mobile station enters the release substate with a power-down indication.

- The mobile station performs the acknowledgment procedures as specified in Section 5.3.2. If an acknowledgment failure is declared, the mobile station disables its transmitter and enters the system determination substate of the mobile station initialization state.

- If the mobile station receives a message which is included in the following list and every message field value is within its permissible range, the mobile station processes the message as described below and in accordance with the message's action time.

    1. Alert with information message
       The mobile station resets the substate timer for T_{53m} seconds. If the alert with information message does not contain a signal information record, the mobile station should use the standard alert.

    2. Analog handoff direction message
       The base station directs the mobile station to perform a CDMA to analog handoff by sending an analog handoff direction message. At the action time specified by the analog handoff direction message, the mobile station disables its transmitter. The mobile station then enables its transmitter on the analog voice channel within T_{63m} = 0.1 seconds. T_{63m} denotes the maximum time to execute a CDMA to analog handoff. The mobile station processes the message and enters the waiting for answer task.

    3. Audit order

    4. Authentication challenge message
       The base station generates the 24-bit random challenge value and sends it to the mobile station in the authentication challenge message on either the paging channel or the forward traffic channel. Upon receipt of the authentication challenge message, the mobile station initializes the authentication algorithm. The mobile station processes the message and responds within T_{32m} seconds.

    5. Base station acknowledgment order

    6. Base station challenge confirmation order
       The mobile station processes the message and responds with an SSD update confirmation order or SSD update rejection order within T_{32m} seconds.

    7. Data burst message

    8. Handoff direction message
       If the mobile station receives the handoff direction message, then the mobile station processes it as described below. When the message takes effect, the mobile station updates the active set in accordance with the handoff direction message and discontinues use of all forward traffic channels associated with pilots not listed in the handoff direction message. If the received current traffic channel frame offset is not equal to the stored current traffic channel frame offset, it changes the frame offset on both the forward traffic channel and the reverse traffic channel.

9. In-traffic system parameters message

   The mobile station stores the following parameters from the in-traffic system parameters message:

   •System and network identifications
   • Search window size for the active set, for the candidate set, for the neighbor set, and for the remaining set
   • Pilot detection threshold
   • Pilot drop threshold
   •Active set versus candidate set

10. Local control order

11. Lock until power-cycled order

    The mobile station disables its transmitter and records the reason for the lock until power-control order in the mobile station's semi-permanent memory. The mobile station enters the system determination substate of the mobile station initialization state, and will not enter the system access state again until after the next mobile station power-up or until it has received an unlock order. This requirement takes precedence over any other mobile station requirement specifying entry to the system access state.

12. Long code transition request order

    If the long code transition request order requests a transition to the private long code, and the mobile station is able to generate the private long code, and the mobile station accepts the request, the mobile station sends a long code transition response order 00000011 within T_{56m} seconds.

    If the long code transition request order requests a transition to the public long code and the mobile station accepts the request, the mobile station sends a long code transition response order 00000010 within T_{56m} seconds.

13. Maintenance order

    The mobile station resets the substate timer for T_{56m} seconds.

14. Maintenance required order

    The mobile station records the reason for the maintenance required order in the mobile station's semi-permanent memory. The mobile station remains in the unlocked condition. The mobile station should notify the user of the maintenance required condition.

15. Message encryption mode order

    In an effort to enhance the authentication process and to protect sensitive subscriber information, a method is provided to encrypt certain fields of selected traffic channel signaling messages. Signaling message encryption is controlled for each call individually. The initial encryption mode for the call is established by the value of the encrypt-mode field in the channel assignment message. If the encrypt mode is set to

'0000', message encryption is off. To turn encryption on after channel assignment, the base station sends the handoff direction message with the encrypt-mode field set to '0001' to the mobile station.

The base station also sends the message encryption mode order to the mobile station with the encrypt-mode field set to '0001'.

16. Mobile station registered message

The mobile station receives the mobile station registered message on the forward traffic channel when the mobile station is considered registered for the base station whose location and other parameters are included in the message.

17. Neighbor list update message

The mobile station supports a neighbor set size of at least $N_{8m} = 20$ pilots. N_{8m} denotes the minimum supported neighbor set size. When the mobile station is first assigned a forward traffic channel, the mobile station initializes the neighbor set to contain the pilots specified in the most recently received neighbor list message. The mobile station maintains a counter for each pilot in neighbor set. The mobile station initializes this counter to zero when it moves the pilot from the active set or the candidate set to the neighbor set. The mobile station initializes this counter to the neighbor set maximum age for retention in the set when it moves the pilot from the remaining set to the neighbor set. The mobile station increments the neighbor list age for each pilot in the neighbor set upon receipt of a neighbor list update message.

18. Parameter update order

The mobile station increments the call history parameter in the mobile station's semi-permanent memory. The mobile station sends a parameter update confirmation order within $T_{56m} = 0.2$ seconds. The mobile station sets the field of the parameter update confirmation order to the same values as the field of the parameter update order.

19. Pilot measurement request order

If the mobile station receives the pilot measurement request order, then the mobile station sends, within T_{56m} seconds, a pilot strength measurement message.

20. Power control parameters message

The mobile station stores the following parameters from the power control parameters message.

- Power control reporting threshold
- Power control reporting frame count
- Threshold report mode indicator
- Periodic report mode indicator
- Power report delay

The mobile station sets the total frames received and the bad frames received count to zero.

21. Release order

The mobile station enters the release substate with a base station release indication.

22. Retrieve parameters message

The mobile station sends, within T_{56m} seconds, a parameters response message.

23. Service option control order

If there is an active service option, the mobile station interprets the message action time of the service option control order in accordance with the requirements for the active service option and processes the service option control order in accordance with those requirements; otherwise, the mobile station sends a mobile station reject order 00000001 within T_{56m} seconds.

24. Service option request order

When the mobile station receives a service option request order, it performs the following actions:

- If the mobile station accepts the requested service option, the mobile station sets the service option requests number to NULL and sends a service option response order accepting the requested service option within $T_{58m} = 5$ seconds. The mobile station interprets the message action time of the service option request order in accordance with the requirements for the requested service option and begins using the requested service option in accordance with those requirements. The mobile station sets the active service option number to the requested service option number when the service option becomes active.

- If the mobile station does not accept the requested service option and has an alternative service option to request, the mobile station sets the service option request number to the alternative service option number and sends a service option request order requesting the alternative service option within T_{58m} seconds.

- If the mobile station does not accept the requested service option and does not have an alternative service option to request, the mobile station sets the service option request number to NULL and sends a service option response order to reject the request within T_{58m} seconds. The mobile station continues to process primary traffic as it did prior to receiving the service option request order and remains in the current state.

25. Service option response order

When the mobile station receives a service option response order, it performs the following actions:

- If the service option number specified in the order is equal to the service option request number, the mobile station sets the service option request number to NULL. The mobile station interprets the message action time of the service option response order in accordance with the requirements for the specified service option, and begins using the specified service option in accordance with those requirements. The mobile station sets the active service option number to the specified service option number when the service option becomes active.

- If the order indicates a service option rejection, the mobile station sets the service option request number to NULL. The mobile station continues to process primary traffic as it did prior to receiving the service option response order and remains in the current state.

- If the order does not indicate a service option rejection and the service option specified in the order is not equal to the stored service option request number, the mobile station sets the service option request number to NULL and sends a mobile station reject order 00000100 within T_{58m} seconds. The mobile station continues to process primary traffic as it did prior to receiving the service option response order and remains in the current state.

26. Set parameters message

If the mobile station can set all of the parameters specified by the parameter-ID fields in the message, the mobile station will set them; otherwise, the mobile station sends, within T_{56m} seconds, a mobile station reject order.

27. SSD update message

The mobile station processes the message and responds with a base station challenge order within T_{32m} seconds.

28. Status request order

The mobile station sends, within T_{56m} seconds, a status message.

- If the mobile station receives any other message with a message type specified in Table 5.2, it will process all layer 2 fields of the message. If the mobile station receives a message that is not included in the above list or cannot be processed, the mobile station discards the message and sends a mobile station reject order within T_{56m} seconds.

5.4.8   Conversation Substate

In this substate, the mobile station's primary traffic service option application exchanges primary traffic bits with the base station. While in the conversation substate, the mobile station performs the following actions:

- The mobile station performs forward traffic channel supervision. If a loss of the forward traffic channel is declared, the mobile station enters the system determination substate of the mobile station initialization state.

- The mobile station adjusts its transmit power and performs the forward traffic channel power control as specified in Section 5.4.1.

- The mobile station performs handoff processing as specified in Section 5.2.4.

- If an active service option number is not equal to NULL, the mobile station processes the received primary traffic bits in accordance with the requirements for the active service option; otherwise, the mobile station discards the received primary traffic bits.

- If there is an active service option (i.e., an active service option number is not equal to NULL), the mobile station transmits primary traffic bits in accordance with the requirements for the active service option; otherwise, the mobile station transmits null traffic channel data.

- The mobile station performs registration timer maintenance.

- If the mobile station originated the call and did not send all the dialed digits in the origination message, the mobile station sends the remaining dialed digits to the base station in the origination continuation message. The mobile station sends the origination continuation message as a message requiring acknowledgment within $T_{54m} = 0.2$ seconds after entering the conversation substate.

- If the mobile station is directed by the user to transmit a message, the mobile station will send a data burst message.

- If the mobile station is directed by the user to request a service option, the mobile station performs a service option request initialization specifying the requested service option number, and sends a service request order.

- If an active service option number is not equal to NULL, the mobile station may send a service option control order to invoke a service option specific function in accordance with the requirements for the service option.

- If the mobile station is directed by the user to request a private long code transition and has the long code mask, the mobile station sends a long code transition request order 00000001 as a message requiring acknowledgment.

- If the mobile station is directed by the user to request a public long code transition, the mobile station sends a long code transition order 00000000 as a message requiring acknowledgment.

- If the mobile station is directed by the user to issue a flash, the mobile station will build a flash with information message with the collected digits contained in a keypad facility information record and send the message to the base station as a message requiring acknowledgment.

- If the mobile station is directed by the user to send burst dual-tone multifrequency (DTMF) digits, the mobile station will build a send burst DTMF message with the dialed digits and send the message as a message requiring acknowledgment.

- If the user directs the mobile station to send a continuous DTMF digit, the mobile station will build a continuous DTMF tone order with the dialed digit and send the order as a message requiring acknowledgment. When the user directs the mobile station to cease sending continuous DTMF digit, the mobile station will send a continuous DTMF tone order 11111111 as a message requiring acknowledgment.

- If the user directs the mobile station to operate in analog mode, the mobile station will send the request analog service order as a message requiring acknowledgment.

- If the mobile station is directed by the user to disconnect the call, the mobile station enters the release substate with a mobile station release indication.

- If the mobile station is directed by the user to power down, the mobile station enters the release substate with a power-down indication.

- The mobile station will perform the acknowledgment procedures. If an acknowledgment failure is declared, the mobile station disables its transmitter and enters the system determination substate of the mobile station initialization state.

- If the mobile station receives a message which is included in the following list and every message field value is within its permissible range, the mobile station will process the message as described below and in accordance with the message's action time.

   1. Alert with information message
      If the message contains a signal information record with the signal type field set to '01' or '10', or if the message does not contain a signal information record, the mobile station enters the waiting for mobile station answer substate. If the alert with information message does not contain a signal information record, the mobile station should use the standard alert.

   2. Analog handoff direction message
      The mobile station will process the message and enter the conversation task with a handoff from CDMA indication.

   3. Audit order

   4. Authentication challenge message
      The mobile station will process the message and respond within T_{32m} seconds. The base station generates the 24-bit random challenge value and sends it to the mobile station in the authentication challenge message on either the paging channel or the forward traffic channel. Upon receipt of the authentication challenge message, the mobile station initializes the authentication algorithm.

   5. Base station acknowledgment order

   6. Base station challenge confirmation order
      The mobile station processes the message and responds with an SSD update confirmation order or an SSD update rejection order within T_{32m} seconds.

   7. Continuous DTMF tone order
      Support of this order by the mobile station is optional.

   8. Data burst message

9. Flash with information message

10. Handoff direction message

    If the mobile station receives the handoff direction message, then the mobile station will process the message as described below.

    - If the received traffic channel frame offset is not equal to the stored traffic channel frame offset and the use action time indicator (1 bit) equals '0', then the mobile station will process the message at the first 80 ms boundary, relative to the system time, occurring at least 80 ms after the end of the frame containing the last bit of the message.
    - When the message takes effect, the mobile station updates the active set in accordance with the handoff direction message and discontinues use of all forward traffic channels associated with pilots not listed in the handoff direction message.
    - If the received traffic channel frame offset is not equal to the current (stored) traffic channel frame offset, it changes the frame offset on both the forward traffic channel and the reverse traffic channel.
    - If the reset layer 2 indicator (1 bit) is equal to '1', it resets the acknowledgment procedures and resets the forward traffic channel power control counters. If this field is set to '1', the mobile station resets its layer 2 sequence numbers up on executing the handoff.

11. In-Traffic system parameters message

    The mobile station stores the following parameters from the in-traffic system parameters message:

    - System and network identifications
    - Search window size for the active set, candidate set, neighbor set, and remaining set.
    - Pilot detection threshold
    - Pilot drop threshold
    - Active set versus candidate set comparison threshold
    - Drop timer value
    - Maximum age for retention of neighbor set members

12. Local control order

13. Lock until power-cycled order

    The mobile station disables its transmitter and records the reason for the lock until power-cycled order in the mobile station's semi-permanent memory. The mobile station should notify the user of the locked condition. The mobile station enters the system determination substate of the mobile station initialization state, and will not enter the system access state again until after the next mobile station power-up or until it has received an unlock order. The requirement takes precedence over any other mobile station requirement specifying entry to the system access state.

14. Long code transition request order
    - If the long code transition request order requests a transition to the public long code and the mobile station accepts the request, the mobile station sends a long code transition response order 00000010 within T_{56m} seconds. The mobile station should indicate to the user that the voice privacy mode is inactive. If the long code transition request order requests a public long code transition, and the mobile station does not accept the request, the mobile station sends a long code transition response order 00000011 within T_{56m} seconds.
    - If the long code transition request order requests a transition to the private long code, and the mobile station is able to generate the private long code, and the mobile station accepts the request, the mobile station sends a long code transition response order 00000011 within T_{56m} seconds. The mobile station uses the private long code on both the forward traffic channel and the reverse traffic channel. If the use action time indicator (1 bit) equals '0', the mobile station begins using the private long code at the first 80 ms boundary (relative to the start of the system time) after N_{4m} frames from the end of the response transmission. The mobile station should indicate to the user that the voice privacy mode is active.
    - If the long code transition request order requests a private long code transition, and the mobile station is not able to generate the private long code or the mobile station does not accept the request, the mobile station sends a long code transition response order 00000010 within T_{56m} seconds.

15. Maintenance order
    The mobile station enters the waiting for mobile station answer substate.

16. Maintenance required order
    The mobile station records the reason for the maintenance required order in the mobile station's semi-permanent memory. The mobile station remains in the unlocked condition. The mobile station should notify the user of the maintenance required condition.

17. Message encryption mode order
    In an effort to enhance the authentication process and to protect sensitive subscriber information, a method is provided to encrypt certain fields of selected traffic channel signaling messages. Messages are not encrypted if authentication is not performed. Signaling message encryption is controlled for each call individually. The initial encryption mode for the call is established by the value of the message encryption mode field (4 bits) in the channel assignment message. If the encryption mode is set to 0000, message encryption is disabled, and all messages sent on the forward and reverse traffic channels are unencrypted. To turn encryption on after channel assignment, the base station sends one of the following messages to the mobile station:
    - Handoff direction message with the message encryption mode field set to 0001.

- Message encryption mode order with the encryption mode field set to 0001.
- Analog handoff direction message with the message encryption mode field (1 bit) set to '1'. When set to '0', encryption of the message on the analog voice channels is off. When set to '1', analog message encryption is enabled.

18. Mobile station registered message

The mobile station receives the mobile station registered message on the forward traffic channel when the mobile station is considered registered for the base station whose location and other parameters are included in the message. The mobile station stores the parameters and performs the actions in accordance with the specified rule.

19. Neighbor list update message

The mobile station supports a neighbor set size of at least $N_{8m} = 20$ pilots. When the mobile station is first assigned a forward traffic channel, the mobile station initiates the neighbor set to contain the pilots specified in the most recently received neighbor list message.

If the mobile station receives a neighbor list update message, it performs the following:

- Increments the neighbor list age for each pilot in the neighbor set.
- Deletes from the neighbor set all pilots whose neighbor list age exceeds the neighbor set maximum age for retention in the set.
- Adds to the neighbor set each pilot named in the message, if it is not already a pilot of the candidate set or neighbor set. If the mobile station can store in the neighbor set only k additional pilots and more than k new pilots were sent in the neighbor list update message, the mobile station stores the first k new pilots listed in the message.

20. Parameter update order

The mobile station increments the call history parameter (6 bits) in the mobile station's semi-permanent memory. The mobile station then sends a parameter update confirmation order within T_{56m} seconds. The mobile station sets the order qualification code field of the parameter update confirmation order to the same value as the order qualification code field of the parameter update order.

21. Pilot measurement request order

The mobile station sends, within T_{56m} seconds, a pilot strength measurement message.

22. Power control parameters message

The mobile station stores the following parameters from the power control parameters message:

- Power control reporting threshold
- Power control reporting frame count
- Threshold report mode indicator

> • Periodic report mode indicator
> • Power report delay

The mobile station sets the total frames received and the bad frames count to zero.

23. Release order

The mobile station enters the release substate with a base station release indication.

24. Retrieve parameters message

The mobile station sends, within T_{56m} seconds, a parameters response message.

25. Send burst DTMF message

Support of this order by the mobile station is optional.

26. Service option control order

If there is an active service option, the mobile station interprets the message action time of the service option control order in accordance with the requirements for the active service option and processes the service option control order in accordance with those requirements; otherwise, the mobile station sends a mobile station reject order 0000001 within T_{56m} seconds.

27. Service option request order

When the mobile station receives a service option request order, it performs the following:

> • If the mobile station accepts the requested service option, the mobile station sets the service option request number to NULL and sends a service option response order accepting the requested service within T_{58m} seconds.
> • If the mobile station does not accept the requested service option and has an alternative service option to request, the mobile station sets the service option request number to the alternative service option number and sends a service option request order requesting the alternative service option within T_{58m} seconds.
> • If the mobile station does not accept the requested service option and does not have an alternative service option to request, the mobile station sets the service option request number to NULL and sends a service option response order to reject the request within T_{58m} seconds. The mobile station continues to process primary traffic as it did prior to receiving the service option request order and remains in the current state.

28. Service option response order

When the mobile station receives a service option response order, it performs the following actions:

> • If the service option number specified in the order is equal to the service option request number, the mobile station sets that request number to NULL. The mobile station interprets the message action time of the service option request order in accordance with the requirements for the specified service

option, and begins using the specified service option in accordance with those requirements. The mobile station sets the active service option number to the specified service option number when the service option becomes active.

- If the order indicates a service option rejection, the mobile station sets the service option request number to NULL. The mobile station continues to process primary traffic as it did prior to receiving the service option response order and remains in the current state.

- If the order does not indicate a service option rejection and the service option specified in the order is not equal to the service option request number, the mobile station sets the service option request number to NULL and sends a mobile station reject order 00000100 within T_{58m} seconds. The mobile station continues to process primary traffic as it did prior to receiving the service option response order and remains in the current state.

29. Set parameters message

   If the mobile station can set all of the parameters specified by the parameter-ID fields in the message, the mobile station will set them; otherwise, the mobile station sends, within T_{56m} seconds, a mobile station reject order.

30. SSD update message

   The mobile station processes the message and responds with a base station challenge order within T_{32m} seconds.

31. Status reject order

   The mobile station sends, within T_{56m} seconds, a status message.

If the mobile station received any other message such as the order message, it will process all layer 2 fields of the message. If the mobile station receives a message that is not included in the forward traffic channel messages, the mobile station discards the message and sends a mobile station reject order within T_{56m} seconds.

5.4.9   Release Substate

In this substate, the mobile station confirms the call disconnect. Upon entering the release substate, the mobile station confirms the following actions:

- The mobile station sets the substate timer for T_{55m} seconds.

- If the mobile station enters the release substate with a power-down indication, the mobile station sends a release order 00000001 and performs power-down registration procedures.

- If the mobile station enters the release substate with a mobile station release indication, the mobile station sends a release order 00000000.

- If the mobile station enters the release substate with a base station release indication, the mobile station sends a release order 00000000. The mobile station disables its transmitter and enters the system determination substate of the mobile station initialization state.

While in the release substate, the mobile station performs the following actions:

- If the substate timer for T_{55m} seconds expires, the mobile station disables its transmitter and enters the system determination substate of the mobile station initialization state.
- The mobile station performs forward traffic channel supervision. If a loss of the forward traffic channel is declared, the mobile station enters the system determination substate of the mobile station initialization state.
- The mobile station adjusts its transmit power and performs the forward traffic channel power control.
- The mobile station performs handoff processing.
- The mobile station transmits null traffic channel data on the reverse traffic channel.
- The mobile station performs registration timer maintenance.
- The mobile station performs the acknowledgment procedures. If an acknowledgment failure is declared, the mobile station disables its transmitter and enters the system determination substate of the mobile station initialization state.
- If the mobile station receives a message which is included in the following list and every message field value is within its permissible range, the mobile station processes the message as described below and in accordance with the message's action time.

    1. Alert with information message
       The mobile station enters the waiting for mobile station answer substate. If the alert with information message does not contain a single information record, the mobile station should use the standard alert which is defined as the signal type '01', the pitch of alerting signal '00' and the signal code '000001'. The information record allows the network to convey information to a user by means of tones and other alerting signals.
    2. Base station acknowledgment order
    3. Data burst message
    4. Handoff direction message
       If the mobile station receives the handoff direction message, then the mobile station processes the message as described below.

        - If the received traffic channel frame offset is not equal to the stored traffic channel frame offset and the use action time indicator equals '0', then the message takes effect on the first 80 ms boundary (relative to the system time) occurring at least 80 ms after the end of the frame containing the last bit of the message.
        - Otherwise, the message will take effect at the action time of the message.
       When the message takes effect, the mobile station performs the following actions:

- Updates the active set, candidate set, and neighbor set in accordance with the handoff direction message.
- Discontinues use of all forward traffic channels associated with pilots not listed in the handoff direction message.
- If the received traffic channel frame offset is not equal to the current traffic channel frame offset, it changes the frame offset on both the forward traffic channel and the reverse traffic channel.
- The reset acknowledgment procedures command is used to reset acknowledgment processing in the mobile station. To direct the mobile station to reset its acknowledgment procedures, the base station sets this field to '1'. Otherwise, the base station sets this field to '0'. If this field is equal to '1', it resets the forward traffic channel power control counters.
- Uses the long code mask specified by the voice privacy and indicates to the user the voice privacy mode status.
- Processes the encrypt-mode field. In an effort to enhance the authentication process and to protect sensitive subscriber information (such as PINs), a method is provided to encrypt certain fields of selected traffic channel signaling messages.

5. In-Traffic system parameters message

The mobile station stores the following parameters from the in-traffic system parameters message:

- System and network identifications
- Search window size for the active set, candidate set, neighbor set, and remaining set
- Pilot detection threshold
- Pilot drop threshold
- Active set versus candidate set comparison threshold
- Drop timer value
- Maximum age for retention of neighbor set members

The mobile station should indicate to the user whether the mobile station is roaming.

6. Local control order

7. Lock until power-cycled order

The mobile station disables its transmitter and records the reason for the lock until power-cycled order in the mobile station's semi-permanent memory. The mobile station should notify the user of the locked condition. The mobile station enters the system determination substate of the mobile station initialization state, and will not enter the system access state again until after the next mobile station power-up or until it has received an unlock order. This requirement takes precedence over any other mobile station requirement specifying entry to the system access state.

8. Maintenance required order

   The mobile station records the reason for the maintenance required order in the mobile station's semi-permanent memory. The mobile station remains in the unlocked condition. The mobile station should notify the user of the maintenance required condition.

9. Mobile station registered message

   The mobile station receives the mobile station registered message on the forward traffic channel when the mobile station is considered registered for the base station whose location and other parameters are included in the message:

   - System and network identifications
   - Registration zone
   - Number of registration zones to be retained
   - Zone timer length
   - Multiple SID storage indicator
   - Multiple NID storage indicator
   - Base station latitude
   - Base station longitude
   - Registration distance

   The mobile station performs the following actions:

   - Sets the first-idle ID status to enabled.
   - Adds the registration zone number of the base station, SID, NID to the zone list if not already in the list.
   - Disables the zone list entry timer for the entry of the zone list containing the registration zone number, SID, and NID. For any other entry of the zone list whose entry timer is not active, it enables the entry timer with the duration specified by the zone timer length.
   - If the zone list contains more than the number of registration zones to be retained, it deletes the excess entries according to the specified rules.
   - Adds the stored value of SID and NID to the SID-NID list if not already in the list.
   - Disables the SID/NID list entry timer for the entry of SID-NID-list containing SID and NID. For any other entry of SID-NID list whose entry timer is not active, it enables the entry timer with the specified duration.
   - If SID-NID list contains more than $N_{10m} = 4$ entries, it deletes the excess entries according to the specified rule of zone-based registration. N_{10m} denotes SID/NID list size.
   - If the multiple SID storage indicator is equal to '0' and the registration SID/NID list contains entries with different SIDs, it deletes the excess entries according to the rules specified in zone-base registration.

- If the multiple NID storage indicator is equal to '0' and the registration SID/NID list contains more than one entry for any SID, it deletes the excess entries according to the rules specified in zone-based registration.
- Sets the stored location of last registration distance to the base station's location and sets the stored registration distance to the base station's registration distance.
- Updates its roaming status and sets the mobile station termination indicator as specified in roaming types. The mobile station should indicate to the user whether the mobile station is roaming.

10. Neighbor list update message

The mobile station supports a neighbor set size of at least $N_{8m} = 20$ pilots. When the mobile station is first assigned a forward traffic channel, the mobile station initializes the neighbor set to contain the pilots specified in the most recently received neighbor list message. The mobile station contains a counter (neighbor list age) for each pilot in the neighbor set. The mobile station initializes this counter to zero when it moves the pilot from the active set or the candidate set to the neighbor set. The mobile station initializes this counter to the neighbor set maximum age for retention in the set when it moves the pilot from the remaining set to the neighbor set. The mobile station increments the neighbor list age for each pilot in the neighbor set upon receipt of a neighbor list update message.

If the mobile station receives a neighbor list update message, it will perform the following actions:

- Increments the neighbor list age for each pilot in the neighbor set.
- Deletes from the neighbor set all pilots whose neighbor list age exceeds the neighbor set maximum age for retention in the set.
- Adds to the neighbor set each pilot named in the message, if it is not already a pilot of the candidate set or neighbor set. If the mobile station can store in the neighbor set only k additional pilots and more than k new pilots were sent in the neighbor list update message, the mobile station stores the first k new pilots listed in the message.

11. Power control parameters message

The mobile station stores the following parameters from the power control parameters message:

- Power control reporting threshold
- Power control reporting frame count
- Threshold report mode indicator
- Periodic report mode indicator
- Power report delay

The mobile station sets the total frames received and the received bad frames to zero.

12. Release order

The mobile station disables its transmitter. If the mobile station enters the release order with a power-down indication, the mobile station may power down; otherwise, the mobile station enters the system determination substate of the mobile station initialization state.

13. Retrieve parameters message

The mobile station sends, within T_{56m} seconds, a parameters response message.

14. Service option control order

If there is an active service option (i.e., the active service option number is not equal to NULL), the mobile station interprets the message action time of the service option control order in accordance with the requirements for the active service option and processes the service option control order in accordance with those requirements; otherwise, the mobile station sends a mobile station reject order 00000001 within T_{56m} seconds.

15. Status request order

The mobile station sends, within T_{56m} seconds, a status message.

If the mobile station receives any other message with a message type specified in the forward traffic channel message body formats, it processes all layer 2 fields of the message. If the mobile station receives a message that is not included in the above list or cannot be processed, the mobile station discards the message and sends a mobile station reject order within $T_{56m} = 0.2$ seconds. T_{56m} denotes the default maximum time to respond to a received message or order on the forward traffic channel.

Base Station Call Processing

This chapter describes base station call processing based on TIA/EIA/IS-95. Call processing refers to the message flow protocols between the base station and the mobile station.

Base station call processing consists of the following types of processing.

- Pilot and Sync Channel Processing

  During this processing, the base station transmits the pilot channel and sync channel which the mobile station uses to acquire and synchronize to the CDMA system while the mobile station is in the mobile station initialization state.

- Paging Channel Processing

  During this processing, the base station transmits the paging channel which the mobile station monitors to receive messages while the mobile station is in the mobile station idle state and the system access state.

- Access Channel Processing

  During this processing, the base station monitors the access channel to receive messages which the mobile station sends while the mobile station is in the system access state.

- Traffic Channel Processing

  During this processing, the base station uses the forward and reverse traffic channels to communicate with the mobile station while the mobile station is in the mobile station control on the traffic channel state.

6.1 PILOT AND SYNC CHANNEL PROCESSING

1. The pilot channel is a reference channel which the mobile station uses for acquisition, timing, and as a phase reference for coherent demodulation. The base station continually transmits a pilot channel for every CDMA channel supported by the base station.

2. The sync channel provides the mobile station with system configuration and timing information. The base station transmits at most one sync channel for each supported CDMA channel. If the base station supports the primary CDMA channel, the base station will transmit a sync channel on the primary CDMA channel. If the base station does not support the primary CDMA channel, the base station will transmit a sync channel on the secondary CDMA channel. The base station continually sends the sync channel message on each sync channel that the base station transmits.

6.2 PAGING CHANNEL PROCESSING

The base station transmits the paging channel which the mobile station monitors to receive messages while the mobile station is in the mobile station idle state and the system access state. The base station may transmit up to seven paging channels on each supported CDMA channel. For each supported CDMA channel for which the base station transmits a sync channel, the base station transmits at least one paging channel.

For each paging channel that the base station transmits, the base station continually sends valid paging channel messages, which may include the null message. The base station will not send any message which ends in a paging channel slot other than the paging channel slot in which the message begins, or the paging channel slot following the paging channel slot in which the message begins.

6.2.1 Paging Channel Procedures

1. To determine the mobile station's assigned CDMA channel, the base station uses the same hash function which is used by the mobile station to select one out of N available resources with the following inputs:
   • Mobile station's identification number (MIN)
   • Number of CDMA channels on which the base station transmits paging channels

2. To determine the mobile station's assigned paging channel, the base station uses the hash function with the following inputs:
   • Mobile station identification number (MIN)
   • Number of paging channels which the base station transmits on the mobile station's assigned CDMA channel

3. To determine the assigned paging channel slots for a mobile station with a given cycle index i, the base station selects a number obtained from the hashing function with the following inputs:
   • Mobile station's identification number (MIN)

- Maximum number of paging channel slots (2048)

The assigned paging channel slots for the mobile station are those slots for which
($\lfloor t/4 \rfloor$ — Hash Value used to determine the assigned paging channel slots)
$\text{mod}(16\text{x}T) = 0$
where t is the system time in frames, and T is the slot cycle length in units of 1.28 seconds
given by
$$T = 2^i$$
where i is the slot cycle index.

6.2.2   Message Transmission and Acknowledgment Procedures

The paging channel acknowledgment procedures facilitate the reliable exchange of messages
between the base station and the mobile station on the paging channel and access channel.

The base station uses the fields such as the acknowledgment address type, acknowledgment sequence number, message sequence number, acknowledgment required, and valid
acknowledgment to support this mechanism. These fields are referred to as layer 2 fields, and the
acknowledgment procedures are referred to as layer 2 procedures. All other message fields and
the processing thereof are referred to as pertaining to layer 3.

A paging channel message can be directed either to a specific mobile station, by means of
the address field, or to a specific MIN (page message and slotted page message only). Since
MINs can be active in more than one mobile station, separate acknowledgment and message
sequence numbering procedures are used for each type of message address.

The base station maintains an independent message numbering sequence on the paging
channel for each message address type (i.e., for each allowed value of the address-type field) and
for each address. The records of the message and slotted page message are considered to be
addressed by MIN (as if the address type were equal to '000').

For each message address type, separate message numbering sequences are maintained for
messages requiring acknowledgment and for messages not requiring acknowledgment. Each
base station may maintain the sequence numbers independently of other base stations. For each
new message sent to a message address, the base station will increment the appropriate message
sequence number value, modulo 8.

The base station waits at least $T_{4m} = 2.2$ seconds after transmitting a message sequence
number in a message sent to a message address before using the same message sequence number
in a different message.

The base station may send a message several times to increase the probability of message
reception. The base station may complete all retransmissions of the same message within T_{4m}
seconds after the first transmission, as shown in Fig. 6.1. If the base station sends a message with
the same contents more than T_{4m} seconds after the first transmission, it will use a different message sequence number.

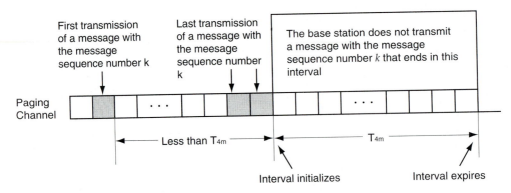

Figure 6.1 Message sequence number reuse

To acknowledge the most recently received access channel message from the mobile station, the base station sets the valid acknowledgment indicator field to '1'. When the base station receives a message with the valid acknowledgment field set to '1', it uses the received acknowledgment address type, acknowledgment sequence number, and mobile station identification fields to determine the message that is being acknowledged. The base station should not retransmit a message requiring acknowledgment after it has received an acknowledgment of the message.

6.2.3   Overhead Information

The base station sends overhead messages to provide the mobile station with the information it needs to operate with the base station.

The base station maintains a configuration sequence number, and increments the configuration sequence number modulo 64 whenever the base station modifies the system parameters message, the neighbor list message, or the CDMA channel list message.

The base station maintains an access configuration sequence number, and increments the access confirmation sequence number modulo 64 whenever the base station modifies the access parameters message.

On each of the paging channels the base station transmits, the base station sends each of the following system overhead messages at least once per $T_{1b} = 1.28$ seconds. T_{1b} denotes the maximum period between subsequent transmissions of an overhead message on the paging channel by the base station.

1. Access Parameters Message

The access parameters message defines the parameters used by mobile stations when transmitting to the base station on an access channel. When the base station sends an access parameters message, it uses the following variable-length message format.

• Message Type (8 bits)

The base station sets this field to 00000010.

- Pilot PN sequence offset index (9 bits)
  The base station sets this field to the pilot PN sequence offset for this base station, in units of 64 PN chips.

- Access parameters message sequence number (6 bits)
  The base station sets this field to the access configuration sequence number.

- Number of access channels (5 bits)
  The base station sets this field to one less than the number of access channels associated with this paging channel.

- Nominal transmit power offset (4 bits)
  The base station sets this field to the correction factor to be used by mobile stations in the open loop power estimate, expressed as a two's complement in units of 1 dB.

- Initial power offset for access (5 bits)
  The base station sets this field to the correction factor to be used by mobile stations in the open loop power estimate for the initial transmission on an access channel, expressed as a two's complement value in units of 1 dB.

- Power increment (3 bits)
  The base station sets this field to the value by which mobile stations are to increase their transmit power between successive access probes in an access probe sequence, in units of 1 dB.

- Number of access probes (4 bits)
  The base station sets this field to one less than the maximum number of access probes that mobile stations are to transmit in a single access probe sequence.

- Maximum access channel message capsule size (3 bits)
  The base station sets this field to a value in the range 0 to 7, three less than the maximum number of access channel frames in an access channel message capsule.

- Access channel preamble length (4 bits)
  The base station sets this field to one less than the number of access channel frames that mobile stations are to transmit in each access channel preamble.

- Persistence value for access overload classes 0 through 9 (6 bits)
  If mobile stations in access overload classes 0 through 9 are permitted to transmit requests on the access channel, the base station sets this field to the persistence value to be used. If such mobile stations are not permitted to transmit requests on the access channel, the base station sets this field to 111111.

- Persistence value for access overload class 10 (Test mobile stations) (3 bits)
  If the mobile stations in access overload class 10 are permitted to transmit requests on the access channel, the base station sets this field to the persistence value to be used. If such mobile stations are not permitted to transmit requests on the access channel, the base station shall set this field to 111.

- Persistence value for access overload class 11 (Energy mobile stations) (3 bits)
  If the mobile stations in access overload class 11 are permitted to transmit requests on the access channel, the base station sets this field to the persistence value to be used. If such mobile stations are not permitted to transmit requests on the access channel, the base station sets this field to 111.

- Persistence value for access overload classes 12, 13, 14, and 15, respectively (3 bits each)
  If the mobile stations in access overload class 12, 13, 14, or 15 are permitted to transmit requests on the access channel, the base station sets this field to the persistence value to be used. If such mobile stations are not permitted to transmit requests on the access channel, the base station sets this field, respectively, to 111.

- Persistence modifier for access channel attempts for message transmissions (3 bits): α
  A mobile station multiplies its transmission probability by $2^{-\alpha}$ for such attempts. The base station then sets this field to the persistence modifier for access channel attempts for message transmissions.

- Persistence modifier for access channel attempts for registrations which are not responses to the registration request order (3 bits)
  A mobile station multiplies its transmission probability by $2^{-(REG\text{-}PSIST)}$ for such attempts. The base station sets this field to the persistence modifier for access channel attempts for registrations which are not responses to the registration request order.

- Time randomization for access channel probes (4 bits): β
  A mobile station delays its transmission from the system time by RN PN chips, where RN is a number determined by hashing between 0 and $2^{\beta} - 1$ PN chips, i.e., the delay in PN chips generated (pseudorandomly) by the mobile station prior to performing an access attempt. The base station then sets this field to a value in the range 0 to 9 inclusive such that the time randomization range is $2^{\beta} - 1$ PN chips.

- Access channel acknowledgment timeout (4 bits)
  The base station sets this field to two less than the length of time mobile stations are to wait after the end of an access channel transmission before determining that the base station did not receive the transmission, in units of 80 ms.

- Access channel probe backoff range (4 bits)
  The base station sets this field to one less than the maximum number of slots mobile stations are to delay due to random backoff between consecutive access probes.

- Access channel probe sequence backoff range (4 bits)
  The base station sets this field to one less than the maximum number of slots mobile stations are to delay due to random backoff between successive access probe sequences and before the first access probe sequence of a response access.

- Maximum number of access probe sequences for an access channel request (4 bits)
  The base station sets this field to the maximum number of access probe sequences mobile stations are to transmit for an access channel request. This field must be greater than 0.

- Maximum number of access probe sequences for an access channel response (4 bits)
  The base station sets this field to the maximum number of access probe sequences mobile stations are to transmit for an access channel response. This field must be greater than 0.

- Authentication mode (2 bits)
  If mobile stations are to include standard authentication data in access channel messages, the base station sets this field to '01'. If mobile stations are not to include authentication data in access channel messages, the base station sets this field to '00'. All other values are reserved.

- Random challenge value (0 or 32 bits)
  If the authentication mode field is set to '01', the base station sets this field to the random challenge value to be used by mobile stations for authentication. If the authentication mode field is set to any other value, the base station omits this field.

- Reserved bits (7 bits)
  The base station always sets this field to 0000000.

2. CDMA channel list message
   When the base station sends a CDMA channel list message, it uses the following variable-length message format.

   - Message Type (8 bits)
     The base station sets this field to 00000100.

   - Pilot PN sequence offset index (9 bits)
     The base station sets this field to the pilot PN sequence offset for this base station, in units of 64 PN chips.

   - Configuration message sequence number (6 bits)
     The base station sets this field to the configuration sequence number.

   - CDMA channel frequency assignment (11 bits)
     The order in which occurrences of this field are included gives the designations of the supported CDMA channels as CDMA channel 1 through CDMA channel N.

     The base station includes one occurrence of this field for each CDMA channel containing a paging channel that is supported by this base station. If the primary CDMA channel is supported by this base station, the base station includes the occurrence of this field first. If the primary CDMA channel is not supported and the secondary CDMA channel is supported, the base station includes the occurrence of this field corresponding to the secondary CDMA channel first.

     The base station sets each occurrence of this field to the CDMA channel number corresponding to the CDMA frequency assignment for that CDMA channel.

   - Reserved bits (0 through 7 bits, as needed)
     The base station adds reserved bits as needed in order to make the length of the entire message equal to an integer number of octets. The base station sets these bits to '0'.

3. Neighbor list message

When the base station sends a neighbor list message, it uses the following variable-length message format.

- Message type (8 bits)

  The base station sets this field to 00000011.

- Pilot PN sequence offset index (9 bits)

  The base station sets this field to the pilot PN sequence offset for this base station, in units of 64 PN chips.

- Configuration message sequence number (6 bits)

  The base station sets this field to the configuration sequence number.

- Pilot PN sequence offset index increment (4 bits)

  A mobile station searches for remaining set pilots at pilot PN sequence index values that are multiples of this value. The base station sets this field to the pilot sequence increment, in units of 64 chips, that mobile stations are to use for searching the remaining set. The base station should set this field to the largest increment such that the pilot PN sequence offsets of all its neighbor base stations are integer multiples of that increment.

  The base station includes one occurrence of the following two-field record for each member which mobile stations are to place in their neighbor sets. The base station may include zero or more occurrences of the following record.

- Neighbor configuration (3 bits)

  The base station sets this field to the value shown in Table 6.1 corresponding to the configuration of this neighbor

Table 6.1   Neighbor Configuration Field

Value (bit)	Neighbor Configuration
000	The neighbor base station has the same configuration as the current base station.
001	The neighbor base station has a different configuration. It does have a primary paging channel on the current CDMA frequency assignment.
010	The neighbor base station does not have a paging channel on the current CDMA frequency assignment. It does have a primary paging channel on the first CDMA channel listed in the CDMA channel list message transmitted by the current base station.
011	The neighbor base station configuration is unknown.
100-111	Reserved.

- Neighbor pilot PN sequence offset index (9 bits)
  The base station sets this field to the pilot PN sequence offset for this neighbor, in units of 64 PN ships.

- Reserved bits (0 through 7 bits, as needed)
  The base station adds reserved bits as needed in order to make the length of the entire message equal to an integer number of octets. The base station sets these bits to '0'.

4. System parameters message
   When the base station sends a system parameters message, it uses the following fixed-length message format.

   - Message Type (8 bits)
     The base station sets this field to 00000001.

   - Pilot PN sequence offset index (9 bits)
     The base station sets this field to the pilot PN sequence offset for this base station, in units of 64 PN chips.

   - Configuration message sequence number (6 bits)
     The base station sets this field to the configuration sequence number.

   - System identification (SID) (16 bits)
     This field serves as a sub-identifier of a system as defined by the owner of the system identification.

     The base station sets this field to the network identification number for this network. The network identification value of 65,535 is reserved.

   - Registration zone (12 bits)
     The base station sets this field to its registration zone number.

   - Number of registration zones to be retained (3 bits)
     The base station sets this field to the number of registration zones the mobile station is to retain for purposes of zone-based registration. If zone-based registration is disabled, the base station sets this field to '000'.

   - Zone timer length (3 bits)
     The base station sets this field to the zone-timer length value shown in Table 6.2 corresponding to the length of the zone registration timer to be used by mobile stations.

Table 6.2   Zone Timer Value

Zone Timer Value (binary)	Timer Length (Minutes)
000	1
001	2
010	5
011	10
100	20
101	30
110	45
111	60

- Multiple SID storage indicator (1 bit)
  If mobile stations may store entries of the SID-NID list containing different SIDs, the base station sets this field to '1'; otherwise, the base station sets this field to '0'.

- Multiple NID storage indicator (1 bit)
  If mobile stations may store multiple entries of the SID-NID list having the same SID (with different NIDs), the base station sets this field to '1'; otherwise, the base station sets this field to '0'.

- Base station identification (16 bits)
  The base station sets this field to its identification number.

- Base station class (4 bits)
  The base station sets this field to the value of '0000' for the public macrocellular system. All other values are reserved.

- Number of paging channels (3 bits)
  The base station sets this field to the number of paging channels on this CDMA channel. This field cannot be set to '000'.

- Maximum slot cycle index (3 bits)
  The base station sets this field to the slot cycle index value corresponding to the maximum slot cycle length permitted.

- Home registration indicator (1 bit)
  If mobile stations that are not roaming and have the mobile station termination indicator equal to '1' are to be enabled for autonomous registrations, the base station sets this field to '1'. If such mobile stations are not to be enabled for autonomous registration, the base station sets this field to '0'.

- SID roamer registration indicator (1 bit)
  If mobile stations that are foreign SID roamers and have the mobile station termination indicator for SID equal to '1' are to be enabled for autonomous registration, the base station sets this field to '1'. If such mobile stations are not to be enabled for autonomous registration, the base station sets this field to '0'.

- NID roamer registration indicator (1 bit)
  If mobile stations that are foreign NID roamers and have the mobile station termination indicator for NID equal to '1' are to be enabled for autonomous registration, the base station sets this field to '1'. If such mobile stations are not to be enabled for autonomous registration, the base station sets this field to '0'.

- Power-up registration indicator (1 bit)
  If mobile stations enabled for autonomous registration are to register immediately after powering on and receiving the system overhead messages, the base station sets this field to '1'; otherwise, the base station sets this field to '0'.

- Power-down registration indicator (1 bit)
  If mobile stations enabled for autonomous registration are to register immediately before powering down, the base station sets this field to '1'; otherwise, the base station sets this field to '0'.

- Parameter-change registration indicator (1 bit)
  If mobile stations are to register on parameter change events, the base station sets this field to '1'. If not, the base station sets this field to '0'.

- Registration period (7 bits)
  If mobile stations are not to perform timer-based registration, the base station sets this field to 0000000. If mobile stations are to perform timer-based registration, the base station sets this field to a value in the range 29 to 85 inclusive, such that the desired timer value is

  $$\lfloor 2^{(\text{Reg-Per})/4} \rfloor \times 0.08 \text{ seconds}$$

- Base station latitude (22 bits)
  The base station sets this field to its latitude in units of 0.25 seconds, expressed as a two's complement assigned number with positive numbers signifying North latitudes.

- Base station longitude (23 bits)
  The base station sets this field to its longitude in units of 0.25 seconds, expressed as a two's complement assigned number with positive numbers signifying East longitude.

- Registration distance (11 bits)
  If mobile stations are to perform distance-based registration, the base station sets this field to the nonzero distance beyond which the mobile station is to reregister. If

mobile stations are not to perform distance-based registration, the base station sets this field to '0'.

- Search window size for the active set and candidate set (4 bits)
  The base station sets this field to the value shown in Table 5.1 corresponding to the search window size to be used by mobile stations for the active set and candidate set.

- Search window size for the neighbor set (4 bits)
  The base station sets this field to the value shown in Table 5.1 corresponding to the search window size to be used by mobile stations for the neighbor set.

- Search window size for the remaining set (4 bits)
  The base station sets this field to the value shown in Table 5.1 corresponding to the search window size to be used by mobile stations for the remaining set.

- Neighbor set maximum age (4 bits)
  The base station sets this field to the maximum age value beyond which mobile stations are to drop members from the neighbor set.

- Power control reporting threshold (5 bits)
  The base station sets this field to the number of bad frames to be received in a measurement period before mobile stations are to generate a power measurement report message. If the base station sets the threshold report mode indicator to '1', it must not set this field to 00000.

- Power control reporting frame count (4 bits) γ:
  The base station sets this field to the value such that the number given by
  $$\lfloor 2^{\gamma/2} \times 5 \text{ frames} \rfloor$$
  is the number of frames over which mobile stations are to count frame errors.

- Threshold report mode indicator (1 bit)
  If mobile stations are to generate threshold power measurement report messages, the base station sets this field to '1'. If mobile stations are not to generate threshold power measurement report messages, the base station sets this field to '0'.

- Periodic power report mode indicator (1 bit)
  If mobile stations are to generate periodic power measurement report messages, the base station sets this field to '1'. If mobile stations are not to generate periodic power measurement report messages, the base station sets this field to '0'.

- Power report delay (5 bits)
  The period that mobile stations wait following a power measurement report message before restarting frame counting for power control purposes is set by this field in units of 4 frames.

- Rescan indicator (1 bit)
  If mobile stations are to reinitialize and reacquire the system upon receiving this message, the base station sets this field to '1'.

6.2.4   Mobile Station Directed Messages

The base station uses the following rules for selecting the paging channel slot in which to send a message to a mobile station:

- If the base station is able to determine that the mobile station is operating in the nonslotted mode, the base station may send the message to the mobile station in any paging channel slot.
- If the base station is able to determine that the mobile station is operating in the slotted mode and is able to determine the mobile station's slot cycle index, the base station sends the message, at least once, as follows:

   1. The base station sends the message in an assigned paging channel slot for the mobile station; and

   2. The base station will not send the message after the last slotted page message sent in that paging channel slot.

- If the base station is not able to determine whether the mobile station is operating in the nonslotted mode, or the base station is not able to determine the mobile station's slot cycle index, the base station will assume that the mobile station is operating in the slotted mode with a slot cycle index which is the smaller of the maximum value of the slot cycle index and 1. The base station sends the message, at least once, as follows:

   1. The base station sends the message in an assigned paging channel slot for the mobile station; and

   2. The base station will not send the message after the last slotted page message sent in that paging channel slot.

The base station sends at least one slotted page message in each paging channel slot. The base station should send messages directed to mobile stations operating in the slotted mode as the first messages in the slot.

The base station may send the following messages directed to a mobile station on the paging channel. If the base station sends a message, the base station must comply with the specified requirements for sending the message, if any.

1. Abbreviated alert order
2. Audit order
3. Authentication challenge message
4. Base station acknowledgment order
5. Base station challenge confirmation order
6. Channel assignment message
7. Data burst message
8. Feature notification message

- Pilot detection threshold (6 bits)
  This value is used by mobile stations to trigger the sending of the pilot strength me surement message initiating the handoff process. The base station sets this field to th pilot detection threshold, expressed as an unsigned binary number equal to $\lfloor -2 \times 1($ $\times \log_{10} E_C/I_0 \rfloor$.

- Pilot drop threshold (6 bits)
  This value is used by mobile stations to start a handoff drop timer for pilots in the active set and candidate set. The base station sets this field to the pilot drop threshold, expressed as an unsigned binary number equal to $\lfloor -2 \times 10 \times \log_{10} E_C/I_0 \rfloor$.

- Active set versus candidate set comparison threshold (4 bits)
  Mobile stations transmit a pilot strength measurement message when the strength of a pilot in the candidate set exceeds that of a pilot in the active set by this margin. The base station sets this field to the threshold candidate set pilot to active set pilot ratio, in units of 0.5 dB.

- Drop timer value (4 bits)
  Timer value after which an action is taken by mobile stations for a pilot that is a member of the active set or candidate set, and whose strength has not become greater than the drop timer value. If the pilot is a member of the action set, a pilot strength measurement message is issued. If the pilot is a member of the candidate set, it will be moved to the neighbor set. The base station sets this field to the drop timer expiration value shown in Table 6.3 corresponding to the drop timer value to be used by mobile stations.

Table 6.3   Handoff Drop Timer Expiration Values

Drop timer value	Timer expiration (seconds)	Drop timer value	Timer expiration (seconds)
0	0.1	8	27
1	1	9	39
2	2	10	55
3	4	11	79
4	6	12	112
5	9	13	159
6	13	14	225
7	19	15	319

The mobile station will indicate the status of the handoff drop timer for all pilots in the active set and candidate set when transmitting a pilot strength measurement message.

- Reserved bits (4 bits)
  The base station sets this field to '0000'.

9. Intercept order

10. Local control order

11. Lock until power-cycled order

12. Maintenance required order

13. Page message: The base station includes both MIN 1 and MIN 2 fields in the message when paging either a foreign SID roamer or a foreign NID roamer.

14. Registration accepted order

15. Registration rejected order

16. Registration request order

17. Release order

18. Reorder order

19. Slotted page message: The base station includes MIN 1 and MIN 2 fields in the message when paging either a foreign SID roamer or a foreign NID roamer.

20. SSD update message

21. Unlock order

6.3 ACCESS CHANNEL PROCESSING

During this processing, the base station monitors the access channel to receive messages which the mobile station sends while the mobile station is in the system access state.

Each access channel is associated with a paging channel. Up to 32 access channels can be associated with a paging channel. The number of access channels associated with a particular paging channel is specified in the access parameters message sent on that paging channel. The base station continually monitors all access channels associated with each paging channel that the base station transmits.

6.3.1   Access Channel Acknowledgment Procedures

The access channel acknowledgment procedures facilitate the reliable exchange of messages between the base station and the mobile station on the paging channel and access channel. The base station uses the fields of the acknowledgment as address type, acknowledgment sequence number, message sequence number, acknowledgment required indicator, and valid acknowledgment to support this mechanism. These fields are referred to as layer 2 fields, and the acknowledgment procedures are referred to as layer 2 procedures. All other message fields and the processing thereof are referred to as pertaining to layer 3.

A message received on the access channel requires acknowledgment if the acknowledgment required field is set to '1'. In this specification, all messages sent on the access channel require acknowledgment. All messages sent on the access channel contain identification data for the mobile station sending the message, and are acknowledged by paging channel messages.

The base station acknowledges a received message by transmitting a message on the paging channel with the acknowledgment sequence number field set equal to the message sequence number field of the received message, and with the valid acknowledgment field set to '1'. A message transmitted with the acknowledgment sequence number and valid acknowledgment fields set in this manner is referred to as including an acknowledgment of the received message.

After receiving a message requiring acknowledgment from a mobile station on the access channel, the base station transmits a message directed to that mobile station, including acknowledgment, on the corresponding paging channel. The acknowledgment is transmitted within the access channel acknowledgment timeout, in units of 80 ms, after receiving the message, where the access channel acknowledgment timeout is the value sent in the access parameters message on the mobile station's assigned paging channel.

When a received message requires acknowledgment and no message directed to the mobile station is available within the access channel acknowledgment timeout x 80 ms after the message is received, the base station transmits a base station acknowledgment order directed to the mobile station, including the acknowledgment.

Whenever a message requiring acknowledgment is received from a mobile station, the base station sets the acknowledgment sequence number field in subsequent paging channel messages directed to that mobile station to the message sequence number specified in the received message. The valid acknowledgment field is set to '1' for the first message with this value of the acknowledgment sequence number sent to the mobile station on the paging channel. For all paging channel messages after the first, directed to the same mobile station and containing the same acknowledgment sequence number field value:

- The base station may set the valid acknowledgment to '1' if the message is sent within T_{4m} seconds after the first message; or
- The base station will set the valid acknowledgment field to '0' if the message is sent more than T_{4m} seconds after the first message.

If the base station performs duplicate message detection using access channel message sequence numbers, it should use the following procedures. The base station should store, for each mobile station that is active on the access channel, a received status indicator for each possible value of the access channel message sequence number field [n], where n is 0 through 7.

The base station should consider a mobile station active on the access channel when it receives an access channel message from the mobile station. The base station should consider the mobile station inactive on the access channel if:

- It has received no message from the mobile station within a time period to be selected by the base station manufacturer; or
- The mobile station has been assigned to a traffic channel; or
- The mobile station has been assigned to the analog system; or
- The base station has received a power-down registration from the mobile station.

When the base station receives an access channel message from an inactive mobile station, it should set the message sequence number received [n] to NO for all values of n from 0 to 7. The base station should then consider the mobile station active on the access channel.

For each active mobile station, the base station should perform the following procedures:

- When a message requiring acknowledgment is received (including a message received while the mobile station was inactive) with the message sequence number, and the message sequence number received is equal to NO, the base station should process the message as a new message. The base station should set the received message sequence number to YES, and should set the received message sequence number = message sequence number + 2 modulo 8 to NO.
- When a message requiring acknowledgment is received with the message sequence number, and the received message sequence number is equal to YES, the base station will acknowledge the message as specified earlier in this section but should not perform any further processing of the message.

6.3.2   Responses to Page Response Message, Origination Message and Registration Message

1. If the base station receives a page response message, the base station should send a channel assignment message or a release order. The base station also may start authentication procedures.

   If the base station sends a channel assignment message, the base station performs the following actions:
   - If the channel assignment message directs the mobile station to a traffic channel, the base station begins traffic channel processing for the mobile station.
   - If the channel assignment message directs the mobile station to an analog voice channel, the base station follows the respective procedure.

2. If the base station receives an origination message, the base station should send a channel assignment message, an intercept order, a reorder order, or a release order. The base station also may commence authentication procedures. If the base station sends a channel assignment message, the base station performs the following actions:
   - If the channel assignment message directs the mobile station to a traffic channel, the base station begins traffic channel processing for the mobile station.
   - If the channel assignment message directs the mobile station to an analog voice channel, the base station follows the corresponding procedure.

   The base station will not set the received response equal to '0' when the channel assignment mode = 001 or 010.

3. If the base station receives a registration message, the base station may send a registration accepted order or a registration reject order. The base station also may start authentication procedures. Authentication is the process by which information is exchanged between a mobile station and a base station for the purpose of confirming the identity of the mobile station. A successful outcome of the authentication process occurs only when it can be demonstrated that the mobile station and base station possess identical sets of shared secret data.

6.4 TRAFFIC CHANNEL PROCESSING

During traffic channel processing, the base station uses the forward and reverse traffic channels to communicate with the mobile station while the mobile station is in the mobile station control on the traffic channel state.

Traffic channel processing consists of the following substates:

- Traffic Channel Initialization Substate
  In this substate, the base station begins transmitting on the forward traffic channel and receiving on the reverse traffic channel.

- Waiting for Order Substate
  In this substate, the base station sends the alert with information message to the mobile station.

- Waiting for Answer Substate
  In this substate, the base station waits for the connect order from the mobile station.

- Conversation Substate
  In this substate, the base station exchanges primary traffic bits with the mobile station's primary service option application.

- Release Substate
  In this substate, the base station disconnects the call.

In the following, it will be shown that the base station performs special functions and actions in one or more of the traffic channel processing substates.

6.4.1 Forward Traffic Channel Power Control

When the base station enables the forward traffic channel power control, the mobile station reports frame error rate statistics to the base station using the power measurement report message. The base station may enable the forward traffic channel power control using the system parameters message sent on the paging channel and the power control parameters message sent on the forward traffic channel. The base station may enable periodic reporting which causes the mobile station to report frame error rate statistics at specified intervals. The base station also may enable threshold reporting which causes the mobile station to report frame error rate statistics when the frame error rate reaches a specified threshold. The base station may use the reported frame error rate statistics to adjust the transmit power of the forward traffic channel.

6.4.2 Service Options

During traffic channel operation, the base station and mobile station may support primary traffic services. Each such service has a set of requirements that govern the way in which the primary traffic bits from forward and reverse traffic channel frames are processed by the base station and mobile station.

Either the base station or mobile station can request a service option. The base station can request a particular service option when paging the mobile station or during traffic channel operation. If the requested service option is acceptable to the mobile station, the base station and mobile station begin using the new service option. If the base station requests a service option that is not acceptable to the mobile station, the mobile station can reject the requested service option or request an alternative option. If the mobile station requests an alternative service option, the base station can accept or reject the mobile station's alternative service option, or request another service option. This process, called service option negotiation, ends when the base station and mobile station find a mutually acceptable service option, or when the base station rejects a service option request from the mobile station or the mobile station rejects a service option request from the base station.

The base station and mobile station use the service option request order either to request a service option or suggest an alternative service option, and the service option response order to accept or reject a service option request. In addition, the base station can request a service option in the page message or the slotted page message, and the mobile station can request a service option in the origination message or the page response message. The base station and mobile station use the service option control order to invoke service option specific functions.

The base station uses a variable service option request to record the number of the service option for which the base station has sent an outstanding request in a service option request order. The service option request number is set to a special value, NULL, when the base station does not have an outstanding service option request. The base station uses another variable to record the number of the service option which is currently active. The current service option request number is set to NULL when there is no active service option.

6.4.3   Service Option Processing

1. Processing for service option requests:
   When processing a service option request in an origination message, a paging response message, or a service option request order, the base station performs the following actions:
   - If the base station accepts the requested service option, the base station sets the service option request number to NULL and sends a service option response order accepting the requested service option within $T_{4b} = 5$ seconds. T_{4b} denotes the maximum time for the base station to respond to a service option request. The base station begins using the requested service option in accordance with the requirements for the requested service option. The base station sets the current service option to the requested service option number when the service option becomes active.
   - If the base station does not accept the requested service option and has an alternative service option to request, the base station sets the service option request number to the alternative service option number and sends a service option request order requesting the alternative service option within T_{4b} seconds.

- If the base station does not accept the requested service option and does not have an alternative service option to request, the base station sets the service option request number to NULL and sends a service option request order to reject the request within T_{4b} seconds. The base station continues to process primary traffic as it did prior to receiving the service option request order and remains in the current state.

2. Processing for service option response order:

   When the base station receives a service option response order, it performs the following actions:

   - If the service option number specified in the order is equal to the service option request number, the base station sets the service option request number to NULL and begins using the specified service option in accordance with the requirements for the service option. The base station sets the current service option to the specified service option number when the service option becomes active.

   - If the order indicates a service option rejection, the base station sets the service option request number to NULL. The base station continues to process primary traffic as it did prior to receiving the service option response order and remains in the current state.

   - If the order does not indicate a service option rejection and the service option specified in the order is not equal to the service option request number, the base station sets the service option request number to NULL, sends a release order 00000010, and enters the release substate.

3. Processing for the received service option control order:

   If there is an active option (the current service option number is not equal to NULL), the base station processes the received service option control order in accordance with the requirements for the active service option.

4. Service option request initialization:

   To perform service option request initialization, the base station sets the service option request number to the specified service option number.

6.4.4   Acknowledgment Procedures

The acknowledgment procedures facilitate the reliable exchange of messages between the mobile station and the base station. The base station uses the fields of the acknowledgment sequence number, the message sequence number and the acknowledgment required to detect duplicate messages and provide a reference for acknowledgments. These message fields are referred to as layer 2 fields, and the acknowledgment procedures are referred to as layer 2 procedures. All other message fields are referred to as layer 3 fields, and the processing of layer 3 fields is referred to as layer 3 processing.

On both the reverse traffic channel and the forward traffic channel, the procedure for messages requiring acknowledgment is a selective repeat scheme in which a message is retransmitted only if an acknowledgment for it is not received.

1. Transmitting messages and receiving acknowledgment:

   A traffic channel message requires acknowledgment when the acknowledgment required field is set to '1'.

   The layer 2 protocol does not guarantee delivery of messages in any order. If the base station requires that the mobile station receives a set of messages in a certain order, the base station must wait for an acknowledgment to each message before transmitting the next message in the set. For messages requiring acknowledgment whose relative ordering is not important, the base station may transmit up to four such messages before receiving an acknowledgment for the first message.

   The base station stores a message sequence number for messages requiring acknowledgment. The base station stores an acknowledgment status indicator for each possible value of the forward traffic channel message sequence number field (acknowledgment waiting [n], where n is 0 through 7). The base station will not send a new message requiring acknowledgment when the acknowledgment waiting (message sequence acknowledgment number + 4, modulo 8) is equal to YES.

   The base station performs the following procedures:

   - When the base station receives a message on the reverse traffic channel, with an acknowledgment sequence number, it sets the acknowledgment waiting (acknowledgment sequence number) to NO.

   - When the base station sends a new message requiring acknowledgment on the forward traffic channel, it sets the acknowledgment waiting (message sequence acknowledgment number) to YES and sets the message sequence number field of the message to the message sequence acknowledgment field. The base station then increments the message sequence acknowledgment number, modulo 8.

   The base station will not retransmit a message for which it has received an acknowledgment.

   If the base station does not receive an acknowledgment after transmitting the message, the base station retransmits the message. If the base station retransmits a message, the base station uses the same message sequence number for the retransmission.

   The base station stores a retransmission counter for each transmitted message requiring acknowledgment. The base station sets the retransmission counter to zero prior to the first transmission of the message. After each transmission of the message, the base station increments the retransmission counter if no acknowledgment is received. The base station will not exceed a maximum number of retransmissions, to be selected by the base station manufacturer. When the retransmission counter is equal to the maximum number of retransmissions, the base station declares an acknowledgment failure.

2. Receiving messages and returning acknowledgments:

   Messages received on the reverse traffic channel contain the message sequence number fields that are incremented by the same rules as messages transmitted on the forward traffic channel. Separate sequence numbers are maintained for reverse traffic channel messages that require acknowledgment and for messages that do not require acknowledgment. The base station acknowledges a received message by transmitting a message with the acknowledgment sequence number field set equal to the message sequence number field of the received message. A message transmitted with the acknowledgment sequence number field set in this manner is referred to as including an acknowledgment of the received message.

Whenever a message requiring acknowledgment is received, the base station sets the acknowledgment sequence number field of subsequent forward traffic channel messages to the message sequence number field of the received message. If no message has been received, the base station sets this field to '111'.

After receiving a message requiring acknowledgment, the base station transmits a message including an acknowledgment within $T_{1m} = 0.4$ seconds as shown in Fig. 5.8. T_{1m} denotes the maximum time the mobile station waits for an acknowledgment.

When a received message requires acknowledgment and no message is available within T_{1m} seconds after the message is received, the base station transmits a base station acknowledgment order including the acknowledgment.

For duplicate message detection, the base station will store a received status indicator for each possible value of the RTC message sequence number field (message sequence number received [n], where n is 0 through 7). The base station performs the following procedures.

- When a message requiring acknowledgment is received with a message sequence number, and the message sequence number received is equal to NO, the base station processes the message as a new message. The base station then sets the message sequence number received to YES, and sets the received message sequence number [message sequence number + 4, modulo 8] to NO.

- When a message requiring acknowledgment is received with the message sequence number, and the received message sequence number is equal to YES, the base station acknowledges the message but will not perform any further processing of the message.

3. Messages not requiring acknowledgment:

A traffic channel message does not require acknowledgment when the acknowledgment required indicator field is set to '0'. The base station stores a message sequence number of messages not requiring acknowledgment. For each new message sent that does not require acknowledgment, the base station sets the message sequence number field of the message to the no acknowledgment message sequence and then increments the no acknowledgment message, modulo 8.

If the base station transmits the same message not requiring acknowledgment more than once, it uses the same message sequence number for all transmissions. The base station completes all retransmissions of the same message within $T_{3m} = 0.32$ seconds after the first transmission, as shown in Fig. 6.2. The base station will wait at least T_{3m} seconds after the last transmission of a message not requiring acknowledgment before transmitting another message not requiring acknowledgment that has the same message sequence number, as shown in Fig. 6.2. T_{3m} designates the period in which two messages received by the mobile station on the forward traffic channel, not requiring an acknowledgment, and carrying the same sequence numbers, are considered duplicates.

4. Acknowledgment procedures reset:

The base station resets the acknowledgment procedures as follows:

- Message sequence number reset

  If the acknowledgment waiting [n] is equal to YES for any n, the base station should save the corresponding messages and retransmit them after completing the reset of the acknowledgment procedures. For each such message, the base station sets the retransmission counter to zero.

- The base station sets the message sequence acknowledgment to 0, the message sequence not requiring acknowledgment to 0, and sets the acknowledgment waiting [n] to NO for all values of n from 0 to 7.

- Acknowledgment sequence number reset:

  The base station sets the acknowledgment sequence number field of all FTC messages to '111' until the first message requiring acknowledgment is received.

- Duplicate detection reset:

  The base station sets the received message sequence number [n] to NO for all values of n from 0 to 7.

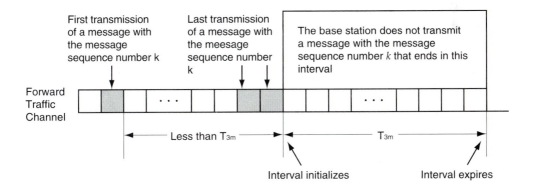

Figure 6.2 Time requirement for the base station not to reuse a message sequence number

6.4.5   Message action times

A forward traffic channel (FTC) message without a use time field or with a use time field set to '0' has an implicit action time. A message with its use time field set to '1' has an explicit action time which is specified in the action time field of the message. A message with a future action time is called a pending message.

Unless otherwise specified, a message having an implicit action time must take effect no later than the first 80 ms boundary, relative to the system time, occurring at least 80 ms after the end of the frame containing the last bit of the message. A message with an explicit action time takes effect when the system time, in units of 80 ms, modulo 64 becomes equal to the message's action time field. The difference in time between the action time and the end of the frame containing the last bit of the message is at least 80 ms.

The base station will support one pending message at any given time, not including pending service option control orders. The number of pending service option control orders that the base station is required to support is specific to the service option.

6.4.6   Long Code Transition Request Processing

If a request for voice privacy is specified in the origination message or page response message, the base station may send a long code transition request order 00000001 requesting a transition to the private long code.

The base station processes the long code transition request order as follows:
- If the long code transition request order requests a transition to the private long code and the base station accepts the request, the base station sends a long code transition request order 00000001. If the base station does not accept the private long code transition request, the base station sends a long code transition request order 00000000.
- If the long code transition request order requests a transition to the public long code and the base station accepts the request, the base station sends a long code transition request order 00000000. If the base station does not accept the public long code transition request, the base station sends a long code transition request order 00000001.

The base station processes the long code transition response order as follows:
- If the long code transition response order indicates that the mobile station accepts the long code transition requested in the long code transition request order sent by the base station, the base station uses the requested long code mask on both the forward traffic channel and the reverse traffic channel. If the base station did not specify an explicit action time in the long code transition request order, the base station should begin using the requested long code mask at the first 80 ms boundary, relative to the start of the system time, after $N_{4m}=20$ frames after the last frame in which any portion of the long code transition response order was received.

6.5 TRAFFIC CHANNEL INITIALIZATION SUBSTATE

In this substate, the base station begins transmitting on the forward traffic channel and acquires the reverse traffic channel.

Upon entering the traffic channel initialization substate, the base station performs the following actions:

- The base station resets the message acknowledgment procedures.
- The base station sets the active service option number to NULL to indicate that there is no active service option.
- The base station performs a service option request initialization specifying NULL as the service option number.
- The base station sets its forward and reverse traffic channel long code masks to the public long code mask.
- The base station sets its forward and reverse traffic channel frame offsets to the frame offset assigned to the mobile station.

While in the traffic channel initialization substate, the base station performs the following actions:

- The base station transmits null traffic channel data.
- The base station performs the message acknowledgment procedures.
- If the base station acquires the reverse traffic channel, the base station sends a base station acknowledgment order. The base station should send the base station acknowledgment order as a message requiring acknowledgment. If the call is a mobile station terminated call, the base station then enters the waiting for order substate. If the call is a mobile station originated call, the base station enters the conversation substate.
- If the base station fails to acquire the reverse traffic channel, the base station either retransmits the channel assignment message on the paging channel and remains in the traffic channel initialization substate, or the base station disables transmission on the forward traffic channel and discontinues the traffic channel processing for the mobile station.

6.6 WAITING FOR ORDER SUBSTATE

In this substate, the base station sends an alert with information message to the mobile station.

Upon entering the waiting for order substate, the base station performs the following actions:

- The base station processes the service option request specified in the page response message as specified in Section 6.4.2.

While in the waiting for order substate, the base station performs the following actions:
- The base station transmits the power control subchannel.

- If there is an active service option (the active service option number is not equal to NULL), the base station processes the received primary traffic bits in accordance with the requirements for the active service option; otherwise, the base station discards the received primary traffic bits.
- If there is an active service option (the active service option number is not equal to NULL), the base station transmits primary traffic bits in accordance with the requirements for the active service option; otherwise, the base station transmits null traffic channel data.
- The base station performs the message acknowledgment procedures.
- If the base station declares a loss of reverse traffic channel continuity, the base station sends a release order to the mobile station and enters the release substate.
- The base station may perform the forward traffic power control.
- The base station may request a service option as specified in Section 6.4.2. To do so, the base station performs a service option request initialization specifying the requested service option number, and sends a service option request order.
- If there is an active service option (the active service option number is not equal to NULL), the base station may send a service option control order to invoke a service option specific function in accordance with the requirements for the active service option.
- The base station may request a long code transition either autonomously or in response to a request for voice privacy specified in the origination message or page response message.
- The base station may perform authentication procedures.
- The base station may send the following messages. If the base station sends a message, the base station will comply with the specified requirements for sending the message, if any.

1. Alert with information message
   The base station then enters the waiting for answer substate.
2. Analog handoff direction message
   The base station then enters the waiting for order task.
3. Audit order
   The audit order can be sent on either the paging channel or on the forward traffic channel. The base station uses the following fixed-length format for the order-specific fields:
   - Order code (6 bits)
     The base station sets this field to 000110.
   - Order qualification code (8 bits)
     The base station sets this field to 00000000.
4. Authentication challenge message
   When the base station sends an authentication challenge message on the paging channel, it uses the following fixed-length message format:

Field	Length (bits)
Message type (00001010)	8
Acknowledgment sequence number	3
Message sequence number	3
Acknowledgment required indicator	1
Valid acknowledgment indicator	1
Address type	3
Address field length	4
Mobile station address	8xAddress field length
Random challenge data	24
Reserved bits	3

5. Base station acknowledgment order

The base station acknowledgment order can be sent on either the paging channel or on the forward traffic channel. The base station uses the following fixed-length format for the order-specific fields.

- Order code (6 bits)

The base station sets this field to 010000.

- Order qualification code (8 bits)

The base station sets this field to 00000000.

6. Base station challenge confirmation order

The base station challenge confirmation order can be sent on either the paging channel or on the forward traffic channel. The base station uses the following fixed-length format for the order-specific fields.

- Order code (6 bits)

The base station sets this field to 000010.

- Order qualification code (8 bits)

The base station sets this field to 00000000.

- Challenge response (18 bits)
- Reserved bits (6 bits)

The base station sets this field to 000000.

7. Data burst message

When the base station sends a data burst message on the paging channel, it uses the following variable-length message format:

Field	Length (bits)
Message type (00001001)	8
Acknowledgment sequence number	3
Message sequence number	3
Acknowledgment required indicator	1
Valid acknowledgment indicator	1
Address type	3
Address field length	4
Mobile station address	8xAddress field length
Message number	8
Data burst type	6
Number of messages in the data burst stream	8
Number of characters in this message	8
Character	8
Reserved bits (00000)	5

8. Handoff direction message

When the base station sends a handoff direction message, it uses the following variable-length message format:

Field	Length (bits)
Message type (00000101)	8
Acknowledgment sequence number	3
Message sequence number	3
Acknowledgment required indicator	1
Message encryption indicator	2
Use action time indicator	1
Action time	6
Handoff direction message sequence number	2
Search window size for the active set and candidate set	4
Pilot detection threshold	6
Pilot drop threshold	6
Active set versus candidate set comparison threshold	4
Drop time value	4
Frame offset	4
Private long code mask indicator	1

Reset acknowledgment procedures command	1
Message encryption mode	2
Alternate frequency assignment indicator	1
Frequency assignment for CDMA channel	0 or 11
Reserved bits	0-7 (as needed)

One or more occurrences of the following record:

Field	Length (bits)
Pilot PN sequence offset index	9
Power control symbol combining indicator	1
Code channel index	8

9. In-traffic system parameters message

When the base station sends an in-traffic system parameters message, it uses the following fixed-length message format.

Message type (00000111)	8
Acknowledgment sequence number	3
Message sequence number	3
Acknowledgment required indicator	1
Message encryption indicator	2
System identification (SID)	15
Network identification (NID)	16
Search window size for the active set and candidate set	4
Search window size for the neighbor set	4
Search window size for the remaining set	4
Pilot detection threshold	6
Pilot drop threshold	6
Active set versus candidate set comparison threshold	4
Drop timer value	4
Maximum age for retention of neighbor set members	4
Reserved bits	4

10. Local control order

    The local control order can be sent on either the paging channel or on the forward traffic channel. The base station uses the following fixed-length format for the order-specific fields.

    • Order code (6 bits)

       The base station sets this field to 0 1 1 1 1 0.

    • Order qualification code (8 bits)

       The specific order is designated by n n n n n n n n as determined by each system.

11. Lock until power-cycled order

    The lock until power-cycled order can be sent on either the paging channel or the forward traffic channel. The base station uses the following fixed-length format for the order-specific fields.

    • Order code (6 bits)

       The base station sets this field to 0 1 0 0 1 0.

    • Order qualification code (8 bits)

       The base station sets this field to 0 0 0 1 n n n n, where n n n n is the lock reason.

12. Long code transition request order

    The long code transition request order may be sent only on the forward traffic channel.

    • Order code (6 bits)

       The base station sets this field to 0 1 0 1 1 1.

    • Order qualification code (8 bits)

       The base station sets this field to 0 0 0 0 0 0 0 0.

13. Maintenance order

    The base station enters the waiting for answer substate.

14. Maintenance required order

    The maintenance required order can be sent on either the paging channel or the forward traffic channel. The base station uses the following fixed-length format for the order-specific fields.

    • Order code (6 bits)

       The base station sets this field to 0 1 0 0 1 0.

    • Order qualification code (8 bits)

       The base station sets this field to 0 0 1 0 n n n n, where n n n n is the maintenance reason.

15. Message encryption mode order

    The message encryption order may be sent only on the forward traffic channel.

    • Order code (6 bits)

       The base station sets this field to 0 0 0 0 1 1.

- Order qualification code (8 bits)

  The base station sets this field to $0\,0\,0\,0\,0\,0\,n\,n$, where $n\,n$ is the mode per the following table:

Encrypt-mode field (binary)	Encryption mode used
00	Encryption disable
01	Encryption call control message

The base station sets this field to the encrypt-mode value shown in the above table corresponding to the encrypting mode that is to be used for messages sent on the forward and reverse traffic channels.

16. Mobile station registered message
17. Neighbor list update order
18. Parameter update order

    The parameter update order may be sent only on the forward traffic channel. The base station uses the following fixed-length format for the order-specific fields.

    - Order code (6 bits)

      The base station sets this field to $0\,0\,0\,1\,0\,1$.

    - Order qualification code (8 bits)

      The base station sets this field to $0\,0\,0\,0\,n\,n\,n\,n$, where $n\,n\,n\,n$ is the request number.

19. Pilot measurement request order

    The pilot measurement request order can be sent only on the forward traffic channel. The base station uses the following fixed-length format for the order-specific fields.

    - Order code (6 bits)

      The base station sets this field to $0\,1\,0\,0\,0\,1$.

    - Order qualification code (8 bits)

      The base station sets this field to $0\,0\,0\,0\,0\,0\,0\,0$.

20. Power control parameters message
21. Release order

    The release order can be sent on either the paging channel or the forward traffic channel. The base station uses the following fixed-length format for the order-specific fields.

    - Order code (6 bits)

      The base station sets this to $0\,1\,0\,1\,0\,1$.

    - Order qualification code (8 bits)

      $0\,0\,0\,0\,0\,0\,0\,0$ for no reason given.

      $0\,0\,0\,0\,0\,0\,1\,0$ for indicating that the requested service option is rejected.

    The base station then enters the release substate.

22. Retrieve parameters message

23. Service option control order

The service option control order may be sent only on the forward traffic channel. The base station uses the following fixed-length format for the order-specific fields.

- Order code (6 bits)
  The base station sets this field to 0 1 1 1 0 1 .

- Order qualification code (8 bits)
  The base station sets this field to n n n n n n n n .

  The specific control is designated by n n n n n n n n as determined by each service option.

24. Service option request order

The service option request order may be sent only on the forward traffic channel. The base station uses the following fixed-length format for the order-specific fields.

- Order code (6 bits)
  The base station sets this field to 0 1 0 0 1 1 .

- Order qualification code (8 bits)
  The base station sets this field to 0 0 0 0 0 0 0 0 .

25. Service option response order

The service option response order can be sent only on the forward traffic channel. The base station uses the following fixed-length format for the order-specific fields.

- Order code (6 bits)
  The base station sets this field to 0 1 0 1 0 0 .

- Order qualification code (8 bits)
  The base station sets this field to 0 0 0 0 0 0 0 0 .

- Service option
  The base station sets this field to the service option code shown in TSB 58 corresponding to the accepted service option, or to 0 0 0 0 0 0 0 0 0 0 0 0 0 0 0 0 to reject the last service option requested by the mobile station.

26. Set parameters message

27. SSD update message

When the base station sends an SSD update message on the paging channel, it uses the following fixed-length message format

Field	Length (bits)
Message type (00001011)	8
Acknowledgment sequence number	3
Message sequence number	3
Acknowledgment required indicator	1
Valid acknowledgment indicator	1
Address type	3
Address field length	4
Mobile station address	8 x Address field length
Random data for the computation of SSD	56
Reserved bits (000)	3

28. Status request order

The status request order can be sent only on the forward traffic channel. The order qualification code field of the status request order specifies the information record to be returned by the mobile station in the status message. The base station uses the following variable-length format for the order-specific fields.

Order specific field	Length (bits)
Order qualification code	8
System identification (SID)	0 or 15
Network identification (NID)	0 or 16
Reserved bits	0 or 1 (as needed)

- Order qualification code (8 bits)

The base station sets this field to the order qualification code corresponding to the information record type to be returned by the mobile station in the status message as shown in the following table.

Information record requested	Order qualification code (binary)
Identification	00000110
Call mode	00000111
Terminal information	00001000
MIN information	00001001
Security status	00001010

- System identification (SID)

  If the order qualification code field is set to $0\,0\,0\,0\,0\,1\,1\,0$, the base station sets this field to the system identification number for this cellular system; otherwise, the base station omits this field.

- Network identification (NID)

  This field serves as a subidentifier of a system as defined by the owner of the system identification. If the order qualification code field is set to $0\,0\,0\,0\,0\,1\,1\,0$, the base station sets this field to the network identification number for this network. The NID value of 65,535 is reserved. Otherwise, the base station omits this field.

- Reserved bits

  The base station adds reserved bits as needed in order to make the length of the order-specific fields equal to an integer number of octets. The base station sets these bits to '0'.

  If the base station receives one of the following autonomous messages from the mobile station, the base station processes the message according to the specified requirements, if any:

1. Data burst message
2. Handoff completion message

   The base station continues transmission to the mobile station on a forward traffic channel removed from the active set until it receives the handoff completion message from the mobile station or determines that the call has been released. The base station discontinues transmission to the mobile station on a forward traffic channel removed from the active set after it receives the handoff completion message.

3. Long code transition request order

   If a request for voice privacy is specified in the origination message or page response message, the base station may send a long code transition request order requesting a transition to the private long code.

4. Parameter update confirmation order
5. Pilot strength measurement message

   The base station should use the pilot strength measurements in the pilot strength measurement message to determine a new active set.

   The base station may respond to a pilot strength measurement message received from the mobile station by sending the handoff direction message.

6. Power measurement report message

   When the base station enables the forward traffic channel power control, the mobile station reports frame error rate statistics to the base station using the power measurement report message.

7. Release order

The base station sends the mobile station a release order, within $T_{2b} = 0.8$ seconds, and enters the release substate, or the base station sends an alert with information message, within T_{2b} seconds, and enters the waiting for answer substate. T_{2b} denotes the maximum time for the base station to send a release order after receiving a release order.

8. Request analog service order

The base station may respond with an analog handoff direction message.

9. Service option control order

If there is an active option, the base station processes the received service option control order in accordance with the requirements for the active service option.

10. Service option request order

When processing a service option request in a service option request order, the base station performs the following. If the base station does not accept the requested service option and has an alternative service option to request, the base station sets the service option request number to the alternative service option number and sends a service option request order requesting the alternative service option within $T_{4b} = 5$ seconds. T_{4b} denotes the maximum time for the base station to respond to a service option request.

11. Service option response order

When the base station receives a service option response order, it performs the following. If the order indicates a service option rejection, the base station sets the service option request number to NULL. The base station continues to process primary traffic as it did prior to receiving the service option response order and remains in the current state.

6.7 WAITING FOR ANSWER SUBSTATE

In this substate, the base station waits for a connect order from the mobile station. While in the waiting for answer substate, the base station performs the following:

- The base station transmits the power control substate.
- If there is an active service option (a service option current value is not equal to NULL), the base station processes the received primary traffic bits in accordance with the requirements for the active service option; otherwise, the base station discards the received primary traffic bits.
- If there is an active service option (a service option current value is not equal to NULL), the base station will transmit primary traffic bits in accordance with the requirements for the active service option; otherwise, the base station transmits null traffic channel data.
- The base station performs the message acknowledgment procedures.
- If the base station declares a loss of reverse traffic channel continuity, the base station should send a release order to the mobile station. If the base station sends a release order, the base station then enters the release substate.
- The base station may perform the forward traffic channel power control.

- The base station may request a service option. To do so, the base station performs a service option request initialization specifying the requested service option number, and sends a service option request order, i.e., the requested service option number.
- If there is an active service option (the current service option number is not equal to NULL), the base station may send a service option control order (order qualification code) to invoke a service option specific function in accordance with the requirements for the active service option.
- The base station may request a long code transition either autonomously or in response to a request for voice privacy specified in the origination message or page response message.
- The base station may perform authentication procedures.
- The base station may send the following messages which were already discussed in the Section 6.6. If the base station sends a message, the base station will comply with the specified requirements for sending the message, if any.

1. Alert with information message

2. Analog handoff direction message

3. Audit order

4. Authentication challenge message

5. Base station acknowledgment order

6. Base station challenge confirmation order

7. Data burst message

8. Handoff direction message

9. In-traffic system parameters message.

10. Local control order

11. Lock until power-cycled order

12. Long code transition request order

13. Maintenance order

14. Maintenance request order

15. Message encryption mode order

16. Mobile station registered message

17. Neighbor list update message

18. Parameter update order

19. Pilot measurement request order

20. Power control parameters message

21. Release order

22. Retrieve parameters message

23. Service option control order

24. Service option request order

25. Service option response order

26. Set parameters message

27. SSD update message

28. Status request order

- If the base station receives one of the following autonomous messages from the mobile station, the base station processes the message according to the specified requirements, if any.

1. Connect order

   The base station enters the conversation substate.

2. Data burst message

3. Handoff completion message

   The base station continues transmission to the mobile station on a forward traffic channel removed from the active set until it receives the handoff completion message from the mobile station or determines that the call has been released.

   The base station discontinues transmission to the mobile station on a forward traffic channel removed from the active set after it receives the handoff completion message.

4. Long code transition request order

   If a request for voice privacy is specified in the origination message or page response message, the base station may send a long code transition request order requesting a transition to the private long code.

5. Parameter update confirmation order

   The parameter update confirmation order may be sent only on the forward traffic channel. The base station uses the following fixed-length codes.
   - Order code (6 bits)
     The base station sets this field to 0 0 0 1 0 1.

- Order qualification code (8 bits)

  The base station sets this field to 0 0 0 0 n n n n , where n n n n is the request number.

6. Pilot strength measurement message

   The base station should use the pilot strength measurements in the pilot strength measurement message to determine a new active set. The base station may respond to a pilot strength measurement message received from the mobile station by sending the handoff direction message.

7. Power measurement report message

   When the base station enables the forward traffic channel power control, the mobile station reports frame error rate statistics to the base station using the power measurement report message.

8. Release order

   The base station sends the mobile station a release order, within T_{2b} seconds, and enters the release substate, or the base station sends an alert with information message, within T_{2b} seconds, and enters the waiting for answer substate.

9. Request analog service order

   The base station may respond with an analog handoff direction message.

10. Service option control order

    If there is an active service option, the base station processes the received service option control order in accordance with the requirements for the active service option.

11. Service option request order

    If the base station does not accept the requested service option and has an alternative service option to request, the base station sets the service option request number to the alternative service option number and sends a service option request order requesting the alternative service option within T_{4b} seconds.

12. Service option response order

    When the base station receives a service option response order, it performs the following:

If the order indicates a service option rejection, the base station sets the service option request number to NULL. The base station continues to process primary traffic as it did prior to receiving the service option response order and remains in the current state.

6.8 CONVERSATION SUBSTATE

In this substate, the base station exchanges primary traffic bits with the mobile station's primary traffic service option application. Upon entering the conversation substate, the base station performs the following:

- If the call is mobile station originated, the base station processes the service option request specified in the origination message.

While in the conversation substate, the base station performs the following:

- The base station transmits the power control subchannel.
- If there is an active service option (the service option current number is not equal to NULL), the base station processes the received primary traffic bits in accordance with the requirements for the active service option; otherwise, the base station discards the received primary traffic bits.
- If there is an active service option (the service option current number is not equal to NULL), the base station transmits primary bits in accordance with the requirements for the active service option; otherwise, the base station transmits null traffic channel data.
- The base station performs the message acknowledgment procedures.
- If the base station declares a loss of reverse traffic channel continuity, the base station sends a release order to the mobile station. If the base station sends a release order, the base station enters the release substate.
- The base station may perform forward traffic channel power control.
- The base station may request a service option. To do so, the base station performs a service option request initialization specifying the requested service option number, and sends a service option request order (requested service option number).
- If there is an active service option (the service option current number is not equal to NULL), the base station may send a service option control order (order qualification code) to invoke a service option specific function in accordance with the requirements for the active service option.
- The base station may request a long code transition either autonomously or in response to a request for voice privacy specified in the origination message or page response message.
- The base station may perform authentication procedures.

• The base station may send the following messages. If the base station sends a message, the base station must comply with the specified requirements for sending the message, if any.

1. Alert with information message

   If the message contains a signal information record with the signal-type field set to '01' or '10', or if the message does not contain a signal information record, the base station enters the waiting for answer substate.

2. Analog handoff direction message

   The base station then enters the conversation task.

3. Audit order

4. Authentication challenge message

5. Base station acknowledgment order

6. Base station challenge confirmation order

7. Continuous DTMF tone order

8. Data burst message

9. Flash with information message

10. Handoff direction message

11. In-traffic system parameters message

12. Local control order

13. Lock until power-cycled order

14. Long code transition request order

15. Maintenance order

    The base station enters the waiting for answer substate.

16. Maintenance required order

17. Message encryption mode order

18. Mobile station registered message

19. Neighbor list update message

20. Parameter update order

21. Pilot measurement request order

22. Power control parameters message

23. Release order

    The base station enters the release substate.

24. Retrieve parameters message

25. Send burst DTMF message

26. Service option control order

27. Service option request order

28. Service option response order

29. Set parameters message

30. SSD update message

31. Status request order

• If the base station receives one of the following autonomous messages from the mobile station, it processes the message according to the specified requirements, if any.

1. Continuous DTMF tone order

2. Data burst message

3. Flash with information message

4. Handoff completion message

5. Long code transition request order

6. Origination continuation message

7. Parameter update confirmation order

8. Pilot strength measurement message

9. Power measurement report message

10. Release order

    The base station sends the mobile station a release order, within T_{2b} seconds, and enters the release substate, or the base station sends an alert with information message, within T_{2b} seconds, and enters the waiting for answer substate.

11. Request analog service order

    The base station may respond with an analog handoff direction message.

12. Send burst DTMF message

13. Service option control order

14. Service option request order

15. Service option response order

6.9 RELEASE SUBSTATE

In this substate, the base station disconnects the call. While in the release substate, the base station performs the following actions:

- The base station transmits the power control subchannel.
- The base station transmits null traffic channel data for at least $T_{3b} = 0.3$ seconds. After this interval, the base station should stop transmitting on the forward traffic channel. T_{3b} denotes the minimum time the base station continues to transmit on a code channel after sending or receiving a release order.
- The base station performs the message acknowledgment procedures.
- The base station may perform forward traffic channel power control.
- The base station may send the following messages. If the base station sends a message, the base station complies with the specified requirements for sending the message, if any.

1. Alert with information message

   If the message contains a signal information record with the signal-type field set to '01' or '10', or if the message does not contain a signal information record, the base station will enter the waiting for answer substate.

2. Audit order

3. Base station acknowledgment order

4. Data burst message

5. Handoff direction message

6. In-traffic system parameters message

7. Local control order

8. Lock until power-cycled order

9. Maintenance order

10. Maintenance required order

11. Mobile station registered message

12. Neighbor list update message

13. Parameter update order

14. Power control parameters message

15. Release order

16. Retrieve parameters message

17. Service option control order

18. Status request order

- If the base station receives one of the following autonomous messages from the mobile station, the base station processes the message according to the specified requirements, if any.

   1. Connect order

   2. Continuous DTMF tone order

      Dual-Tone Multifrequency (DTMF) designates method of signaling by the simultaneous transmission of two tones, one from a group of low frequencies and another from a group of high frequencies. Each group of frequencies consists of four frequencies.

3. Data burst message

4. Flash with information message

5. Handoff completion message

6. Pilot strength measurement message

7. Power measurement report message

8. Long code transition request order

9. Origination continuation message

10. Release order

11. Request analog service order

12. Send burst DTMF message

13. Service option control order

14. Service option request order

15. Service option response order

6.10 REGISTRATION

Registration is the process by which a mobile station notifies the base station of its location, status, identification, slot cycle, and other characteristics. The base station can make use of location information to efficiently page the mobile station when establishing a mobile-terminated call. Registration also provides the mobile station's slot-cycle-index parameter so that the base station can determine which paging channel slots a mobile station operating in the slotted mode is monitoring. Registration also provides the station class mark and protocol revision number so that the base station knows the capabilities of the mobile station.

The CDMA system supports nine different forms of registration.

1. Power-up registration

    The mobile station registers when it powers on, switches from using the alternate serving system, or switches from using the analog system.

2. Power-down registration

    The mobile station registers when it powers off if previously registered in the current serving system.

3. Timer-based registration

    The mobile station registers when a timer expires.

4. Distance-based registration

   The mobile station registers when the distance between the current base station and the base station in which it last registered exceeds a threshold.

5. Zone-based registration

   The mobile station registers when it enters a new zone.

6. Parameter-change registration

   The mobile station registers when certain of its stored parameters change.

7. Ordered registration

   The mobile station registers when the base station requests it.

8. Implicit registration

   When a mobile station successfully sends an origination message or page response message, the base station can infer the mobile station's location. This is considered an implicit registration.

9. Traffic channel registration

   Whenever the base station has registration information for a mobile station that has been assigned to a traffic channel, the base station can notify the mobile station that it is registered.

   The first five forms of registration, as a group, are called autonomous registration and are conditioned, in part, by roaming status and by indicators contained in the system parameters message. The base station may initiate ordered registration through an order message.

   While a mobile station is assigned a traffic channel, the base station may obtain registration information by using the status request order to obtain status messages from the mobile station. The base station may notify the mobile station that it is registered through the mobile station registered message.

6.10.1 Registration on the Paging and Access Channels

The base station specifies the forms of registration that are enabled, the corresponding registration parameters, and the roaming status conditions for which registration is enabled in the system parameters message. If any of the autonomous registration forms are enabled, the base station should also enable parameter-based registration.

The base station should process an origination message or page response message sent on the access channel as an implicit registration of the mobile station sending the message. The base station can obtain complete registration information about the mobile station at any time by sending a registration request order to the mobile station.

6.10.2 Registration on the Traffic Channels

The base station can obtain registration information from a mobile station on the traffic channels by means of the status request order. When the base station has registration data for a mobile station, the base station may send a mobile station registered message to the mobile station, specifying the base station's registration system, zone and location information.

6.11 HANDOFF PROCEDURES

This section describes the mobile station requirements during handoff. The base station supports the following three handoff procedures:

- Soft handoff
  A handoff in which a new base station commences communications with the mobile station without interrupting the communications with the old base station. The base station can direct the mobile station to perform a soft handoff only when all forward traffic channels assigned to the mobile station have identical frequency assignments. Soft handoff provides diversity of forward traffic channels and reverse traffic channel paths on the boundaries between base stations.

- CDMA to CDMA hard handoff
  A handoff in which the base station directs the mobile station to transition between disjoint sets of base stations, different frequency assignments, or different frame offsets.

- CDMA to analog handoff
  A handoff in which the base station directs the mobile station from a forward traffic channel to an analog voice channel.

The active set contains the pilots associated with the forward traffic channels assigned to the mobile station. The base station informs the mobile station of the contents of the active set using the channel assignment message and handoff direction message.

6.11.1 Overhead Information

The base station sends the following messages governing the pilot search procedures performed by the mobile station:

- System parameters message
  The base station sends handoff related parameters on the paging channel in the system parameters message.

- In-traffic system parameters message
  The base station may revise handoff related parameters for a mobile station operating on the traffic channel by sending the in-traffic system parameters message.

- Neighbor list message
  The base station sends a neighbor list on the paging channel, in the neighbor list message.

The base station must not specify more than $N_{8m} = 20$ pilots in the neighbor list message.

- Neighbor list update message

  The base station may revise the neighbor list for a mobile station operating on the traffic channel by sending a neighbor list update message. The base station will not include a pilot that is a member of the mobile station's active set in a neighbor list update message. The base station will not specify more than N_{8m} pilots in the neighbor list update message. The base station should list the pilots in the neighbor list update message in descending priority order. The mobile station maintains a counter, neighbor list age, for each pilot in the neighbor set. The mobile station initializes this counter to zero when it moves the pilot from the active set or candidate set to the neighbor set. The mobile station initializes this counter to the neighbor set maximum age for retention in the set when it moves the pilot from the remaining set to the neighbor set. The mobile station increments the neighbor list age for each pilot in the neighbor set upon receipt of a neighbor list message.

- Handoff direction message

  The base station may also modify the values of the following parameters through the handoff direction message:

  1. Search window size for the active set and candidate set

     The base station sets this field to the window size parameter corresponding to the number of PN chips that the mobile station is to reach for pilots in the active set and candidate set.

  2. Pilot detection threshold

     This value is used by the mobile station to trigger the sending of the pilot strength measurement message initiating the handoff process.

  3. Pilot drop threshold

     This value is used by the mobile station to trigger the sending of the pilot strength measurement message transmitting the handoff process and to more pilots from the candidate set to the neighbor set.

  4. Active set versus candidate set comparison threshold

     The mobile station transmits a pilot strength measurement message when the strength of a pilot in the candidate set exceeds that of a pilot in the active set by this margin. The base station sets this field to the threshold candidate set pilot to active set pilot ratio, in units of 0.5 dB.

  5. Drop timer value

     Timer value after which an action is taken by the mobile station for a pilot that is a member of the active set or candidate set, and where strength has not become greater than the drop timer value. If the pilot is a member of the active set, a pilot strength measurement message is issued. If the pilot is a member of the candidate set, it will be moved to the neighbor set.

6.11.2 Call Processing During Handoff

1. Processing the pilot strength measurement message

The base station should use the pilot strength measurements in the pilot strength measurement message to determine a new active set.

The base station may also use the PN phase measurements in the pilot strength measurement message to estimate the propagation delay to the mobile station. This estimate can be used to reduce the RTC acquisition time.

The base station responds to a pilot strength measurement message received from the mobile station by sending the handoff direction message.

2. Processing the handoff direction message

The base station retains a handoff direction message sequence number. The sequence number is initialized to zero prior to the transmission of the first handoff direction message to the mobile station. The base station increments the sequence number modulo 4 each time the base station modifies the pilot list sent to the mobile station in a handoff direction message.

Following a hard handoff, the base station should set the handoff direction message (HDM) sequence number to the value of the last HDM sequence field of the handoff completion message and should use the pilot order contained in the handoff completion message to interpret the contents of subsequent power measurement report messages.

The base station sets the contents of a handoff direction message according to the following rules:

- A handoff direction message lists no more than $N_{6m} = 6$ pilots in the new active set. N_{6m} represents the supported traffic channel active set size.
- A handoff direction message identifies the identical power control subchannels.
- When the CDMA frequency assignment is not changed, the handoff direction message will not change the code channel associated with an active set pilot that remains in the new active set.
- The base station specifies the long code mask to be used on the new forward traffic channel by using the private long code mask field of the handoff direction message. The base station may change the contents of this field only for CDMA to CDMA hard handoffs. If a change of long code mask is specified and the base station does not specify an explicit action time in the handoff direction message, the base station will begin using the new long code mask on the first 80 ms boundary (relative to the system time) occurring at least 80 ms after the end of the frame containing the last bit of the message.

- For CDMA to CDMA hard handoffs, the base station may require the mobile station to perform a reset of the acknowledgment procedures by using the reset layer 2 field of the handoff direction message. If the base station requires the mobile station to reset the acknowledgment procedures, the base station will also reset the acknowledgment procedures.
- For CDMA to CDMA hard handoffs, the base station may alter the frame offset by setting the frame offset field to a new value. If the base station specifies a new frame offset and does not specify an explicit action time, the base station will change its forward and reverse traffic channel frame offsets at the second 80 ms boundary (relative to the system time) after the end of transmission of the handoff direction message, unless the end of transmission of the message coincides with an 80 ms boundary, in which case the change in frame offsets will occur 80 ms after the end of transmission.

3. Transmitting during handoff

The base station will continue transmission to the mobile station on a forward traffic channel removed from the active set until it receives the handoff completion message from the mobile station or determines that the call has been released. The base station discontinues transmission to the mobile station on a forward traffic channel removed from the active set after it receives the handoff completion message.

4. Ordering pilot measurements from the mobile station

The base station may direct the mobile station to send a pilot strength measurement message by sending a pilot measurement request order.

6.11.3 Active Set Maintenance

The base station maintains an active set for each mobile station under its control as follows:

- When the base station sends the channel assignment message it initializes the active set to contain only the pilot associated with the assigned forward traffic channel.
- When the base station sends a handoff direction message, it adds to the active set, before the action time of the message, all pilots named in the message if they are not already in the active set.
- The base station deletes the pilots that were not named in the most recent handoff direction message from the active set upon receipt of the handoff completion message.

6.11.4 Soft Handoff

The base station should use soft handoff when directing a mobile station from one forward traffic channel to another forward traffic channel having the same frequency assignment.

1. Receiving during soft handoff

   Each base station in the active set demodulates the reverse traffic channel. The base station should provide diversity combining of the demodulated signals obtained by each base station in the active set.

2. Transmitting during soft handoff

   The base station begins transmitting identical modulation symbols on all forward traffic channels specified in a handoff direction message (with the possible exception of the power control subchannel) by the action time of the message.

The base station transmits identical power control symbols on all identical power control subchannels that were identified as such in the last handoff direction message.

The base station uses the same long code mask on the reverse traffic channel and on all forward traffic channels whose associated pilots are in the active set.

6.11.5 CDMA to Analog Hard Handoff

The base station may direct the mobile station to perform a handoff from the CDMA system to the analog system by sending an analog handoff direction message.

Example 6.1   The following simple call flow illustrates the mobile station origination example

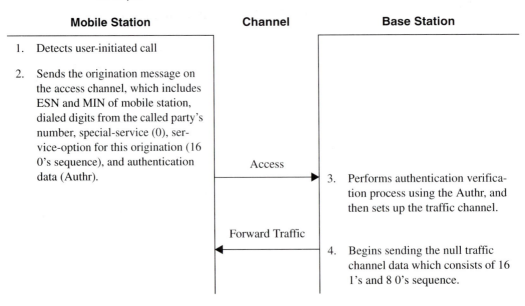

Example 6.1 The following simple call flow illustrates the mobile station origination example *(continued)*

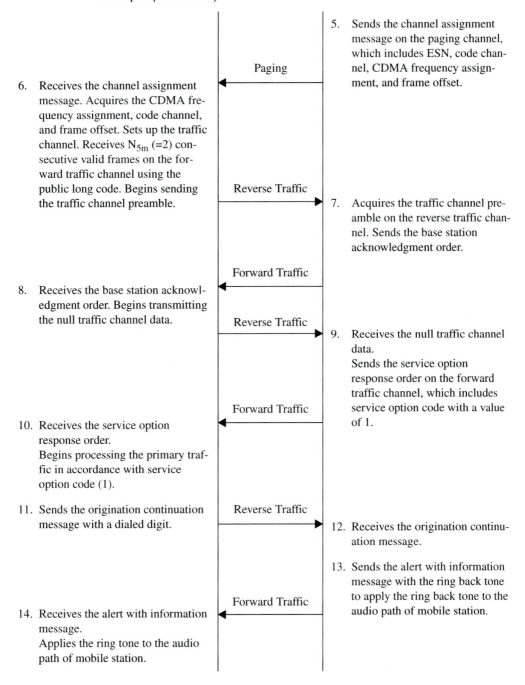

5. Sends the channel assignment message on the paging channel, which includes ESN, code channel, CDMA frequency assignment, and frame offset.

Paging

6. Receives the channel assignment message. Acquires the CDMA frequency assignment, code channel, and frame offset. Sets up the traffic channel. Receives N_{5m} (=2) consecutive valid frames on the forward traffic channel using the public long code. Begins sending the traffic channel preamble.

Reverse Traffic

7. Acquires the traffic channel preamble on the reverse traffic channel. Sends the base station acknowledgment order.

Forward Traffic

8. Receives the base station acknowledgment order. Begins transmitting the null traffic channel data.

Reverse Traffic

9. Receives the null traffic channel data.
Sends the service option response order on the forward traffic channel, which includes service option code with a value of 1.

Forward Traffic

10. Receives the service option response order.
Begins processing the primary traffic in accordance with service option code (1).

11. Sends the origination continuation message with a dialed digit.

Reverse Traffic

12. Receives the origination continuation message.

13. Sends the alert with information message with the ring back tone to apply the ring back tone to the audio path of mobile station.

Forward Traffic

14. Receives the alert with information message.
Applies the ring tone to the audio path of mobile station.

Example 6.1   The following simple call flow illustrates the mobile station origination
example *(continued)*

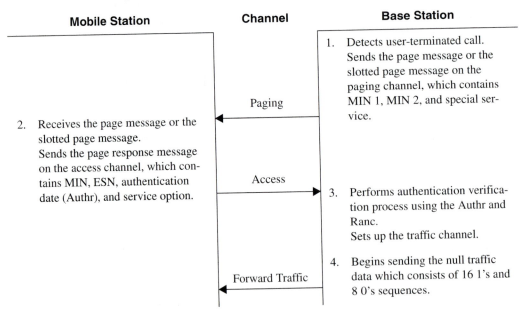

15. Detects the hook-off state of the
    receiver user.
    Sends the alert with information
    message with the ring back tone
    off.

Forward Traffic

16. Receives the alert with information
    message.
    Removes the ring back tone from
    the audio path of mobile station.

17. Begins conversation between users.

Example 6.2   The following simple call flow illustrates the mobile station termination
example

Mobile Station	**Channel**	**Base Station**

1. Detects user-terminated call.
   Sends the page message or the
   slotted page message on the
   paging channel, which contains
   MIN 1, MIN 2, and special service.

Paging

2. Receives the page message or the
   slotted page message.
   Sends the page response message
   on the access channel, which contains MIN, ESN, authentication
   date (Authr), and service option.

Access

3. Performs authentication verifica-
   tion process using the Authr and
   Ranc.
   Sets up the traffic channel.

4. Begins sending the null traffic
   data which consists of 16 1's and
   8 0's sequences.

Forward Traffic

Example 6.2 The following simple call flow illustrates the mobile station termination example

5. Sends the channel assignment message on the paging channel, which contains ESN, CDMA code channel, and frame offset.

Paging

6. Receives the channel assignment message.
   Acquires the CDMA code channel and frame offset.
   Sets up the traffic channel using the public long code. Receives N_{5m} (=2) consecutive valid frames on the forward traffic channel.
   Begins sending the traffic channel preamble.

Reverse Traffic

7. Acquires the traffic channel preamble on the traffic channel.
   Sends the base station acknowledgment order.

Forward Traffic

8. Receives the base station acknowledgment order. Begins transmitting the null traffic channel data.

Reverse Traffic

9. Receives the null traffic channel data.
   Sends the service option response order on the forward traffic channel, which includes the service option code with a value of 1.

Forward Traffic

10. Receives the service option response order.
    Begins processing the primary traffic in accordance with service option code (1).

11. Sends the alert with information with ring to apply the ring to the mobile station.

Forward Traffic

12. Receives the alert with information.
    Starts ringing to the mobile station.
    Detects the hook-off state of the user, i.e., user answers call.
    Stops ringing.
    Sends the connect order on reverse traffic channel.

Reverse Traffic

13. Receives the connect order.

14. Begins conversation between users.

Table 6.3   The following call flow shows the mobile station initiated call disconnect example.

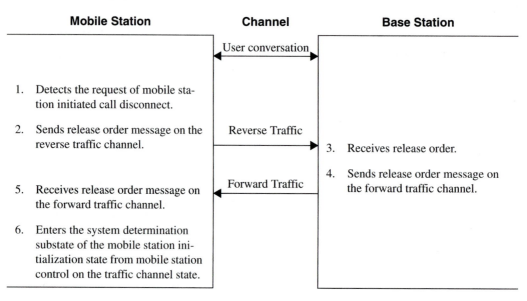

Mobile Station	Channel	Base Station
	User conversation	
1. Detects the request of mobile station initiated call disconnect.		
2. Sends release order message on the reverse traffic channel.	Reverse Traffic	3. Receives release order.
5. Receives release order message on the forward traffic channel.	Forward Traffic	4. Sends release order message on the forward traffic channel.
6. Enters the system determination substate of the mobile station initialization state from mobile station control on the traffic channel state.		

Brief Survey of One-way Hash Functions and Message Digests

This chapter provides a variety of hash functions which are suitable for security techniques, particularly applicable to the provision of authentication and digital signature.

A hash function is a one-way function. A one-way hash function is also known as a message digest function. Therefore, we may use the terms hash code and message digest interchangeably. A hash function accepts a variable-size message M as input and converts it to a fixed-size message digest H(M) as output. In general, H(M) is much smaller than M; for example H(M) might be 18, 64, or 128 bits, whereas M might be a megabyte or more.

For message authentication, the message digest function often plays an important role, so it is useful to understand the properties described in the following:

1. A hash function H is applicable to any data block of a variable-length message M.
2. A hash function H produces a fixed-size hash code H(M).
3. H(M) should be computed for any given M for the user's specific security requirements.
4. It must be computationally infeasible to find a message M which has a given prespecified message digest H(M).
5. For any given message M, it must be computationally infeasible to find another message M' for which H(M) = H(M'). In other words, it is computationally infeasible to find any two distinct messages M and M', $M \neq M'$, which map to the same message digests H(M) = H(M'). A hash function satisfying this property is called a collision-resistant hash function.

7.1 One-Way Function for Crypto-Algorithm

Some techniques for designing crypto-algorithms are listed in the following:

- Discrete logarithm problem

  Given x, modular exponentiation $g^x \bmod p$ is easy to calculate, but the inverse operation to compute x is much harder.

- Prime factorization problem

  It is easy to multiply the two large prime numbers to get a composite, but it is harder to factor the product and to recover the two large primes.

- Trapdoor knapsack problem

  Knapsack systems were proposed by Merkle and Hellman and have been suggested for use in designing public key cryptosystems. The Merkle-Hellman approach was to employ a superincreasing sequence.

  A superincreasing sequence $\{K_i\}$, $1 \le i \le n$, is defined as a sequence in which every term is greater than the sum of all previous terms, namely,

  $$K_i > K_1 + K_2 + \cdots + K_{i-1},\ 1 \le i \le n$$

  Given a pile of binary values x_i (0 or 1), $1 \le i \le n$, and a sum $Y = \{y_i\}$ such that

  $$Y = x_1 K_1 + x_2 K_2 + \cdots + x_n K_n$$

  The sum Y resulted from the product of a block of x_i equal in length to the number of items K_i in the pile is easy to compute, but recovery of x_i from Y and K_i is hard to solve.

  The associated knapsack problem is to find the binary values $\{x_i\}$ from the sum sequence $Y = \{y_i\}$. The solution of a superincreasing knapsack can be found as follows:

  If $y_n < K_n$, then set $x_n = 0$ and $y_{n-1} = y_n$.

  If $y_n > K_n$, then set $x_n = 1$ and $y_{n-1} = y_n - K_n$.

  Using the computed value of y_{n-1}, x_{n-1} and y_{n-2} can be found in a similar fashion. The recovery procedure continues until $X = (x_1, x_2, \cdots, x_n)$ has been completed.

  Disguised knapsack problems such as multiplicative trapdoor knapsack and multiple iterative knapsack may be proved to be secure, but unfortunatley the trapdoor knapsack above and most of its variations have been broken. However, the theoretical idea behind the Merkle-Hellman knapsack algorithm has historical value.

- Modular square root problems

  If n is the product of two primes, then the ability to calculate square root mod n will be computationally equivalent to the ability to factor n. User A (who knows the prime factors of n) can easily compute the square root of a number mod n, but for any other user, the computation has been proven to be as hard as computing the prime factors of n.

- Quadrature residuosity problem

  If m is prime and $a < m$, then a is a quadrature residue mod m

  if $x^2 = a \pmod{m}$ for some x. If $x^2 = a \pmod{m}$, gcd $(a, m) = 1$, has a solution, then a is called a quadratic residue. Quadrature residuosity problem is that given x and m, it is easy to compute x^2 mod m, but given a and m, it is very difficult to compute x from a and m.

7.2 MESSAGE DIGEST ALGORITHMS FOR AUTHENTICATION DATA

The purpose of this section is to provide a variety of hash functions which are suitable for security techniques. The following specified hash functions are applicable to the provision of authentication.

1. MD2 Algorithm

   The input to MD2 is a message whose length is an integral number of 8-bit bytes and produces a 128-bit message digest.

   The message is padded to be a multiple of the 16-byte checksum which is appended to the end. The message is processed with 16 bytes at a time to produce an intermediate message digest. Each intermediate value of the digest depends on the previous intermediate value and the 16 bytes of the original message being processed. The structure of MD2 is similar to MD4 and MD5, but it is slower and less secure.

2. MD4 Algorithm

   MD4 was designed, by Rivest in 1990, to be a 32-bit word-oriented scheme. MD2 requires the message to be an integral number of bytes, while MD4 can handle messages with an arbitrary number of bits. Hence, MD4 can be computed faster on 32-bit machines than a byte-oriented MD2 scheme.

   MD4 takes an input message of arbitrary length and produces an output 128-bit message digest, in such a way that it is computationally infeasible to produce two messages having the same message digest, or to produce any message having a given prespecified target message digest.

   The MD4 algorithm is intended for digital signature applications, where a large file must be compressed in a secure manner before being encrypted with a secret key under a public-key cryptosystem.

3. MD5 Algorithm

   The MD4 message digest algorithm is a one-way hash function which produces a 128-bit hash. When the MD4 algorithm was introduced, several researchers attacked the first few rounds of MD4. Rivest subsequently strengthened the algorithm which is now called MD5. But MD5 is slightly slower than MD4.

   MD5 is actually an improved version of MD4, but it also produces a 128-bit hash value for the 512-bit input blocks, divided into sixteen 32-bit subblocks. The output of MD5 is a set of four blocks which concatenate to form a single 128-bit message digest.

4. Secure Hash Algorithm (SHA)

SHA like MD5 is a variation of MD4, jointly designed by NIST and NSA. FIPS proposed a Secure Hash Standard (SHS) specified as a Secure Hash Algorithm (SHA), which ensures the security of the Digital Signature Algorithm (DSA). However, the SHA is usable whenever a secure hash function is required.

5. Snefru

Snefru, designed by Merkle is a one-way hash function which hashes arbitrary-length messages into either 128-bit or 256-bit values. The message is broken into a 512-m in length. The algorithm hashes a 512-bit value into an m-bit value. A hash function H was designed so that it was easy to compute the hash of an input but computationally infeasible to compute an input that generates a specific hash value. Biham and Shamir demonstrated the insecurity of two-pass snefru by using differential cryptanalysis.

6. Kerberos

Kerberos was originally developed at MIT in 1988 as an authentication service able to restrict access to authorized users and to authenticate requests for service.

Kerberos relies exclusively on conventional encryption, making no use of public key encryption.

Version 4 and Version 5 of Kerberos are conceptually similar, but Version 4 (1988) is the most widely used version and works only with TCP/IP networks, while Version 5 (1994) corrects some of the security deficiencies of Version 4 and has been issued as a draft Internet Standard (RFC 510).

An implementation of Kerberos consists of a Key Distribution Center (KDC) and a library of subroutines.

7. X. 509

X. 509 [CCIT87] was initially issued in 1988 and is based on the public-key cryptography and digital signature. The digital signature scheme is to require the use of a hash function. X. 509 is an important standard because it is expected that the X. 509 directory service will become widely used.

The X. 509 standard does not dictate a specific hash algorithm but recommends RSA. However, the standard was subsequently revised to address some security concerns, and a revised recommendation was issued in 1993.

8. DM Algorithm

Hash code computation based on the DM algorithm was proposed independently by Davies and Meyer in 1985. The message to be hashed is first divided into fixed length blocks: $M_1, M_2, \cdots, M_t$. The message block M_i, $(1 \leq i \leq t)$, is used as the encryption key. The previous message block is encrypted using that key and then EX-ORed with itself. The result becomes the input to the next round. Thus, at the end, the length of message becomes the last M_i. Let H_i $(1 \leq i \leq t)$ be defined as $H_i = E_{Mi}(H_{i-1}) \oplus H_{i-1}$, $1 \leq i \leq t$. Using the DES algorithm, the message M is first partitioned into 56-bit DES key blocks. The first block M_1 becomes the 56-bit DES key. The input to the DES algorithm is an initializing vector H_0. The ciphertext output of the DES algorithm is H_1 which is EX-ORed $E_{M1}(H_0)$ with H_0 and becomes the input to the hash of the second block. The next 56-bit message block M_2 becomes the new DES key. H_1 is the plaintext input to the algorithm and H_2 becomes the ciphertext output. $H_2 = E_{M2}(H_1) \oplus H_1$ is the input to the hash of the third block, and so on.

9. Modified CBC Algorithm

There are many ways to consider the computation of hash code. Modified CBS scheme is one of them. The conceptual idea of this modified CBC algorithm is somewhat similar to the DM scheme described above. The padded message to be hashed is divided into t blocks, $M_1, M_2, \cdots, M_t$ where M_i, $1 \le i \le t$, is 64 bits long each. Let H_0 be an initializing vector (IV). Then the DES-like multiple encryption steps are shown as follows:

$$H_1 = E_{H_0 \oplus M_1}(H_0) \oplus H_0$$
$$H_2 = E_{H_1 \oplus M_2}(H_1) \oplus H_1$$

$$H_t = E_{H_{t-1} \oplus M_t}(H_{t-1}) \oplus H_{t-1}$$

where $H_i \oplus M_t$ denotes the enciphering key.

Message digest algorithms are much more useful for designing either public key or secret key cryptosystems. Computation of a signature on a long message is often impractical because of slow speed and poor performance. The message can then be compressed into a message digest of a small size. Therefore, instead of computing a signature over an entire long message, a signature over the compressed digest could be used.

CHAPTER 8

Authentication, Secrecy, and Identification

$\mathbf{D}$ata security embraces the problems of both secrecy (or privacy) and authentication. The privacy problem is concerned with the task of preventing an opponent from extracting message information from the channel. Therefore, the information message or data must be protected to ensure its secrecy and integrity and then the message will be accepted as genuine. Authentication deals with the problem of preventing the injection of false data into the channel or altering of messages that change their meaning.

The mobile station should operate in conjunction with the base station to authenticate the identity of the mobile station. Authentication is the process by which information is exchanged between a mobile station and base station for the purpose of confirming the identity of the mobile station. A successful outcome of the authentication process occurs only when it can be demonstrated that the mobile station and base station process identical sets of shared secret data. For example, in a CDMA authentication protocol, a mobile station and base station each have matching sealed authenticators (i.e., identical shared secret data), actually a short message digest of symbols produced and distributed by the authentication algorithm.

8.1 MOBILE STATION IDENTIFICATION NUMBER

The mobile station identification number (MSIN or MIN) is defined according to the International Mobile Station Identity (IMSI) in the ITU-T Recommendation E.212. The IMSI is structured as shown in Fig. 8.1.

IMSI may not exceed 15 digits. IMSI consists of two parts. The first part is the Mobile Country Code (MCC) with 3 digits. The next part is the National Mobile Station Identity (NMSI). In this standard, IMSI consists of an MCC of 3 digits, an MNC of 2 digits and an MSIN of 10 digits.

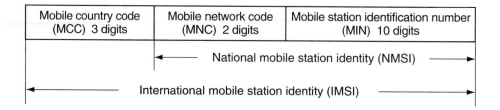

Figure 8.1 International mobile station identity (IMSI)

The mobile station identification number (MSIN)[1], as shown in Fig. 8.2, is a 34-bit binary number which is derived from a 10-digit directory telephone number according to the following procedure.

Suppose a 10-digit telephone number is $D_1D_2D_3$-$D_4D_5D_6$-$D_7D_8D_9D_{10}$, where $D_1D_2D_3$ is an area code, $D_4D_5D_6$ denotes a switching station, and $D_7D_8D_9D_{10}$ represents an individual phone number.

1. The first three digits are mapped into 10 bits (corresponding MSIN2) by the following coding algorithm:
   • Represent the 3-digit field as $D_1D_2D_3$ with the digit 0 having the value 10.
   • Compute $100 D_1 + 10 D_2 + D_3 - 111$.
   • Convert the result computed above to binary using a standard decimal-to-binary conversion (see Table 8.1).
2. The second three digits are mapped into the 10 most significant bits of MSIN1 by the coding algorithm described in (1) above.
3. The last four digits are mapped into the 14 least-significant bits of MSIN1 as follows:
   • The thousand digit should be mapped into four bits by a Binary-Coded-Decimal (BCD) conversion, as specified in Table 8.1.
   • The last three digits are mapped into 10 bits by the coding algorithm described in (1).

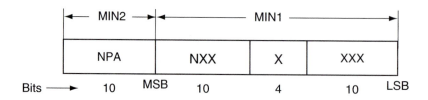

Figure 8.2 Mobile station identification number (MIN)

---

1.   MSIN or MIN is often used interchangeably.

Table 8.1   Decimal-to-Binary Conversion and BCD Mapping

Decimal-to-Binary Conversion		Thousand-Digit BCD Mapping Procedure	
Decimal Number	Binary Number	Decimal Number	Binary Number
1	0000000001	1	0001
2	0000000010	2	0010
3	0000000011	3	0011
4	0000000100	4	0100
5	0000000101	5	0101
.	.	6	0110
.	.	7	0111
.	.	8	1000
998	1111100110	9	1001
999	1111100111	0	1010

Example 8.1   Suppose the 10-digit directory telephone number 290-453-7186 is encoded into MSIN2 and MSIN1 using the procedure just described.

1. MSIN2 computation: The 10-bit MSIN2 is derived from the three digits (i.e., 290) of the telephone number.
   - $D_1 = 2$, $D_2 = 9$, and $D_3 = 0$
   - $100D_1 + 10D_2 + D_3 - 111 = 100(2) + 10(9) + 0 - 111 = 179$
   - 179 in binary is '00 1011 0011'.
   Therefore, MSIN2 is '00 1011 0011'.
2. MSIN1 computation: The 10 most significant bits of MSIN1 are derived from the second three digits (i.e., 453) of the telephone number.
   - $D_1 = 4$, $D_2 = 5$, and $D_3 = 3$
   - 453 - 111 = 342
   - 342 in binary is '0101 0101 10'.
3. BCD conversion of D_7: The next four most significant bits of MSIN1 are derived from the thousands digit (i.e., 7) of the telephone number by BCD conversion:
   - 7 in BCD is '0111'.
4. Computation of the last three digits (i.e., $D_8D_9D_{10}$) of MSIN1: The 10 least significant bits of MSIN1 are derived from the last three digits (i.e., 186) of the telephone number.
   - $D_1 = 1$, $D_2 = 8$, and $D_3 = 6$
   - $100D_1 + 10D_2 + D_3 - 111 = 100(1) + 10(8) + 6 - 111 = 75$
   - 75 in binary is '00 0100 1011'.
   Therefore, MSIN1 is '0101 0101 1001 1100 0100 1011'.

Referring to Fig. 8.1, the MNC binary mapping is defined as follows:

1. Represent the 2-digit mobile network code D_1D_2, with a digit equal to zero being given the value of ten.
2. Compute $10D_1 + D_2 - 11$.
3. Convert the result in step (2) to binary using a standard decimal-to-binary conversion (see Table 8.1).

Next, the MCC binary mapping is defined as follows:

1. Represent the 3-digit mobile country code as $D_1D_2D_3$ with the digit equal to zero being given the value ten.
2. Compute $100D_1 + 10D_2 + D_3 - 111$.
3. Convert the result computed in step (2) to binary using a standard decimal-to-binary conversion.

8.2 ELECTRONIC SERIAL NUMBER (ESN)

The ESN is a 32-bit binary number that uniquely identifies the mobile station to any cellular system. The bit allocation of the ESN is shown in Fig. 8.3.

The ESN must be factory-set and not readily altered in the field. Modification of the ESN requires a special facility not normally available to subscribers. The circuitry that provides the ESN must be isolated from fraudulent contact and tampering. Electronic storage devices mounted in sockets or connected with a cable are deemed not to comply with this requirement. Attempts to change the ESN circuitry must render the mobile station inoperative.

At the time of issuance of initial type acceptance, the manufacturer is assigned a manufacturer's code within the eight most-significant bits (i.e., bit 31 through bit 24) of the 32-bit serial number. Bits 23 through 18 are reserved (initially all zero), and bits 17 through 0 are uniquely assigned by each manufacturer. When a manufacturer has used substantially all possible combinations of serial numbers within bits 17 through 0, the manufacturer may submit notification to the FCC of the U. S. government. The FCC will allocate the next sequential binary number within the reserved block (i.e., bits 23 through 18).

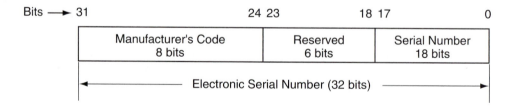

Figure 8.3 The ESN bit allocation

8.3 AUTHENTICATION

Authentication is the process by which information is exchanged between a mobile station and base station for the purpose of confirming the identity of the mobile station. A successful outcome of the authentication process occurs only when it can be demonstrated that the mobile station and base station possess identical sets of shared secret data (SSD). SSD is a 128-bit pattern stored in the mobile station (in semi-permanent memory) and readily available to the base station. SSD-A is used to support the authentication procedures and SSD-B is used to support voice privacy and message encryption.

In this standard, two main authentication procedures are defined: Global Challenge and Unique Challenge.

Global challenges are initiated over the paging and access channels. The mobile (or personal) station may initiate the following authentication procedures over the access channel:

- Authentication at registrations
- Authentication at originations
- Authentication at page responses

On the other hand, unique challenges will be initiated over the forward and reverse traffic channels (the signaling channel for W-CDMA), or over the paging and access channels. The base station initiates the unique challenge-response procedure according to the following conditions:

- Authentication at registration has failed.
- Authentication at origination has failed.
- Authentication at page response has failed.
- At any time after channel assignment.

Table 8.2 summarizes the setting of the input parameters of the authentication algorithm for each of its uses in this standard

Table 8.2 Authentication Algorithm Input Parameters

Procedure	Rand-Challenge	ESN	AUTH-Data	SSD-AUTH	Save-registers
Registration	RAND	ESN	MSIN1	SSD-A	False
Unique challenge	256 x RANDU + (8 LSBs of MSIN2)	ESN	MSIN1	SSD-A	False
Originations	RAND	ESN	MSIN1	SSD-A	True
Terminations	RAND	ESN	MSIN1	SSD-A	True
Base station challenge	RANDBS	ESN	MSIN1	SSD-A-NEW	False

RAND: Authentication random challenge value (0 or 32 bits)

RANDU: Unique random variable (24 bits)

RANDBS: Random challenge data (32 bits)

8.3.1   Shared Secret Data (SSD)

SSD is a 128-bit pattern stored in semi-permanent memory in the mobile station. As depicted in Fig. 8.4, SSD is partitioned into two distinct subsets. Each subset is used to support a different process.

SSD-A is used to support the authentication procedures, and SSD-B is used to support voice privacy and message confidentiality. SSD is generated according to the procedure specified later in Section 8.12.

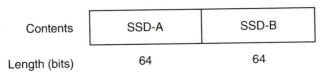

SSD-A: For authentication procedures
SSD-B: For voice privacy and message encryption

Figure 8.4 Partitioning of SSD

8.3.2   Random Challenge Memory (RAND)

RAND is a 32-bit value held in the mobile station. When received on the CDMA paging channel, it is equal to the RAND value received in the last access parameters message. The access parameters message defines the parameters used by mobile stations when transmitting to the base station on an access channel.

RAND is used in conjunction with SSD-A and other parameters, as appropriate, to authenticate mobile station originations, terminations, and registrations.

8.3.3   Call History Parameter (COUNT)

The call history parameter defines a modulo-64 event count held in the mobile station. COUNT is updated by the mobile station when a parameter update order is received on the CDMA forward traffic channel.

8.4 AUTHENTICATION OF MOBILE STATION REGISTRATIONS

When the authentication field of the access parameters message is set to '01', and the mobile station attempts to register by sending a registration request message on the access channel, the following authentication-related procedures are performed:

The mobile station will:
- set the input parameters of the authentication algorithm procedure as illustrated in Fig. 8.5;
- set the save-registers input parameter to FALSE;
- execute the authentication algorithm procedure;
- set the AUTHR field of the registration message equal to the 18 bits of the Authentication Algorithm output; and
- send the authentication data (AUTHR) with the random challenge value (RANDC: eight most significant bits of RAND) and the call history parameter (COUNT) to the base station via the authentication response message.

The base station will:
- compare the received value of RANDC to the most significant eight bits of its internally stored value of RAND;
- compare the received value of COUNT with its internally stored value associated with the received MSIN/ESN;
- compute the value of AUTHR in the same manner as the mobile station, except using the internally stored value of SSD-A; and
- compare the value for AUTHR computed internally with the value of AUTHR received from the mobile station.

If any of the comparisons by the base station fail, the base station may deem the registration attempt unsuccessful, initiate the unique challenge response procedure, or commence the process of updating the SSD.

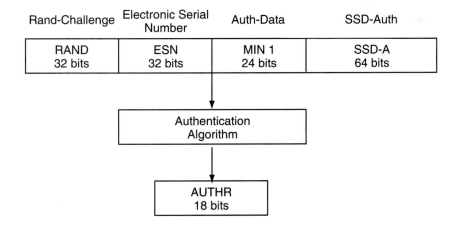

Figure 8.5 Computation of AUTHR for authentication
of mobile station registrations

8.4.1   Common Authentication Fields

Most access channel messages share the same four fields related to authentication:

1. Authentication mode (AUTH-MODE)

   If authentication is required by the base station and authentication information is available, the mobile station sets this field to '01'. All other values are reserved.

2. Authentication data (AUTHR)

   If the authentication mode field is set to '01', the mobile station sets this field as specified in Section 8.3. If the authentication mode field is set to any other value, the mobile station omits this field.

3. Random challenge value (RANDC)

   If the authentication mode field is set to '01', the mobile station sets this field as specified in Section 8.3. If the authentication mode field is set to any other value, the mobile station omits this field.

4. Call history parameter (COUNT)

   If the authentication mode field is set to '01', the mobile station sets this field to the current value of the COUNT parameter stored in the semi-permanent memory. If the authentication mode field is set to any other value, the mobile station omits this field.

8.4.2   Registration Message

This section specifies the contents of the registration message body that may be sent on the access channel. When the mobile station sends a registration message, it uses the following variable-length message format:

Field	Length(bits)	Field	Length (bits)
Message type (00000001)	8	Authentication mode	2
Acknowledgment sequence number	3	Authentication data	0 or 18
Message sequence number	3	Random challenge value	0 or 8
Acknowledgment required indicator	1	Call history parameter	0 or 6
Valid acknowledgment indicator	1	Registration type	4
Acknowledgment address type	3	Slot cycle index	3
Mobile station identifier field type	3	Protocol revision of MS	8
MS identifier field length	4	Station class mark	8
Mobile station identifier	8 x MSID-LEN	Mobile terminated calls accepted indicator	1
		Reserved bits	6

Among the variable-length fields contained in the registration message body, the following are explained further:

- Mobile station identification field type (MSID-TYPE)
  The mobile station sets this field to the value shown in Table 8.3 corresponding to the identifier type contained in the MSID field.

Table 8.3   Address Type

Description	MSID-Type (binary)	MSID-LEN (octets)
MSIN and ESN	000	9
All other MSID-TYPE values are reserved		

- Mobile station identifier field length (MSID-LEN)
  The mobile station sets this field to the number of octets in the MSID field.

- Mobile station identifier (MSID)
  The mobile station sets this field to the mobile station identifier, using the identifier type specified in the MSID-TYPE field. If the MSID-TYPE is equal to '000', the MSID field consists of the following subfields:

Subfield	Length (bits)
Mobile station identification number 1 (MSIN 1)	24
Mobile station identification number 2 (MSIN 2)	10
Mobile station's electronic serial number (ESN)	32
Reserved (000000)	6

8.4.3   Access Channel Message Body Format

The messages sent on the access channel are summarized in Table 8.4.

Table 8.4   Access Channel Messages

Message Name	Message Type (binary)
Registration message	00000001
Order message	00000010
Data burst message	00000011
Origination message	00000100
Page response message	00000101
Authentication challenge response message	00000110

8.5 UNIQUE CHALLENGE-RESPONSE PROCEDURES

The unique challenge-response procedure is initiated by the base station and can be carried out either on the paging and access channels, or on the forward and reverse traffic channels. The procedure is as follows:

The base station will:

- generate a 24-bit random challenge data (RANDU) and send it to the mobile station via a unique (or authentication) challenge message on the paging channel of the forward traffic channel (in case of W-CDMA, the signaling channel of the forward traffic channel);
- initialize the authentication algorithm, as illustrated in Fig. 8.6;
- set AUTHU (Authentication Challenge Response) equal to the 18 bits of the authentication output.

Upon receipt of the unique challenge request message, the mobile station will:

- set input parameters of the authentication algorithm procedure, as illustrated in Fig. 8.6;
- set the save-register input parameter to FALSE;
- compute AUTHU as described above using the received RANDU and its internally stored values for the remaining input parameters; and
- send AUTHU to the base station via the unique challenge response messages.

Upon receipt of the unique challenge response from the mobile station, the base station compares the received value for AUTHU to that generated or stored internally. If the comparison fails, the base station may deny further access attempts by the mobile station, drop the call in progress, or initiate the process of updating the shared secret data (SSD).

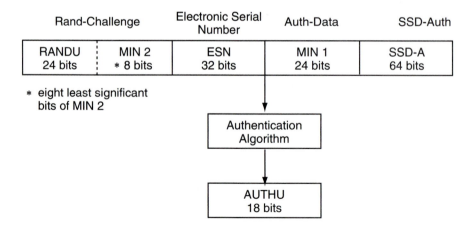

Figure 8.6   Computation of AUTHU for the unique
challenge-reponse procedure

8.5.1   Authentication Challenge Message

When the base station sends an authentication challenge message on the paging channel, it uses the following fixed-length message format:

Field	Length (bits)
Message type (MSG-TYPE) '00001010'	8
Acknowledgment sequence number (ACK-SEQ)	3
Message sequence number (MSG-SEQ)	3
Acknowledgment required indicator (ACK-REQ)	1
Valid acknowledgment indicator (VALID-ACK)	1
Address type (ADDR-TYPE)	3
Address field length (ADDR-LEN)	4
Mobile station address (ADDRESS)	8 x ADDR-LEN
Random challenge data (RANDU)	24
Reserved bits (RESERVED) '000'	3

8.5.2   Authentication Challenge Response Message

When the mobile station sends an authentication challenge response message on the access channel, it uses the following fixed-length message format:

Field	Length (bits)
Message type (MSG-TYPE) '00000110'	8
Acknowledgment sequence number (ACK-SEQ)	3
Message sequence number (MSG-SEQ)	3
Acknowledgment required indicator (ACK-REQ)	1
Valid acknowledgment indicator (VALID-ACK)	1
Acknowledgment address type (ACK-TYPE)	3
Mobile station identifier field type (MSID-TYPE)	3
Mobile station identifier field length (MSID-LEN)	4
Mobile station identifier (MSID)	8 x MSID-LEN
Reserved bits (RESERVED) These bits take the place of the AUTH-MODE field. The mobile station sets this field to '00'.	2
Authentication challenge response (AUTHU)	18
Reserved bits (RESERVED) '0000'	4

8.6 AUTHENTICATION OF MOBILE STATION ORIGINATIONS

When the authentication indicator on the authentication identification information element is set to '01', and the mobile station attempts to originate a call by sending an origination message on the access channel, the following authentication-related procedures are performed:

The mobile station will:

- set the input parameters of the authentification algorithm procedure, as illustrated in Fig. 8.7;
- set the save-registers input parameter to TRUE;
- execute the authentication algorithm procedure;
- set the authentication response (AUTHR) equal to the 18 bits of the authentication algorithm output; and
- send AUTHR together with RANDC (eight most significant bits of RAND) and COUNT to the base station via the authentication response message. (see Section 8.5.2).

The base station will:

- compare the received value of RANDC to the most significant eight bits of its internally stored value of RAND;
- compare the received value of COUNT with its internally stored value associated with the received MSIN/ESN;
- compute the value of AUTH in the same manner as the mobile station, except use the internally stored value of SSD-A; and
- compare the value of AUTHR computed internally with the value of AUTH received from the mobile station.

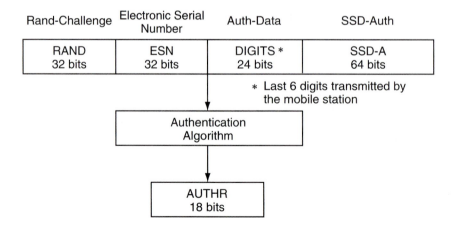

Figure 8.7   Computation of AUTHR for authentication
of mobile station originations

If the comparisons executed at the base station are successful, the appropriate channel assignment procedures are commenced. Once assigned to a forward traffic channel, the base station may, at the discretion of the system operator, issue a parameter update order to the mobile station, updating the value of COUNT stored in the mobile station's semi-permanent memory.

If any of the comparisons by the base station fail, the base station may deny service, initiate the unique challenge response procedure, or commence the process of updating the SSD.

8.7 AUTHENTICATION OF MOBILE STATION TERMINATIONS

If AUTH (authentication mode, 2 bits) is equal to '01', authentication data are included in access channel messages where appropriate.

When the AUTH field of the access parameters message sent on the paging channel is set to '01' (standard authentication mode), and the mobile station responds to a page (by sending a page response message on the access channel), the following authentication procedures are performed:

The mobile station will:

- set the input parameters of the authentification algorithm procedure as illustrated in Fig. 8.8;
- set the save-registers input parameter to TRUE;
- execute the authentication algorithm;
- set AUTHR equal to the 18 bits of the authentication algorithm output that are used to fill the AUTHR field of the page response message; and
- send AUTHR together with RANDC (eight most significant bits of RAND) and COUNT to the base station via the page (authentication) response message.

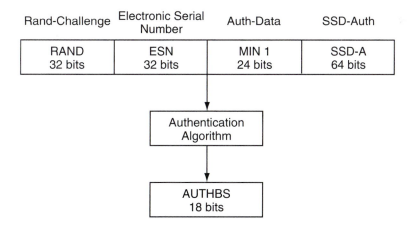

Figure 8.8   Computation of AUTHR for authentication
of mobile station terminations

The base station will:

- compare the received value of RANDC to the eight most significant bits of its internally stored value of RAND;
- compare the received value of COUNT with its internally stored value associated with the received MSIN/ESN;
- compute AUTHR in the same manner as the mobile station, but using its internally stored value of SSD-A; and
- compare the value of AUTHR computed internally with the value of AUTHR received from the mobile station.

If the comparisons executed at the base station are successful, the appropriate channel assignment procedures are commenced. Once assigned to a forward traffic channel, the base station may issue a parameter update order to the mobile station, updating the value of COUNT stored in the mobile station's semi-permanent memory.

If any of the comparisons by the base station fail, the base station may deny services, initiate the unique challenge response procedure, or commence the process of updating the SSD.

The access parameters message defines the parameters used by mobile stations when transmitting to the base station on an access channel. For the reader's reference, note that (1) if mobile stations are to include standard authentication data in access channel messages, the base station must set this field to '01'. If mobile stations are not to include authentication data in access channel messages, the base station sets this filed to '00'; (2) if the AUTH field is set to '01', the base station sets this field to the random challenge value (RAND) to be used by mobile stations for authentication. If the AUTH field is set to any other value, the base station omits this field.

8.8 SSD UPDATE

SSD is updated using the SSD generation procedure, initialized with mobile station specific information, random data, and the mobile station's A-key.

The A-key is:

- 64 bits long;
- assigned to the mobile station;
- stored in the mobile station's permanent security and identification memory; and
- known only to the mobile station and to its associated Home Location Register/Authentication Center (HLR/AC).

NOTES:

1. The last item in the above list is intended to enhance the security of the mobile station's secret data by eliminating the need to pass the A-key itself from base station to base sta-

tion as the subscriber roams. As a consequence, SSD updates are carried out only in the mobile station and its associated HLR/AC, not in the serving base station. The serving base station obtains a copy of the SSD computed by the HLR/AC via intersystem communication with the mobile station's HLR/AC.

2. Since the SSD update procedure involves multiple transactions and can be started on one channel and completed on another channel, call processing and signaling text has to be included here for the sake of added clarity.

An A-key must be entered into the mobile station. More specifically, updating the SSD in the mobile station proceeds as illustrated in Fig. 8.9.

The SSD update procedure is performed as follows:

The base station will:

- send an SSD update order on either the paging channel or the forward traffic channel, with the RANDSSD field set to the same 56-bit random number used in the HLR/AC computations, to the mobile station via the SSD update message.

Upon receipt of the SSD update message, the mobile station will:

- set the input parameters of the SSD-generation procedure;
- execute the SSD-generation procedure;
- compute SSD-A-NEW equal to the 64 most significant bits of the SSD-generator output and SSD-B-NEW to the 64 least significant bits of the SSD-generator output (see Fig. 8.10);
- set the input parameters of the authentication algorithm procedure as illustrated in Fig. 8.9;
- execute the authentication algorithm;
- select a 32-bit random number, RANDBS, and send it to the base station in a BS challenge order on the access channel or reverse traffic channel;
- set AUTHBS equal to the 18-bit output resulted from the authentication algorithm; and
- set the save-registers input parameter to FALSE.

Upon receipt of a base station challenge message, the base station will:

- set the input parameters of the authentication algorithm as illustrated in Fig. 8.9, where RANDBS is set to the value received in the base station challenge message;
- set the save-registers input parameter to FALSE;
- execute the authentication algorithm;
- set AUTHBS equal to the 18 bits of the authentication algorithm output (see Fig. 8.11); and
- acknowledge receipt of the base station challenge order by including AUTHBS in the base station challenge confirmation order on the paging channel or forward traffic channel, i.e., the base station sends its computed value of AUTHBS to the mobile station in a base station challenge confirmation order on the paging or forward traffic channel.

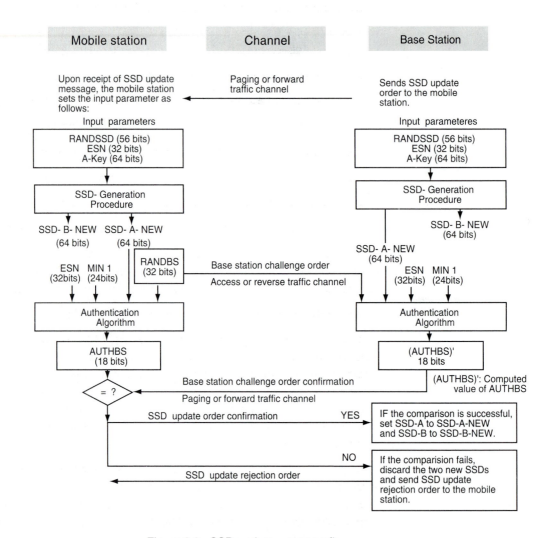

Figure 8.9   SSD update message flow

Upon receipt of the base station challenge confirmation order, the mobile station will:

• compare the received value of AUTHBS to its internally computed value;
• acknowledge receipt of the SSD update order confirmation as follows:
    − if the comparison at the mobile station is successful, execute the SSD update procedure to set SSD-A and SSD-B to SSD-A-NEW and SSD-B-NEW, respectively.

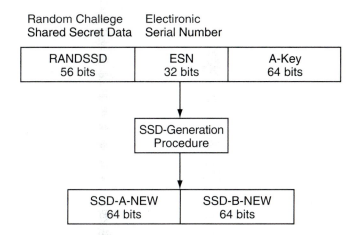

Figure 8.10   Computation of Shared Secret Data (SSD)

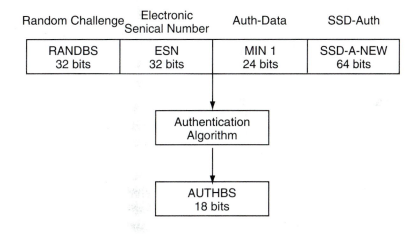

Figure 8.11   Computation of AUTHBS

- send an SSD update confirmation message to the base station with the SSD in the authentication parameter information element, indicating successful completion of the SSD update.
- if the comparison at the mobile station fails, discard SSD-A-NEW and SSD-B-NEW.
- send an SSD update confirmation message to the base station with the SSD in the authentication parameter information element, indicating unsuccessful completion of the SSD update.

In the base station, if the SSD update order confirmation received from the mobile station indicates a success, set SSD-A and SSD-B to the values received from the HLR/AC.

If the mobile station fails to receive the base station challenge order confirmation within the specified seconds after acknowledgment to the base station challenge confirmation was received, the mobile station discards SSD-A-NEW and SSD-B-NEW. The mobile station then terminates the SSD update process.

8.9 SIGNALING MESSAGE ENCRYPTION

In an effort to enhance the authentication process and to protect sensitive subscriber information (such as PINS), a method is available to encrypt certain fields of selected traffic channel signaling messages. However, TIA/EIA/IS-95 neither discusses nor lists messages and fields to be encrypted due to the fact that the availability of encryption algorithm information is entirely governed by the U.S.A. International Traffic and Arms Regulation (ITAR) and the Export Administration Regulations.

Messages are not to be encrypted if authentication is not performed (i.e., AUTH field equal to '00' in the access parameters message).

Signaling message encryption is controlled for each call individually. The initial encryption mode for the call is established by the value of the signaling encryption field in the encryption message at the channel assignment. If the signaling encryption field is set to '00', message encryption is off. To turn encryption on after channel assignment, the base station sends the encryption message with the signaling encryption field set to '01'.

To turn signaling message encryption off, the base station sends an encryption message with the signaling encryption field set to '00'.

Every reverse traffic channel message contains an encryption field which identifies the message encryption mode active at the time the message was created.

8.10 VOICE PRIVACY

Voice privacy is provided in the CDMA system by means of the private long code mask used for PN spreading. The generation and application of the private long code mask is not specified in TIA/EIA/IS-95.

Voice privacy is provided on the traffic channel only. All calls are initiated using the public long code mask for PN spreading. The mobile station user may request voice privacy using the origination message or page response message with voice privacy field set to '01' during call setup. During traffic channel operation, the mobile station may request voice privacy using the encryption message with voice privacy set to '01'.

The transition to private long code mask will not be performed if authentication is not performed (AUTH field set to '00' in the access parameters message or mobile station unable to perform authentication).

To initiate a transition to the private or public long code mask, either the base station or the mobile station sends a long code transition request order on the traffic channel.

The mobile station actions in response to a long code transition request order are specified as follows:

- If the long code transition request order requests a transition to the private long code, and the mobile station accepts the request, the mobile station sends a long code transition response order (00000011) within $T_{56m} = 0.2$ seconds. The mobile station uses the private long code on both the forward traffic channel and the reverse traffic channel. If use-time equals '0', the mobile station begins using the private long code at the first 80 ms boundary (relative to the start of system time) after $N_{4m} = 20$ frames from the end of the response transmission. The mobile station should indicate to the user that the voice privacy mode is active. If the long code transition request order requests a private long code transition, and the mobile station is not able to generate the private long code or the mobile station does not accept the request, the mobile station then sends a long code transition response order (00000010) within T_{56m} seconds.
- If the long code transition request order requests a transition to the public long code and the mobile station accepts the request, the mobile station sends a long code transition response order (00000010) within T_{56m} seconds. The mobile station should indicate to the user that the voice privacy mode is inactive. If the long code transition request order requests a public long code transition, and the mobile station does not accept the request, the mobile station sends a long code transition response order (00000011) within T_{56m} seconds.

The base station actions in response to receipt of a long code transition request order are specified in the following. If a request for voice privacy is specified in the origination message or page response message, the base station may send a long code transition request order (00000001) requesting a transition to the private long code.

The base station processes the long code transition request order as follows:

- If the long code transition request order requests a transition to the private long code and the base station accepts the request, the base station sends a long code transition request order (00000001). If the base station does not accept the private long code transition request, the base station sends a long code transition request order (00000000).
- If the long code transition request order requests a transition to the public long code and the base station accepts the request, the base station sends a long code transition request order (00000000). If the base station does not accept the public long code transition request, the base station sends a long code transition request order (00000001).

The base station processes the long code transition request order as follows:

- If the long code transition response order indicates that the mobile station accepts the long code transition requested in the long code transition request order sent by the base station, the base station uses the requested long code mask on both the forward traffic channel and the reverse traffic channel. If the base station did not specify an explicit action time in the long code transition request order, the base station should begin using the requested long code mask at the first 80 ms boundary (relative to the start of system time) after N_{4m} frames after the last frame in which any portion of the long code transition response order was received.

8.11 AUTHENTICATION ALGORITHM

In computer-communication networks, it is often necessary for communication parties to verify one another's identity. One practical way is the use of cryptographic authentication protocols employing a one-way hash function. In order to authenticate the identity of the mobile station, the mobile station must have operated in conjunction with the base station. Authentication in the CDMA system is the process of confirming the identity of the mobile station by exchanging information between a mobile station and base station.

One possible authentication scheme may be considered for the case of any block cipher. It is possible to use a symmetric block cipher algorithm (such as DES) to compute the 18-bit hash code. If the block algorithm is secure, then it is assumed that the one-way hash function is also secure.

The 152-bit message block M is input into the authentication algorithm device. Using DES, M is first broken down into the 64-bit blocks such that $M = M_1 M_2 M_3 \cdots$. The first message block M_1 becomes the DES key. Division into 64-bit blocks can be accomplished by mapping the 152-bit message value onto the 192-bit value by padding with 40 bit zeros. Appropriate padding is needed to force the message to conveniently divide into certain fixed lengths. Our authentication scheme is to generate an 18-bit authentication data from a message length of 192 bits.

Some authentic computations for mobile station registrations, unique challenge-response procedures, originations, terminations, and SSD-update are considered, in that order, in the following discussion.

Several techniques will be proposed for the key generation in this section. Unfortunately, the design for key generation is not as easy as one may expect from the standpoint of easy implementation and secure hash functions. In the following, few attempts are made to devise the key generation structure that occupies the gray block in Fig. 8.12.

8.11.1   Key Generation Technique (I) and AUTHR Computation

Suppose M_1, M_2, M_3 are the decomposition of a 176-bit value into 48 bits, 64 bits, and 64 bits, respectively. The 176-bit value is composed of 152-bit message length and 24-bit padding. M_1=48 bits will be used as input to the key generation scheme, (i.e., the gray block in Fig. 8.12).

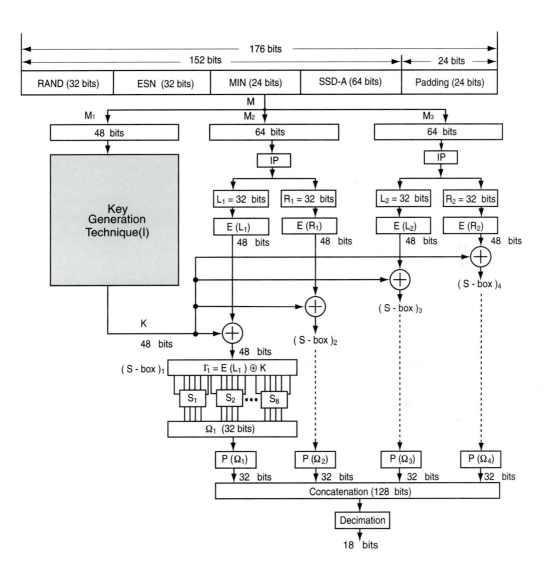

Figure 8.12 Computation of AUTHR (18 bits) for a mobile
station at registrations or terminations

Let us arrange the 48-bit input into a 6 x 8 array, as shown below:

Table 8.5   A 6 x 8 Array for Key Computation

Input (column by column)
⇩

1	7	13	19	25	31	37	43
2	8	14	20	26	32	38	44
3	9	15	21	27	33	39	45
4	10	16	22	28	34	40	46
5	11	17	23	29	35	41	47
6	12	18	24	30	36	42	48

Column-wise permutation

7	31	1	43	13	37	25	19
8	32	2	44	14	38	26	20
9	33	3	45	15	39	27	21
10	34	4	46	16	40	28	22
11	35	5	47	17	41	29	23
12	36	6	48	18	42	30	24

Row-wise permutation

11	35	5	47	17	41	29	23
7	31	1	43	13	37	25	19
9	33	3	45	15	39	27	21
12	36	6	48	18	42	30	24
8	32	2	44	14	38	26	20
10	34	4	46	16	40	28	22

⇨   Output (row by row)

A 48-bit key generation from M_1 is computed as follows:

11	35	5	47	17	41	29	23	7	31	1	43	13	37	25	19
9	33	3	45	15	39	27	21	12	36	6	48	18	42	30	24
8	32	2	44	14	38	26	20	10	34	4	46	16	40	28	22

Example 8.2   First, consider the key generation by column/row permutations described in this section. Assume that the key data is 16c27a415f39 (hexadecimal) = 0001 0110 1100 0010 0111 1010 0100 0001 0101 1111 0011 1001 (binary).

Using Table 8.5, the key data will be executed as follows:

1. Input the key data column by column.

    ⇩

0101 0011

0001 1111

0111 0011

1100 0110

0001 0000

1010 0101

2. Execute column-wise permutation

1001 0101

0101 0111

1001 1101

1110 0100

0000 0001

0111 1000

3. Execute row-wise permutation

0000 0001

1001 0101

1001 1101 ⇨ 01959d7857e4

0111 1000

0101 0111

1110 0100

Thus, the 48-bit key output is obtained writing out row by row as follows:

    0000 0001 1001 0101 1001 1101 0111 1000 0101 0111 1110 0100

    (binary) = 01959d7857e4 (hexadecimal)

This 48-bit key K was computed by implementing the key generation technique (I), which is applicable to the gray block in Fig. 8.12.

Example 8.3   Now, using Fig. 8.12, the 18-bit message digest for AUTHR is derived by hashing the 176-bit padded message as described below:

Assume that the M_2 data block (64 bits) is 1 7 b 4 3 9 a 1 2 f 5 1 c 5 a 8 . This M_2 data is first subjected to an initial permutation (IP) to make it split into two blocks L_1 (left) and R_1 (right) where each block consists of 32 bits as indicated in Table 8.6.

When M_2 hexadecimal data are converted into the binary values first and arranged according to Table 8.6, the initial permuted sequence can be found as follows: Given M_2 as

    17b439a1 → 0001 0111 1011 0100 0011 1001 1010 0001

    2f51c5a8 → 0010 1111 0101 0001 1100 0101 1010 1000

Executing the initial permutation according to Table 8.6 yields:

Table 8.6   Initial Permutation (IP)

	58	50	42	34	26	18	10	2
L_1	60	52	44	36	28	20	12	4
	62	54	46	38	30	22	14	6
	64	56	48	40	32	24	16	8
	57	49	41	33	25	17	9	1
R_1	59	51	43	35	27	19	11	3
	61	53	45	37	29	21	13	5
	63	55	47	39	31	23	15	7

L_1 (32 bits):	0110	0000	0010	0111	0101	0011	0111	1101
	6	0	2	7	5	3	7	d

R_1 (32 bits):	1100	1010	1001	1110	1001	0100	0001	0001
	c	a	9	e	9	4	1	1

Thus, the initial permuted sequence is expressed as:

6 0 2 7 5 3 7 d (L_1) ; c a 9 e 9 4 1 1 (R_1)

This 64-bit permuted sequence consists of a 32-bit block L_1, and another 32-bit block R_1, as obtained above.

These L_1 and R_1 are expanded from 32 bits to 48 bits, respectively, according to Table 8.7.

Table 8.7   Bit Expansion Table

32	1	2	3	4	5
4	5	6	7	8	9
5	9	10	11	12	13
12	13	14	15	16	17
16	17	18	19	20	21
20	21	22	23	24	25
24	25	26	27	28	29
28	29	30	31	32	1

$E(L_1)$ denotes a function which takes a block of 32 bits as its input and yields a block of 48 bits as its output, as shown below:

$$E(L_1) = b\,0\,0\,1\,0\,e\,a\,a\,6\,b\,f\,a$$

Once $E(L_1)$ is computed, it is added bit by bit to the key K such that

$$\Gamma_1 = E(L_1) \oplus K$$
$$= (b0010eaa6bfa) \oplus (01959d7857e4)$$
$$= (b19493d23c1e)$$

This 48-bit input Γ_1 to the (S-box)$_1$ is passed through a nonlinear substitution to form the 32-bit output.

The 48-bit vector Γ_1 is used as argument in the substitution operation of (S-box)$_1$ from S_1 through S_8. Each S_i, $1 \le i \le 8$, is a matrix of four rows and 16 columns as shown Table 8.8. The input to each S_i is composed of 6 bits. Its first and last bits represent the row number of S_i and the middle 4 bits produce a column number. For example, for an input (010011) to S_i, denotes as S_i^{01} (1001), the row number is 01, i.e., row 1, and the column number is determined by 1001, i.e., column 9.

Table 8.8   Primitive S-box Functions

S_1

14	4	13	1	2	15	11	8	3	10	6	12	5	9	0	7
0	15	7	4	14	2	13	1	10	6	12	11	9	5	3	8
4	1	14	8	13	6	2	11	15	12	9	7	3	10	5	0
15	12	8	2	4	9	1	7	5	11	3	14	10	0	6	13

S_2

15	1	8	14	6	11	3	4	9	7	2	13	12	0	5	10
3	13	4	7	15	2	8	14	12	0	1	10	6	9	11	5
0	14	7	11	10	4	13	1	5	8	12	6	9	3	2	15
13	8	10	1	3	15	4	2	11	6	7	12	0	5	14	9

S_3

10	0	9	14	6	3	15	5	1	13	12	7	11	4	2	8
13	7	0	9	3	4	6	10	2	8	5	14	12	11	15	1
13	6	4	9	8	15	3	0	11	1	2	12	5	10	14	7
1	10	13	0	6	9	8	7	4	15	14	3	11	5	2	12

S_4

7	13	14	3	0	6	9	10	1	2	8	5	11	12	4	15
13	8	11	5	6	15	0	3	4	7	2	12	1	10	14	9
10	6	9	0	12	11	7	13	15	1	3	14	5	2	8	4
3	15	0	6	10	1	13	8	9	4	5	11	12	7	2	14

S_5

2	12	4	1	7	10	11	6	8	5	3	15	13	0	14	9
14	11	2	12	4	7	13	1	5	0	15	10	3	9	8	6
4	2	1	11	10	13	7	8	15	9	12	5	6	3	0	14
11	8	12	7	1	14	2	13	6	15	0	9	10	4	5	3

S_6

12	1	10	15	9	2	6	8	0	13	3	4	14	7	5	11
10	15	4	2	7	12	9	5	6	1	13	14	0	11	3	8
9	14	15	5	2	8	12	3	7	0	4	10	1	13	11	6
4	3	2	12	9	5	15	10	11	14	1	7	6	0	8	13

S_7

4	11	2	14	15	0	8	13	3	12	9	7	5	10	6	1
13	0	11	7	4	9	1	10	14	3	5	12	2	15	8	6
1	4	11	13	12	3	7	14	10	15	6	8	0	5	9	2
6	11	13	8	1	4	10	7	9	5	0	15	14	2	3	12

S_8

13	2	8	4	6	15	11	1	10	9	3	14	5	0	12	7
1	15	13	8	10	3	7	4	12	5	6	11	0	14	9	2
7	11	4	1	9	12	14	2	0	6	10	13	15	3	5	8
2	1	14	7	4	10	8	13	15	12	9	0	3	5	6	11

If the S-box input, $\Gamma_1 = b\,1\,9\,4\,9\,3\,d\,2\,3\,c\,1\,e$, is expressed in the binary notation, it gives:

$$101100011001010010010011110100100011110000011110$$

Grouping this 48-bit Γ_1 into sets of 6 bits leads to easy computation of substitution operations from S_1 through S_8 as follows:

$$S_1^{10}(0110) = S_1^1(6) = 2 = 0010$$

$$S_2^{01}(1100) = S_2^1(12) = 6 = 0110$$

$$S_3^{00}(1001) = S_3^0(9) = d = 1101$$

$$S_4^{01}(1001) = S_4^1(9) = 7 = 0111$$

$$S_5^{10}(1010) = S_5^2(10) = c = 1100$$

$$S_6^{11}(0001) = S_6^3(1) = 3 = 0011$$

$$S_7^{10}(1000) = S_7^2(8) = a = 1010$$

$$S_8^{00}(1111) = S_8^0(15) = 7 = 0111$$

Concatenating all of these 4-bit S_i, $1 \le i \le 8$, yields the 32-bit Ω_1:

$\Omega_1 \quad = 0010\ 0110\ 1101\ 0111\ 1100\ 0011\ 1010\ 0111$

$\qquad = 26d7c3a7$

This is the output of (S-box)$_1$.

Using Table 8.9, the 32-bit permuted output $P(\Omega_1)$ is produced by permuting the bits of Ω_1 as follows: $P(\Omega_1)$ is obtained from the input Ω_1 by taking the 16th bit of Ω_1 as the first bit of $P(\Omega_1)$, the seventh bit of Ω_1 as the second bit of $P(\Omega_1)$, and so on until the 25th bit of Ω_1 is taken as the 32nd bit of $P(\Omega_1)$.

Table 8.9   Permutation Function P

16	7	20	21
29	12	28	17
1	15	23	26
5	18	31	10
2	8	24	14
32	27	3	9
19	13	30	6
22	11	4	25

Thus the result of $P(\Omega_1)$ is represented as

$P(\Omega_1) \quad = 1100\ 0101\ 0110\ 0111\ 0011\ 1111\ 0011\ 0001$

$\qquad = c5673f31$

Next, consider R_1 = 32 bits of M_2. If R_1 expands to 48 bits by using Table 8.7, we have

$E(R_1)$ = e554fd4a80a3

EX-ORing $E(R_1)$ with the K, the 48-bit input Γ_2 to (S-box)$_2$ is computed as

$\Gamma_2 = E(R_1) \oplus K$

= (e554fd4a80a3) $\oplus$ (01959d7857e4)

= e4c16032d747

= 1110 0100 1100 0001 0110 0000 0011 0010 1101 0111 0100 0111

Division of Γ_2 into 8 sets of 6 bits yields (S-box)$_2$ as shown below:

i	Input to S_i, Γ_2	S_i, $1 \le i \le 8$		Output of S_i Ω_2
		Row	Column	
1	111001	3	12	a
2	001100	0	6	3
3	000101	1	2	0
4	100000	2	0	a
5	001100	0	6	b
6	101101	3	6	f
7	011101	1	14	8
8	000111	1	3	8

Thus, the (S-box)$_2$ output will be

Ω_2 = a30abf88

Using Table 8.9, the data sequence by permutation becomes

$P(\Omega_2)$ = 79c062c9

Thus far, two permutation data $P(_1\Omega)$ and $P(\Omega_2)$ corresponding to the 64-bit M_2 block have been completely computed.

Suppose the third data block M_3 (64 bits) is 51cb36af43000000. Using Table 8.6, the data sequence by initial permutation of M_3 becomes 13050c1ba0c0a1e, where L_2 = 13050c1b and R_2 = 0a0c0a1e. Both L_2 (left half) and R_2 (right half) of M_3 are each expanded from 32 bits to 48 bits according to Table 8.7 as shown below:

$E(L_2)$ = 8a680a8580f6

As you see, the 32-bit L_2 or R_2 can be spread out and scrambled into 48 bits with the E-table for expansion.

The expanded data $E(L_2)$ or $E(R_2)$ is EX-ORed with the key data K such that

$\Gamma_3 = E(L_2) \oplus K$

= (8a680a8580f6) $\oplus$ (01959d7857e4)

= (8bfd97fdd712)

= (1000 1011 1111 1101 1001 0111 1111 1101 1101 0111 0001 0010)

This 48-bit Γ_3 is the input data to (S-box)$_3$.

The (S-box) $_3$ operation from S_1 through S_8 is executed as follows:

Input to S_i	$S_{i,}$ $1 \leq i \leq 8$		Output from S_i
Γ_3	Row	Column	Ω_3
S_1^{10}	2	1	1
S_2^{11}	3	15	9
S_3^{10}	2	11	12
S_4^{01}	1	11	12
S_5^{11}	3	15	3
S_6^{01}	1	14	3
S_7^{00}	0	14	6
S_8^{00}	0	9	9

The output data from (S-box)$_3$ is written as
$$\Omega_3 = 19cc3369$$
The permutation function $P(\Omega_3)$ yields a 32-bit output from a 32-bit input by permuting the bits of Ω_3 according to Table 8.9.

Hence $P(\Omega_3)$ resulting from permutation is
$$P(\Omega_3) = 28397dc2$$
Finally, consider the right half $R_2 = 0a0c0a1e$ which resulted from the initial permutation (Table 8.6) of the 64-bit M_3. The 48-bit expanded data, $E(R_2)$, from the 32-bit R_2 is
$$E(R_2) = 0540580540fc$$
The (S-box) $_4$ input data Γ_4 can be computed from EX-ORing $E(R_2)$ with K such that
$$\Gamma_4 = E(R_2) \oplus K$$
$$= 04d5c57d1718$$
where K is the key data sequence.

The $(S\text{-box})_4$ substitution operation from S_1 through S_8 is executed as shown in the following table:

Input to S_i	$S_i, 1 \leq i \leq 8$		Output from S_i
Γ_4	Row	Column	Ω_4
S_1^{10}	1	0	0
S_2^{01}	1	6	8
S_3^{01}	1	11	14
S_4^{01}	1	2	11
S_5^{01}	1	15	6
S_6^{01}	1	8	6
S_7^{00}	0	14	6
S_8^{00}	0	12	5

Thus, the output data Ω_4 from $(S\text{-box})_4$ is computed as
$$\Omega_4 = 08\text{eb}6665$$
Using Table 8.9, the permutation of Ω_4 can be shown as
$$P(\Omega_4) = 807\text{d}0\text{dec}$$
Thus, we have completely computed all four permutations, i.e., $P(\Omega_1)$, $P(\Omega_2)$, $P(\Omega_3)$, and $P(\Omega_4)$. The 128-bit final output data $P(\Omega)$ can be obtained from the concatenation process such that
$$P(\Omega) = P(\Omega_1) \| P(\Omega_2) \| P(\Omega_3) \| P(\Omega_4).$$
$$\begin{aligned} P(\Omega) &= (\text{c}5673\text{f}31, 79\text{e}062\text{c}9, 28397\text{dc}2, 807\text{d}0\text{dec}) \\ &= (1100\ 0101\ 0110\ 0111\ 0011\ 1111\ 0011\ 0001 \\ &\quad 0111\ 1001\ 1110\ 0000\ 0111\ 0010\ 1100\ 1001 \\ &\quad 0010\ 1000\ 0011\ 1001\ 0111\ 1101\ 1100\ 0010 \\ &\quad 1000\ 0000\ 0111\ 1101\ 0000\ 1101\ 1110\ 1100) \end{aligned}$$
Finally, the 18-bit authentication data (AUTHR) is computed by taking the decimation process from the 128-bit $P(\Omega)$ by picking 1 bit per every 7 bits:
$$\text{AUTHR} : 011111000011000101$$

8.11.2 Key Generation Technique (II) and AUTHR Computation

The 152-bit message length is expanded into the 192-bit padded message value by adding 40-bit padding to the message in order to divide it into three 64-bit blocks, $M_i = 64$ bits, $1 \le i \le 3$.

The message M is evenly partitioned into 64-bit blocks

$$M_1, M_2, M_3$$

where $M_1 = 64$ bits must be hashed for the generation of the 48-bit encryption key. Thus, the first message block M_1 becomes the DES key which is applicable to the gray block in Fig. 8.14. Figure 8.13 illustrates this technique for key generation.

Utilizing Fig. 8.13 for the DES key computation, we provide Tables 8.10, 8.11, and 8.12 for the permuted choices 1 and 2 and for the number of left shifts

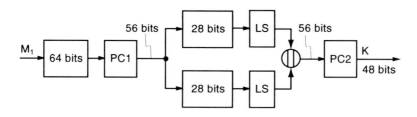

PC1: Permuted choice 1
PC2: Permuted choice 2
⓪ : Concatenation
LS: Left shift once

Figure 8.13 Key generation based on DES key schedule

Table 8.10   Permuted Choice 1 (PC-1)

57	49	41	33	25	17	9	1	58	50	42	34	26	18
10	2	59	51	43	35	27	19	11	3	60	52	44	36
63	55	47	39	31	23	15	7	62	54	46	38	30	22
14	6	61	53	45	37	29	21	13	5	28	20	12	4

Table 8.11   Permuted Choice 2 (PC-2)

14	17	11	24	1	5	3	28	15	6	21	10
23	19	12	4	26	8	16	7	27	20	13	2
41	52	31	37	47	55	30	51	51	45	33	48
44	49	39	56	34	53	46	50	50	36	29	32

Table 8.12   Number of Left Shifts

Round Number	1	2	3	4	5	6	7	8	9	10	11	12	13	14	15	16
Number of Left Shifts	1	1	2	2	2	2	2	2	1	2	2	2	2	2	2	1

Example 8.4   Referring to Fig. 8.14, the 64-bit DES key M_1 should be hashed into the 48-bit key data as follows: Let the DES key be 7a138b2524af17c3. Using Table 8.10, the PC-1 output data becomes

$$L_0 = a481394 \text{ (left half)} ; R_0 = e778253 \text{ (right half)}$$

From Table 8.12, shifting one place to the left (LS) yields LS data as

$$4902729 \ cef04a7$$

These two LS data are concatenated and transformed into the 48-bit key data according to Table 8.11. Then the 48-bit key data will be

$$K = 058c4517a7a2$$

Example 8.5   Given that $M_2 = $ 17b439a12f51c5a8.

Using Table 8.6, an initial permutation (IP) splits into two blocks L_1 (left) and R_1 (right) where each consists of 32 bits. Hence the initial permuted data is

$$IP(L_1) = 6027537d \ \ IP(R_1) = ca9e9411$$

1. Using Table 8.7, the data expansion $E(L_1)$ becomes

$E(L_1) = $ b0010eaa6bfa

EX-ORing $E(L_1)$ with K, we have

$$\Gamma_1 = E(L_1) \oplus K$$
$$= (b0010eaa6bfa) \oplus (058c4517a7a2)$$
$$= b58a4bbacc58$$

This is the input data to (S-box)$_1$, as shown in Fig. 8.14.

(S-box)$_1$ Operation:

Input	S_1^{11}	S_2^{00}	S_3^{11}	S_4^{01}	S_5^{11}	S_6^{00}	S_7^{11}	S_8^{00}
Row	3	0	3	1	3	0	3	0
Column	6	12	10	5	7	14	8	12
Output	1	12	14	15	13	5	9	5

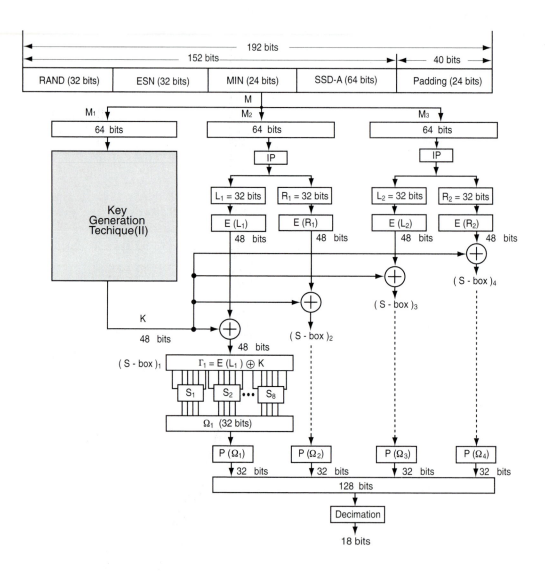

Figure 8.14   Computation of authentications of mobile station at registrations,
             unique challenge-response procedure, originations, and terminations.
             Authentication input parameters:
             · Registrations and terminations are identical.
             · Unique challenge: RAND(32 bits) = RANDU(24 bits) + MIN2 (8 LSBS)
             · Originations: Auth-Data(24 bits) = Digits(Last 6 digits transmitted
               by the mobile station)

Thus, (S-box)$_1$ output will be $\Omega_1 = 1cefd595$.

Using Table 8.9, the permutation function $P(\Omega_1)$ becomes

$$P(\Omega_1) = a34d397f$$

2. Using Table 8.7, the data expansion of R_1 yields

$$E(R_1) = e554fd4a80a3$$

EX-ORing $E(R_1)$ with K becomes the input Γ_2 to (S-box)$_2$:

$$\Gamma_2 = E(R_1) \oplus K$$
$$= (e554fd4a80a3) \oplus (058c4517a7a2)$$
$$= e0d8b85d2701$$

(S-box)$_2$ Operation:

Input	S_1^{10}	S_2^{01}	S_3^{10}	S_4^{10}	S_5^{01}	S_6^{00}	S_7^{00}	S_8^{01}
Row	2	1	2	2	1	0	0	1
Column	12	6	1	12	11	9	14	0
Output	3	8	6	5	10	13	6	1

The output data of (S-box)$_2$ is $\Omega_2 = 3865ad61$.

Using Table 8.9, the permutation function $P(\Omega_2)$ becomes

$$P(\Omega_2) = 91193e8e$$

3. Given that $M_3 = 5\ 1\ c\ b\ 3\ 6\ 0\ 0\ 0\ 0\ 0\ 0\ 0\ 0\ 0\ 0$.

Using Table 8.6, an initial permutation (IP) of M_3 can be obtained as follows:

Left block $L_2 = 03050403$; right block $R_2 = 02040206$

These two blocks are expanded into 48 bits each according to Table 8.7.

$$E(L_2) = 80680a808006$$

EX-ORing $E(L_2)$ with K yields the input data to (S-box)$_3$:

$$\Gamma_2 = E(L_2) \oplus K$$
$$= (80680a808006) \oplus (058c4517a7a2)$$
$$= 85c44f9727a4$$

(S-box)$_3$ Operation:

Input	S_1^{11}	S_2^{00}	S_3^{01}	S_4^{01}	S_5^{11}	S_6^{10}	S_7^{00}	S_8^{10}
Row	3	0	1	1	3	2	0	2
Column	0	15	8	7	2	9	15	2
Output	15	10	2	3	12	0	1	4

Thus, the output data from (S-box)$_3$ can be obtained as

$$\Omega_3 = fa23c914$$

The permutation function $P(\Omega_3)$ is computed according to Table 8.9:

$$P(\Omega_3) = c3cc8226$$

4. Next, consider the case for $R_2 = 02040206$

Following the bit expansion table (Table 8.7), the expansion of R_2 becomes

$$E(R_2) = 00400800400c$$

The input Γ_4 to (S-box)$_4$ is computed as follows:

$\Gamma_4 = E(R_2) \oplus K_1$

$\quad = (00400800400c) \oplus (058c4517a7a2)$

$\quad = 05cc4d17e7ae$

(S-box)$_4$ Operation:

Input	S^{01}_1	S^{00}_2	S^{11}_3	S^{01}_4	S^{01}_5	S^{10}_6	S^{00}_7	S^{10}_8
Row	1	0	3	1	1	2	0	2
Column	0	14	8	6	2	15	15	7
Output	0	5	4	0	2	6	1	2

Hence, the output data from (S-box)$_4$ is

$$\Omega_4 = 05402612$$

Using Table 8.9, the permutation of Ω_4 becomes

$$P(\Omega_4) = 02234098$$

Thus, the output of S-box is a set of four blocks which concatenate to form a single 18-bit hash. Final data output from the S-box is

$$P(\Omega) = P(\Omega_1) \| P(\Omega_2) \| P(\Omega_3) \| P(\Omega_4) = a34d397f\ 91193e8e\ c3cc8226\ 02234098$$

Finally, the 18-bit authentication data (AUTHR) is computed by making use of decimation process from the 128-bit $P(\Omega)$ by picking 1 bit per every 7 bits:

$$AUTHR = 111100001010100100$$

8.11.3 Computation of AUTHR Using Concatenation, Permutation, and Substitutions (S-boxes)

Figure 8.15 illustrates another authentication data computation. The computation procedure for authentication data is described as shown below:

1. Enlarge the 152-bit message into 176 bits by adding 24-bit padding.
2. Divide this enlarged message of 176 bits into three blocks: $M_1 = 48$ bits, $M_2 = 64$ bits, and $M_3 = 64$ bits. Use M_1 as the 48-bit key.
3. Divide the 48-bit key into two halves: $K_l = 24$ bits and $K_r = 24$ bits.
4. Permute M_2 and M_3 to L_0 and R_0, respectively: $L_0 = 64$ bits and $R_0 = 64$ bits.
5. Concatenate K_l with L_0 to produce 88 bits: $L_0 \| K_l \rightarrow 88$ bits
6. Concatenate the result of $L_0 \| K_l$ with R_0 to produce $L_0 \| K_l \| R_0 \rightarrow 152$ bits
7. Concatenate the 152-bit result at step 6 with Kr such that

$L_0 \| K_l \| R_0 \| K_r \rightarrow 176$ bits

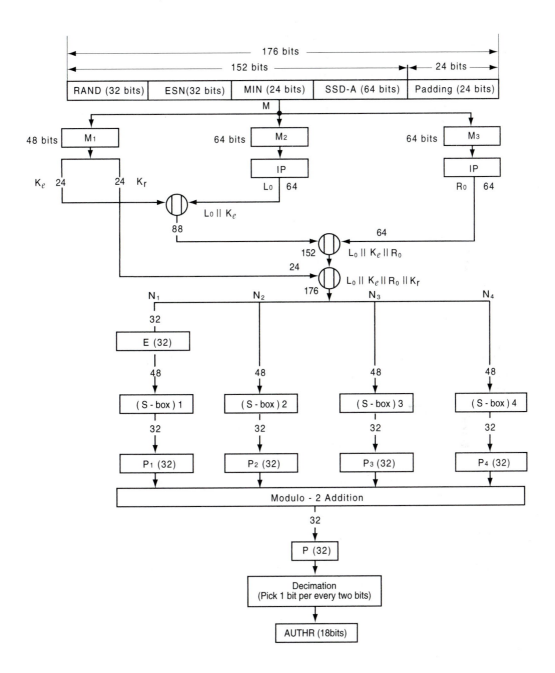

Figure 8.15 Computation of AUTHR for authentication of mobile station's registrations or terminations

8. Divide this 176-bit message into four parts:

N_1 = 32 bits, N_2 = 48 bits, N_3= 48 bits, N_4 = 48 bits

9. Expand N_1 = 32 bits into $E(N_1)$ = 48 bits using Table 8.7.

10. Thus, it can be accomplished into four evenly partitioned blocks of 48 bits each.

11. These four 48-bit blocks are subject to input to four S-boxes.

12. The sum of these four S-box outputs are modulo-2 added, and resulted in 32 bits.

13. Finally, these 32 bits are decimated into AUTHR (18 bits).

Example 8.6   Referring to Fig. 8.15, the 176-bit padded message M is divided into three blocks, i.e., M_1=48 bits, M_2=64 bits, and M_3= 64 bits where M_1 will be used as the key such that K_l = 16c27a(24 bits) and K_r = 415f39 (24 bits).

Assume that

M_2 = 17b439a12f51c5a8 (64 bits)

M_3 = 51cb36af43000000 (64 bits)

Using Table 8.6, the initial permutations of M_2 and M_3 become, respectively,

$IP(M_2)$ = 6027537dca9e9411 = L_o

$IP(M_3)$ = 13050c1b0a0c0a1e = R_o

Data concatenation: $L_o \parallel K_l \parallel R_o \parallel K_r$ (176 bits)

Divide the concatenation data into 4 blocks, i.e., one block of 32 bits and three blocks of 48 bits each, as follows:

6027537d   ca9e941116c2   7a13050c1b0a   0c0a1e415f39

| ←32 bits→ | ← 48 bits → | ← 48 bits → | ← 48 bits→ |

1. Using Table 8.7, the 32-bit block is expanded into a 48-bit block such that

$E(6027537d)$ = b0010eaa6bfa (48 bits)

This is the input data Γ_1 to (S-box) $_1$ from S_1 through S_8, as shown below.

Input to S_i	S_i		Output from S_i
Γ_1	Row	Column	Ω_1
S_1^{10}	2	6	2
S_2^{00}	0	0	15
S_3^{00}	0	2	9
S_4^{00}	0	7	10
S_5^{10}	2	5	13
S_6^{10}	2	3	5
S_7^{11}	3	7	7
S_8^{10}	2	13	3

Thus, the output data from (S-box)$_1$ is $\Omega_1 = $ 2f9ad573.

Using Table 8.9, the permutation function $P(\Omega_1)$ becomes

$$P(\Omega_1) = 6\,7\,5\,e\,6\,f\,5\,8$$

2. Input data to (S-box)$_2$ is $\Gamma_2 = $ ca9e941116c2. In binary notation,

$\Gamma_2 = $ 11001010100111101001010000010001000101101011000010

Partitioning Γ_2 into every 6 bits and feeding them into S_1 through S_8, the output data from (S-box)$_2$ is computed as shown below:

$\Gamma\,2$	S_i's Row and Column		$\Omega\,2$
S_1^{10}	2	9	12
S_2^{11}	3	4	3
S_3^{10}	2	13	10
S_4^{00}	0	10	8
S_5^{00}	0	2	4
S_6^{01}	1	8	6
S_7^{01}	1	13	15
S_8^{00}	0	1	2

Thus, the output Ω_2 of (S-box)$_2$ becomes

$$\Omega_2 = \text{c3a846f2}$$

Using Table 8.9, the permutation of Ω_2 yields

$$P(\Omega_2) = \text{42b6c54d}$$

3. Input data to (S-box)$_3$ is $\Gamma_3 = $ 7a13050c1b0a.

Feeding every 6 bits of Γ_3 into (S-box)$_3$, the output data from (S-box)$_3$ is found to be

$$\Omega_3 = 7\,d\,f\,b\,b\,a\,7\,f$$

The permutation of Ω_3 yields

$$P(\Omega_3) = \text{b}\,f\,7\,b\,c\,f\,f\,6$$

4. Input data to (S-box)$_4$ is $\Gamma_4 = 0\,c\,0\,a\,1\,e\,4\,1\,5\,f\,f\,3\,9$.

Express Γ_4 in binary notation first and break down Γ_4 into 8 blocks. Each block contains 6 bits, which becomes the input to S_i, $1 \leq i \leq 8$. Inputting each 6 bits to (S-box)$_4$ results in the output from (S-box)$_4$, as follows:

$$\Omega_4 = \text{ff858d93}$$

and

$$P(\Omega_4) = \text{d38afb1b}$$

Thus, all output data from (S-box)$_i$ $1 \leq i \leq 4$, becomes

$$675e6f58 \quad 42b6c54d \quad bf7bcff6 \quad d38afb1b$$

Modulo-2 addition of these four groups produces
$$49199ef8$$
and $P(49199ef8) = bf38c449.$

Using Table 8.7, this permuted data is expanded into 48 bits, as shown below:
$$dfe9f1608253$$

Converting these hexadecimal values to the binary values first and deleting the left 6 bits as well as the right 6 bits, the following 36-bit binary sequence remains:
$$111110100111110001011000001000001001$$

If the decimation rule is applied to this sequence, the 18-bit message digest (AUTHR) is obtained by picking 1 bit from every two bits as shown below:
$$AUTHR = (110011101100000001)$$

8.11.4 Computation of AUTHR Using DM Scheme

The DM scheme was proposed independently by Davies and Meyer in 1985 and its conceptual idea is somewhat similar to cipher block chaining (CBC) mode. In this section, AUTHR computation based on the DM scheme is described using the DES block cipher to construct a hash function. The message to be hashed is first divided into fixed length blocks: $M_1, M_2, \cdots, M_t$. The message block M_i, $1 \leq i \leq t$, is used as the encryption key, as illustrated in Fig. 8.16. The previous message block is encrypted using that key and then EX-ORed with itself. The result becomes the input to the next round. Thus, at the end, the length of the message becomes the last M_i. Let H_i ($1 \leq i \leq t$) be defined as $H_i = E_{M_i}(H_{i-1}) \oplus H_{i-1}$, $1 \leq i \leq t$.

Using the DES algorithm, the message M is first partitioned into 56-bit DES key blocks. The first block M_1 becomes the 56-bit DES key. The input to the DES algorithm is an initializing vector H_0. The ciphertext output of the DES algorithm is H_1 which is EX-Ored $E_{M_i}(H_0)$ with H_0 and becomes the input to the hash of the second block. The next 56-bit message block M_2 becomes the new DES key. H_1 is the plaintext input to the algorithm and H_2 becomes the ciphertext output. $H_2 = E_{M_i}(H_1) \oplus H_1$ is the input to the hash of the third block, and so on.

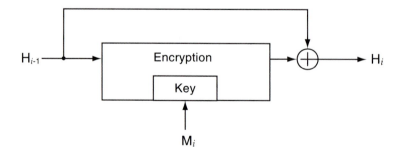

Figure 8.16 Davies-Meyer scheme

In the following, we attempt to design the authentication procedure to compute the 18-bit hash value, based on the DM scheme. The proposed authentication system is illustrated in Fig. 8.17.

The message to be hashed is first partitioned evenly into a series of m-bit blocks. Appropriate padding may be needed to force the message to divide evenly. The n-bit message with k-bit padding is mapped n-bit values onto $(n + k) =$ bit values so that $(n + k) =$ bit messages are divided into a series of m-bit blocks of equal length.

The proposed authentication procedure based on Fig. 8.17 is described below:

1. Transform the 152-bit message into a 192-bit padded message by appending 40-bit padding.
2. Divide the 192-bit padded message into three equal lengths of 64 bits. Each 64-bit message block will be used as the DES key.
3. Choose 64 bits randomly for an initializing vector IV (or H_0), which will act as the plaintext input to the DES algorithm.
4. Compute the ciphertext output H_1 by EX-ORing with H_0 and this becomes the input to the hash of the second block.
5. M_2, the next 64-bit message block, becomes the new DES key.
6. H_2 becomes the ciphertext output, $H_2 = E_{M_2}(H_1) \oplus H_1$, that is the input to the third block.
7. H_3 becomes $H_3 = E_{M_3}(H_2) \oplus H_2$ of the 64-bit value from which the 18-bit message digest (AUTHR data) can be computed.

Example 8.7   The 192-bit padded message is divided into three 64-bit blocks, i.e., M_1, M_2, M_3. These three blocks will be used as DES keys.

Letting

Initializing Vector (IV):

$$H_0(IV) = 67542301efcdab89$$

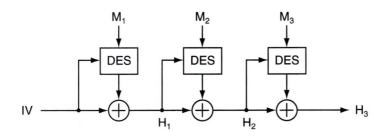

Figure 8.17 The authentication system based on DM scheme

DES Key Data:

$$M_1 = 7a138b2524af17c3$$
$$M_2 = 17b439a12f51c5a8$$
$$M_3 = 51cb360000000000$$

I. First DES operation (see Fig. 8.17)

1. Key schedule

Applying Table 8.10 (PC-1) to the key data M_1, PC-1 output becomes a481394 (left half) and e778253 (right half). These two halves are shifted simultaneously one or two places to the left according to Table 8.12 and then concatenated. The 48-bit key data from round 1 through round 16 is then computed, as shown below:

K_1=058c457c2c45	K2=d805187690e5
K_3=4368a26ecode	K_4=44822dc2aceb
K_5=b8a5a061f48b	K_6 =8398062ebf19
K_7 =d0060bae146b	K_8=282ae27f5472
K_9=619214cadb66	K_{10}=b074284dc82a
K_{11}=0498c614cff8	K_{12}=c00750c47c5c
K_{13}=326072dd9c51	K_{14}=44da15e993fc
K_{15}=ac4520cbe738	K_{16}=07b14291dfab

2. Encryption

H_0 (IV) is encrypted using the key M_1, as follows:
The 64-bit H_0 to be enciphered is

$$H_0 = 67452301efcdab89$$

Using Table 8.6, the initial permuted data is

$$IP = 330033fff055f055$$

where $L_0 = 330033ff$ and $R_0 = f055f055$.

Utilizing Table 8.7, the 48-bit expanded data $E(R_0)$ transformed from the initial permuted data R_0 is computed as

$$E(R_0) = fa02abfa02ab.$$

EX-ORing this expanded data with the key data K_1 yields:

$$\Gamma_1 = E(R_0) \oplus K_1 = (fa02abfa02ab) \oplus (058c457c2c45)$$
$$= ff8eee862eee$$

This data sequence is the input to (S-box)$_1$. From Table 8.8, the output from (S-box)$_1$ is

$$\Omega_1 = d95dbe22.$$

Using Table 8.9, the permutation function of $(S\text{-box})_1$ output produces

$$P\Omega_{(1)} = \text{b5abd4ca}.$$

Finally, the 32-bit R_1 can be obtained by the following formula:

$R_1 = P(\Omega_1) \oplus L_0$

$\quad = (\text{b5abd4ca}) \oplus (330033\text{ff})$

$\quad = 86\text{abc735}$

Thus far, we have completed the computation of first round encryption of H_0.

Computation up to the 16th round would surely involve lengthy calculation. But the computing process of each round is exactly identical. Therefore, the results of all 16 rounds can be summarized as shown below.

Consider the 16-round DES encryption using (H_0, M_1):

$$H_0 = 67452301\text{efcdab89}$$
$$M_1 = 7\text{a}138\text{b}2524\text{af}17\text{c}3$$

Round 1

$K_1 = 058\text{c}457\text{c}2\text{c}45$

$L_0 = 330033\text{ff}, \ R_0 = \text{f}055\text{f}055$

$E(R_0) = \text{fa}02\text{abfa}02\text{ab}$

$\Gamma_1 = E(R_0) \oplus K_1$

$\quad = \text{ff}8\text{eee}862\text{eee}$

$\Omega_1 = \text{d}95\text{dbe}22$

$P(\Omega_1) = \text{b5abd4ca}$

$R_1 = P(\Omega_1) \oplus L_0$

$\quad = 86\text{abe}735$

Round 2

$K_2 = \text{d}805187690\text{e}5$

$R_1 = 86\text{abe}735$

$E(R_1) = \text{c}0\text{d}557\text{f}0\text{e}9\text{ab}$

$\Gamma_2 = E(R_1) \oplus K_2$

$\quad = 18\text{d}04\text{f}86794\text{e}$

$\Omega_2 = 18\text{d}3\text{bcd}1$

$P(\Omega_2) \ \text{b}759098\text{b}$

$R_2 = P(\Omega_2) \oplus L_1 = P(\Omega_2) \oplus R_0$

$\quad = 470\text{cf}9\text{de}$

Round 3

$K_3 = 4368\text{a}26\text{ee}0\text{de}$

$R_2 = \text{f}470\text{cf}9\text{de}$

$E(R_2) = 20\text{e}8597\text{f}3\text{efc}$

$\Gamma_3 = E(R_2) \oplus K_3$

$\quad = 6380\text{fb}11\text{de}22$

$\Omega_3 = 5977430\text{b}$

$P(\Omega_3) = 8\text{c}6\text{ff}806$

$R_3 = P(\Omega_3) \oplus R_1$

$\quad = 0\text{ac}41\text{f}33$

Round 4

$K_4 = 44822\text{dc}2\text{aceb}$

$R_3 = 0\text{ac}41\text{f}33$

$E(R_3) = 8556080\text{fe}9\text{a}6$

$\Gamma_4 = E(R_3) \oplus K_4$

$\quad = \text{c}1\text{d}425\text{cd}454\text{d}$

$\Omega_4 = \text{fb}10\text{f}357$

$P(\Omega_4) \ 67\text{beeaa}2$

$R_4 = P(\Omega_4) \oplus R_2$

$\quad = 20\text{b}2137\text{c}$

Round 5

$$K_5 = \text{b8a5a061f48b}$$
$$R_4 = \text{20b2137c}$$
$$E(R_4) = \text{1015a4a6bf8}$$
$$\Gamma_5 = E(R_4) \oplus K_5$$
$$= \text{a8b0046b9f73}$$
$$\Omega_5 = \text{62ae063c}$$
$$P(\Omega_5) = \text{4a60976c}$$
$$R_5 = P(\Omega_5) \oplus R_3$$
$$= \text{40a4885f}$$

Round 6

$$K_6 = \text{8398062ebf19}$$
$$R_5 = \text{40a4885f}$$
$$E(R_5) = \text{a015094502fe}$$
$$\Gamma_6 = E(R_5) \oplus K_6$$
$$= \text{238d0f6bbde7}$$
$$\Omega_6 = \text{292300f7}$$
$$P(\Omega_6) = \text{825a4e25}$$
$$R_6 = P(\Omega_6) \oplus R_4$$
$$= \text{a2e85d59}$$

Round 7

$$K_7 = \text{d0060bae146b}$$
$$R_6 = \text{a2e85d59}$$
$$E(R_6) = \text{d057502faaf3}$$
$$\Gamma_7 = E(R_6) \oplus K_7$$
$$= \text{00515b81be98}$$
$$\Omega_7 = \text{e40a4b55}$$
$$P(\Omega_7) = \text{12f4aa70}$$
$$R_7 = P(\Omega_7) \oplus R_5$$
$$= \text{05250222f}$$

Round 8

$$K_8 = \text{282ae27f5472}$$
$$R_7 = \text{5250222f}$$
$$E(R_7) = \text{aa42a010415e}$$
$$\Gamma_8 = E(R_7) \oplus K_8$$
$$= \text{8268426f152c}$$
$$\Omega_8 = \text{4b1d9b9e}$$
$$P(\Omega_8) = \text{ff2af061}$$
$$R_8 = P\Omega_{(8)} \oplus R_6$$
$$= \text{5dc2ad38}$$

Round 9

$$K_9 = \text{619214cadb66}$$
$$R_8 = \text{5dc2ad38}$$
$$E(R_8) = \text{2fbe0555a9f0}$$
$$\Gamma_9 = E(R_8) \oplus K_9$$
$$= \text{4e2c119f7296}$$
$$\Omega_9 = \text{6eb4770e}$$
$$P(\Omega_9) = \text{6c2eb3bc}$$
$$R_9 = P(\Omega_9) \oplus R_7$$
$$= \text{3e7e9193}$$

Round 10

$$K_{10} = \text{b074284dc82a}$$
$$R_9 = \text{3e7e9193}$$
$$E(R_9) = \text{9fc3fd4a3ca6}$$
$$\Gamma_{10} = E(R_9) \oplus K_{10}$$
$$= \text{2fb7d507f48c}$$
$$\Omega_{10} = \text{12512edcb}$$
$$P(\Omega_{10}) = \text{1d566a99}$$
$$R_{10} = P(\Omega_{10}) \oplus R_8$$
$$= \text{4094c7a1}$$

Round 11

$$K_{11} = \quad 0498c614cff8$$
$$R_{10} = \quad 4094c7a1$$
$$E(R_{10}) = \quad a014a960fd02$$
$$\Gamma_{11} = \quad E(R_{10}) \oplus K_{11}$$
$$= \quad a48c6f7432fa$$
$$\Omega_{11} = \quad 46488f93$$
$$P(\Omega_{11}) = \quad 5323a859$$
$$R_{11} = \quad P(\Omega_{11}) \oplus R_9$$
$$= \quad 6d5d39ca$$

Round 12

$$K_{12} = \quad c00750c47c5c$$
$$R_{11} = \quad 6d5d39ca$$
$$E(R_{11}) = \quad 35aafa9f3e54$$
$$\Gamma_{12} = \quad E(R_{11}) \oplus K_{12}$$
$$= \quad f5adaa5b4208$$
$$\Omega_{12} = \quad 60cbf4f6$$
$$P(\Omega_{12}) = \quad a35787e9$$
$$R_{12} = \quad P(\Omega_{12}) \oplus R_{10}$$
$$= \quad e3c34048$$

Round 13

$$K_{13} = \quad 326072dd9c51$$
$$R_{12} = \quad e3c34048$$
$$E(R_{12}) = \quad 707e06a00251$$
$$\Gamma_{13} = \quad E(R_{12}) \oplus K_{13}$$
$$= \quad 421e747d9e00$$
$$\Omega_{13} = \quad 3db3600d$$
$$P(\Omega_{13}) = \quad 8c4c4bb6$$
$$R_{13} = \quad P(\Omega_{13}) \oplus R_{11}$$
$$= \quad e111727c$$

Round 14

$$K_{14} = \quad 44da15e993fc$$
$$R_{13} = \quad e111727c$$
$$E(R_{13}) = \quad 7028a2ba43f9$$
$$\Gamma_{14} = \quad E(R_{13}) \oplus K_{14}$$
$$= \quad 34f2b753d005$$
$$\Omega_{14} = \quad de3b384d$$
$$P(\Omega_{14}) = \quad fcd888f6$$
$$R_{14} = \quad P(\Omega_{14}) \oplus R_{12}$$
$$= \quad 1f1bc8be$$

Round 15

$$K_{15} = \quad ac4520cbe738$$
$$R_{14} = \quad 1f1bc8be$$
$$E(R_{14}) = \quad fe8f7e515fc$$
$$\Gamma_{15} = \quad E(R_{14}) \oplus K_{15}$$
$$= \quad a3add72ef2c4$$
$$\Omega_{15} = \quad d33c7a98$$
$$P(\Omega_{15}) = \quad 7ea4d0c7$$
$$R_{15} = \quad P(\Omega_{15}) \oplus R_{13}$$
$$= \quad 9fb5a2bb$$

Round 16

$$K_{16} = \quad 07b14291dfab$$
$$R_{15} = \quad 9fb5a2bb$$
$$E(R_{15}) = \quad cffdabd055f7$$
$$\Gamma_{16} = \quad E(R_{15}) \oplus K_{16}$$
$$= \quad c84ce9418a5c$$
$$\Omega_{16} = \quad c8fa8e1c$$
$$P(\Omega_{16}) = \quad 1fe9816c$$
$$R_{16} = \quad P(\Omega_{16}) \oplus R_{14}$$
$$= \quad 00f249d2$$

The pre-output block ($R_{16} \parallel R_{15}$) is the concatenation of R_{16} with R_{15} ($=L_{16}$). Using Table 8.13, the inverse permutation IP^{-1} applied to the pre-output block ($R_{16} \parallel R_{15}$) computes the output of the first DES algorithm, $E_{M_1}(H_0)$.

$$R_{16} \parallel R_{15} = 00f249d29fb5a2bb$$

Thus, the data from IP^{-1} is

$$IP^{-1} = a69ba086b33a15bb.$$

EX-ORing IP^{-1} data with H_0 becomes the input H_1 to the second DES algorithm such that

$$
\begin{aligned}
H_1 &= IP^{-1} \oplus H_0 \\
&= (a69ba086b33a15bb) \oplus (67452301efcdab89) \\
&= c1de83875cf7be32
\end{aligned}
$$

Table 8.13   Inverse of Initial Permutation, IP^{-1}

40	8	48	16	56	24	64	32
39	7	47	15	55	23	63	31
38	6	46	14	54	22	62	30
37	5	45	13	53	21	61	29
36	4	44	12	52	20	60	28
35	3	43	11	51	19	59	27
34	2	42	10	50	18	58	26
33	1	41	9	49	17	57	25

II. Second DES operation (see Fig. 8.17)

    Key M_2: 17b439a12f51c5a8

    Input H_1: c1de83875cf7be32

  1. Key schedule

Applying Table 8.10 (PC-1) to the key data M_2, PC-1 output becomes ca609e2 (left half) and 1153947 (right half). These two halves are shifted according to Table 8.12 to generate the 48-bit key data from round 1 through round 16, as shown below:

$K_1 = 0975a4d9e052$	$K_2 = 07a44cc18bad$
$K_3 = d005dd21e708$	$K_4 = 4a4cb2161f99$
$K_5 = 55c221b83446$	$K_6 = fca9285b1035$
$K_7 = 8399a6a8c3a6$	$K_8 = 82a60b0768a8$
$K_9 = b822c7106fc3$	$K_{10} = 691e16643815$
$K_{11} = 315628ba8051$	$K_{12} = 64b8c8e700be$
$K_{13} = c011f483e742$	$K_{14} = 12e472411b8b$
$K_{15} = 94ca553ca700$	$K_{16} = ec4d1216107d$

2. Encryption

The input H_1 is encrypted using the key M_2, as shown below:

Using Table 8.6, the initial permuted data is

$$IP = 33f27a2d6fc052ee$$

where $L_0 = 33f27a2d$ and $R_0 = 6fe052ee$

Utilizing Table 8.7, the 48-bit $E(R_0)$ from the 32-bit R_0 is computed as

$$E(R_0) = 35ff002a575c$$

EX-ORing $E(R_0)$ with the key data K_1 is computed as

$$\Gamma_1 = E(R_0) \oplus K_1 = (35ff002a575c) \oplus (0975a4d9e052)$$
$$= 3c8aa4f3b70e$$

This data sequence is the input to $(S\text{-box})_1$. Using Table 8.8, the output Ω_1 from $(S\text{-box})_1$ is computed as

$$\Omega_1 = 16f90061$$

From Table 8.9, the permuted data of Ω_1 becomes

$$P(\Omega_1) = c4110d56$$

Thus, the 32-bit R_1 is obtained by

$$R_1 = P(\Omega_1) \oplus L_0$$
$$= (c4110d56) \oplus (33f27a2d)$$
$$= f7e3777b$$

This is the result of first round encryption of H_1.

The 16-round encryption process using (H_1, M_2) is summarized in the following:

Round 1		**Round 2**	
$K_1 =$	0975a4d9e052	$K_2 =$	07a44cc18bad
$L_0 =$	$33f27a2d, R_0 = 6fe052ee$	$R_1 =$	f7e3777b
$E(R_0) =$	35ff002a575c	$E(R_1) =$	faff06baebf7
$\Gamma_1 =$	$E(R_0) \oplus K_1$	$\Gamma_2 =$	$E(R_1) \oplus K_2$
$=$	3c8aa4f3b70e	$=$	fd5b4a7b605a
$\Omega_1 =$	16f90061	$\Omega_2 =$	d1869ad0
$P(\Omega_1) =$	c4110d56	$P(\Omega_2) =$	33f0d103
$R_1 =$	$P(\Omega_1) \oplus L_0$	$R_2 =$	$P(\Omega_2) \oplus R_0$
$=$	$(c4110d56) \oplus (33f27a2d)$	$=$	5c1083ed
$=$	f7e3777b		

Round 3

$$K_3 = \text{d005dd21e708}$$
$$R_2 = \text{75c1083cd}$$
$$E(R_2) = \text{af80a1407f5a}$$
$$\Gamma_3 = E(R_2) \oplus K_3$$
$$= \text{7f857c619852}$$
$$\Omega_3 = \text{8958d069}$$
$$P(\Omega_3) = \text{2d9d4c40}$$
$$R_3 = P(\Omega_3) \oplus R_1$$
$$= \text{da7e3b3b}$$

Round 4

$$K_4 = \text{4a4cb2161f99}$$
$$R_3 = \text{da7e3b3b}$$
$$E(R_3) = \text{ef43fc1f69f7}$$
$$\Gamma_4 = E(R_3) \oplus K_4$$
$$= \text{a50f4e09766e}$$
$$\Omega_4 = \text{d492ace22}$$
$$P(\Omega_4) = \text{116ec44c}$$
$$R_4 = P(\Omega_4) \oplus R_2$$
$$= \text{4d7e47a1}$$

Round 5

$$K_5 = \text{55c221b83446}$$
$$R_4 = \text{4d7e47a1}$$
$$E(R_4) = \text{a5abfc20fd02}$$
$$\Gamma_5 = E(R_4) \oplus K_5$$
$$= \text{f069dd98c944}$$
$$\Omega_5 = \text{5e0eb6d8}$$
$$P(\Omega_5) = \text{6b7890db}$$
$$R_5 = P(\Omega_5) \oplus R_3$$
$$= \text{b106abe0}$$

Round 6

$$K_6 = \text{fca9285b1035}$$
$$R_5 = \text{b106abe0}$$
$$E(R_5) = \text{5a280d557f01}$$
$$\Gamma_6 = E(R_5) \oplus K_6$$
$$= \text{a681250e6f34}$$
$$\Omega_6 = \text{4a90b59a}$$
$$P(\Omega_6) = \text{6f0aa189}$$
$$R_6 = P(\Omega_6) \oplus R_4$$
$$= \text{2274e628}$$

Round 7

$$K_7 = \text{8399a6a8c3a6}$$
$$R_6 = \text{2274e628}$$
$$E(R_6) = \text{1043a970c150}$$
$$\Gamma_7 = E(R_6) \oplus K_7$$
$$= \text{93da0fd802f6}$$
$$\Omega_7 = \text{e4835c9d}$$
$$P(\Omega_7) = \text{bac48b39}$$
$$R_7 = P(\Omega_7) \oplus R_5$$
$$= \text{bc220d9}$$

Round 8

$$K_8 = \text{82a60b0768a8}$$
$$R_7 = \text{bc220d9}$$
$$E(R_7) = \text{857e041016f2}$$
$$\Gamma_8 = E(R_7) \oplus K_8$$
$$= \text{07d80f177e5a}$$
$$\Omega_8 = \text{04d327e0}$$
$$P(\Omega_8) = \text{84712599}$$
$$R_8 = P(\Omega_8) \oplus R_6$$
$$= \text{a605c3b1}$$

Round 9

$$K_9 = \text{b822c7106fc3}$$
$$R_8 = \text{a605c3b1}$$
$$E(R_8) = \text{d0c00be07da3}$$
$$\Gamma_9 = E(R_8) \oplus K_9$$
$$= \text{68e2eef01260}$$
$$\Omega_9 = \text{94490a47}$$
$$P(\Omega_9) = \text{90b30872}$$
$$R_9 = P(\Omega_9) \oplus R_7$$
$$= \text{9b7128ab}$$

Round 10

$$K_{10} = \text{691e16643815}$$
$$R_9 = \text{9b7128ab}$$
$$E(R_9) = \text{cf6ba2951557}$$
$$\Gamma_{10} = E(R_9) \oplus K_{10}$$
$$= \text{a675b4f12d42}$$
$$\Omega_{10} = \text{41730d02}$$
$$P(\Omega_{10}) = \text{9443e00c}$$
$$R_{10} = P(\Omega_{10}) \oplus R_8$$
$$= \text{324623bd}$$

Round 11

$$K_{11} = \text{315628ba8051}$$
$$R_{10} = \text{324623bd}$$
$$E(R_{10}) = \text{9a420c107dfa}$$
$$\Gamma_{11} = E(R_{10}) \oplus K_{11}$$
$$= \text{ab1424afdab}$$
$$\Omega_{11} = \text{6b19da8a}$$
$$P(\Omega_{11}) = \text{fd2ec241}$$
$$R_{11} = P(\Omega_{11}) \oplus R_9$$
$$= \text{665feaea}$$

Round 12

$$K_{12} = \text{64b8c8e700be}$$
$$R_{11} = \text{665feaea}$$
$$E(R_{11}) = \text{30c2fff55754}$$
$$\Gamma_{12} = E(R_{11}) \oplus K_{12}$$
$$= \text{547a371257ea}$$
$$\Omega_{12} = \text{c78b426c}$$
$$P(\Omega_{12}) = \text{c8f4c570}$$
$$R_{12} = P(\Omega_{12}) \oplus R_{10}$$
$$= \text{fab2e6cd}$$

Round 13

$$K_{13} = \text{c011f483e742}$$
$$R_{12} = \text{fab2e6cd}$$
$$E(R_{12}) = \text{30c2fff55754}$$
$$\Gamma_{13} = E(R_{12}) \oplus K_{13}$$
$$= \text{3f4451f33119}$$
$$\Omega_{13} = \text{1c240e20}$$
$$P(\Omega_{13}) = \text{1028141e}$$
$$R_{13} = P(\Omega_{13}) \oplus R_{11}$$
$$= \text{7677fef4}$$

Round 14

$$K_{14} = \text{12e472411b8b}$$
$$R_{13} = \text{7677fef4}$$
$$E(R_{13}) = \text{3ac3afffd7a8}$$
$$\Gamma_{14} = E(R_{13}) \oplus K_{14}$$
$$= \text{2827ddbecc23}$$
$$\Omega_{14} = \text{f11edca1}$$
$$P(\Omega_{14}) = \text{35c4de4b}$$
$$R_{14} = P(\Omega_{14}) \oplus R_{12}$$
$$= \text{cf763886}$$

Round 15

$K_{15} =$ 94ca553ca700

$R_{14} =$ cf763886

$E(R_{14}) =$ 65ebac1f140d

$\Gamma_{15} =$ $E(R_{14}) \oplus K_{15}$

$=$ f121f923b30d

$\Omega_{15} =$ 579c7087

$P(\Omega_{15}) =$ 6406d9f3

$R_{15} =$ $P(\Omega_{15}) \oplus R_{13}$

$=$ 12712707

Round 16

$K_{16} =$ ec4d1216107d

$R_{15} =$ 12712707

$E(R_{15}) =$ 8a43a290e80e

$\Gamma_{16} =$ $E(R_{15}) \oplus K_{16}$

$=$ 660eb086f873

$\Omega_{16} =$ 90afba6c

$P(\Omega_{16}) =$ b9f015e6

$R_{16} =$ $P(\Omega_{16}) \oplus R_{14}$

$=$ 76862d60

The pre-output block is the concatenation of R_{16} with R_{15} ($= L_{16}$), i.e., $R_{16} \parallel R_{15}$. Using Table 8.13, the inverse of initial permutation IP^{-1} applicable to the pre-output block ($R_{16} \parallel R_{15}$) computes the output of the second DES algorithm, $E_{M_2}(H_1)$.

$$R_{16} \parallel R_{15} = 76862d6012712707$$

The data resulting from IP^{-1} is

$$IP^{-1} = 2eda5e04e06d6110$$

EX-Oring IP^{-1} data with H_1 becomes the input H_2 to the third DES algorithm such that

$H_2 = IP^{-1} \oplus H_1$

$= (2eda5e04e06d6110) \oplus (c1de83875cf7be32)$

$= ef04dd83bc9adf22$

III. Third DES operation (see Fig. 8.17)

Key M_3 : 51cb360000000000

Input H_2 : ef04dd83bc9adf22

1. Key schedule

Applying Table 8.10 PC-1 to the data M_3, the PC-1 output becomes 203040(left half) and 604025(right half). These two halves are shifted according to Table 8.12, generate the 48-bit key data from round 1 through round 16, as shown below:

$K_1 = 80e000209341$

$K_3 = 000702128422$

$K_5 = 601001080c00$

$K_7 = 0180400c6050$

$K_9 = 00409261c000$

$K_{11} = 34010080040a$

$K_{13} = 020001881304$

$K_{15} = 0900141043e0$

$K_2 = 8400066a0008$

$K_4 = 22020000500e$

$K_6 = 0810200410a0$

$K_8 = 800058c00861$

$K_{10} = 044200028b18$

$K_{12} = 021900111410$

$K_{14} = 0820410d0020$

$K_{16} = 014408406804$

2. Encryption

The input H_2 is encrypted using the key M_3, as shown below:

Using Table 8.6, the initial permuted data is

$$IP = 4574574d7d9175e9$$

Utilizing Table 8.7, the 48-bit $E(R_0)$ from the 32-bit R_0 is found as follows:

$$E(R_0) = bfbca2babf52$$

EX-ORing $E(R_0)$ with the key K_1 computes

$$\Gamma_1 = E(R_0) \oplus K_1 = (bfbca2babf52) \oplus (80e000209341)$$
$$= 3f5ca29a2c13$$

This data sequence Γ_1 is the input to $(S\text{-box})_1$. Using Table 8.8, the output Ω_1 from $(S\text{-box})_1$ is computed as

$$\Omega_1 = 1716ea5$$

Using Table 8.9, the permuted data of Ω_1 becomes

$$P(\Omega_1) = 75605cbb$$

Thus, the 32-bit R_1 is computed as

$$R_1 = P(\Omega_1) \oplus L_0$$
$$= (75605cbb) \oplus (4574574d)$$
$$= 30140bf6$$

This is the result of first round encryption of H_2. The 16-round encryption process using (H_2, M_3) is summarized as follows:

Beginning with the 16-round DES encryption, the following data is used:

$$H_2 = ef04dd83bc9adf22$$
$$M_3 = 51cb36000000000$$

Round 1		**Round 2**	
$K_1 =$	80e000209341	$K_2 =$	8400066a0008
$L_0=$	4574574d, $R_0 = 7d9175e9$	$R_1 =$	30140bf6
$E(R_0) =$	bfbca2babf52	$E(R_1) =$	1a00a8057fac
$\Gamma_1 =$	$E(R_0) \oplus K_1$	$\Gamma_2 =$	$E(R_1) \oplus K_2$
$=$	3f5ca29a2c13	$=$	9e00ae6f7fa4
$\Omega_1 =$	1716bea5	$\Omega_2 =$	200d9724
$P(\Omega_1) =$	75605cbb	$P(\Omega_2) =$	a1203668
$R_1 =$	$P(\Omega_1) \oplus L_0$	$R_2 =$	$P(\Omega_2) \oplus R_0$
$=$	30140bf6	$=$	dcb14381

Round 3

$$K_3 = \quad 000702128422$$
$$R_2 = \quad dcb14381$$
$$E(R_2) = \quad ef95a2a07c03$$
$$\Gamma_3 = \quad E(R_2) \oplus K_3$$
$$= \quad ef92a0b2f821$$
$$\Omega_3 = \quad 003a7a12$$
$$P(\Omega_3) = \quad 366600c4$$
$$R_3 = \quad P(\Omega_3) \oplus R_1$$
$$= \quad 06720b32$$

Round 4

$$K_4 = \quad 22020000500e$$
$$R_3 = \quad 06720b32$$
$$E(R_3) = \quad 00c3a40569a4$$
$$\Gamma_4 = \quad E(R_3) \oplus K_4$$
$$= \quad 22c1a40539aa$$
$$\Omega_4 = \quad 2de9e1dc$$
$$P(\Omega_4) = \quad 8b1d63f5$$
$$R_4 = \quad P(\Omega_4) \oplus R_2$$
$$= \quad 57ac2074$$

Round 5

$$K_5 = \quad 601001080c00$$
$$R_4 = \quad 57ac2074$$
$$E(R_4) = \quad 2afd581003a8$$
$$\Gamma_5 = \quad E(R_4) \oplus K_5$$
$$= \quad 4aed59180fa8$$
$$\Omega_5 = \quad a1e11c29$$
$$P(\Omega_5) = \quad b8814f0c$$
$$R_5 = \quad P(\Omega_5) \oplus R_3$$
$$= \quad bef3443e$$

Round 6

$$K_6 = \quad 0810200410a0$$
$$R_5 = \quad bef3443e$$
$$E(R_5) = \quad 5fd7a6a081fd$$
$$\Gamma_6 = \quad E(R_5) \oplus K_6$$
$$= \quad 57c786a4915d$$
$$\Omega_6 = \quad c28317b9$$
$$P(\Omega_6) = \quad eae0ad09$$
$$R_6 = \quad P(\Omega_6) \oplus R_4$$
$$= \quad bd4c8d7d$$

Round 7

$$K_7 = \quad 0180400c6050$$
$$R_6 = \quad bd4c8d7d$$
$$E(R_6) = \quad dfaa5945abfb$$
$$\Gamma_7 = \quad E(R_6) \oplus K_7$$
$$= \quad de2a1949cbab$$
$$\Omega_7 = \quad ee8155ea$$
$$P(\Omega_7) = \quad e89ea719$$
$$R_7 = \quad P(\Omega_7) \oplus R_5$$
$$= \quad 566de327$$

Round 8

$$K_8 = \quad 800058c00861$$
$$R_7 = \quad 566de327$$
$$E(R_7) = \quad aac35bf0690e$$
$$\Gamma_8 = \quad E(R_7) \oplus K_8$$
$$= \quad 2ac30330616f$$
$$\Omega_8 = \quad fdf8bfbd$$
$$P(\Omega_8) = \quad 3fa9efff$$
$$R_8 = \quad P(\Omega_8) \oplus R_6$$
$$= \quad 82e56282$$

Round 9

$$K_9 = \quad 00409261c000$$
$$R_8 = \quad 82e56282$$
$$E(R_8) = \quad 40750ab05405$$
$$\Gamma_9 = \quad E(R_8) \oplus K_9$$
$$= \quad 401798d19405$$
$$\Omega_9 = \quad 338bc03d$$
$$P(\Omega_9) = \quad cb444f62$$
$$R_9 = \quad P(\Omega_9) \oplus R_7$$
$$= \quad 9d29ac45$$

Round 10

$$K_{10} = \quad 044200028b18$$
$$R_9 = \quad 9d29ac45$$
$$E(R_9) = \quad cfa953d5820b$$
$$\Gamma_{10} = \quad E(R_9) \oplus K_{10}$$
$$= \quad cbeb53d70913$$
$$\Omega_{10} = \quad cf8707b5$$
$$P(\Omega_{10}) = \quad c2e8fd39$$
$$R_{10} = \quad P(\Omega_{10}) \oplus R_8$$
$$= \quad 400d9fbb$$

Round 11

$$K_{11} = \quad 34010080040a$$
$$R_{10} = \quad 400d9fbb$$
$$E(R_{10}) = \quad a0005bcffdf6$$
$$\Gamma_{11} = \quad E(R_{10}) \oplus K_{11}$$
$$= \quad 94015b4ff9fc$$
$$\Omega_{11} = \quad 8f0a0d85$$
$$P(\Omega_{11}) = \quad 50c86879$$
$$R_{11} = \quad P(\Omega_{11}) \oplus R_9$$
$$= \quad cde1c43c$$

Round 12

$$K_{12} = \quad 021900111410$$
$$R_{11} = \quad cde1c43c$$
$$E(R_{11}) = \quad 65bf03e081f9$$
$$\Gamma_{12} = \quad E(R_{11}) \oplus K_{12}$$
$$= \quad 67a603f195e9$$
$$\Omega_{12} = \quad 93b800c4$$
$$P(\Omega_{12}) = \quad 44904167$$
$$R_{12} = \quad P(\Omega_{12}) \oplus R_{10}$$
$$= \quad 049ddedc$$

Round 13

$$K_{13} = \quad 020001881304$$
$$R_{12} = \quad 049ddedc$$
$$E(R_{12}) = \quad 0094fbefd6f8$$
$$\Gamma_{13} = \quad E(R_{12}) \oplus K_{13}$$
$$= \quad 0294fa67c5fc$$
$$\Omega_{13} = \quad e3823bc5$$
$$P(\Omega_{13}) = \quad 70f0eba1$$
$$R_{13} = \quad P(\Omega_{13}) \oplus R_{11}$$
$$= \quad bd112f9d$$

Round 14

$$K_{14} = \quad 0820410d0020$$
$$R_{13} = \quad bd112f9d$$
$$E(R_{13}) = \quad dfa8a295fcfb$$
$$\Gamma_{14} = \quad E(R_{13}) \oplus K_{14}$$
$$= \quad d788e398fcdb$$
$$\Omega_{14} = \quad 39afb55e$$
$$P(\Omega_{14}) = \quad ab5a73ee$$
$$R_{14} = \quad P(\Omega_{14}) \oplus R_{12}$$
$$= \quad afc7ad32$$

Round 15	Round 16

$$K_{15} = \text{0900141043e0} \qquad\qquad K_{16} = \text{014408406804}$$
$$R_{14} = \text{afc7ad32} \qquad\qquad R_{15} = \text{e3fc056f}$$
$$E(R_{14}) = \text{55fe0fd5a9a5} \qquad\qquad E(R_{15}) = \text{f07ff800ab5f}$$
$$\Gamma_{15} = E(R_{14}) \oplus K_{15} \qquad\qquad \Gamma_{16} = E(R_{15}) \oplus K_{16}$$
$$= \text{5cfe1bc5ea45} \qquad\qquad = \text{f13bf040c35b}$$
$$\Omega_{15} = \text{be5a6b1d} \qquad\qquad \Omega_{16} = \text{507f861e}$$
$$P(\Omega_{15}) = \text{5eed2af2} \qquad\qquad P(\Omega_{16}) = \text{8f63906e}$$
$$R_{15} = P(\Omega_{15}) \oplus R_{13} \qquad\qquad R_{16} = P(\Omega_{16}) \oplus R_{14}$$
$$= \text{e3fc056f} \qquad\qquad = \text{20a43d5c}$$

Thus, the pre-output block is
$$R_{16} \parallel R_{15} = \text{20a43d5ce3fc056f}$$
Using Table 8.13, the data from IP^{-1} yields
$$IP^{-1} = \text{8e823f2725f6a3b0}$$
EX-ORing this IP^{-1} data with H_2 becomes the third DES output as follows:
$$H_3 = IP^{-1} \oplus H_2$$
$$= (\text{8e823f2725f6a3b0}) \oplus (\text{ef04dd83bc9adf22})$$
$$= \text{6186f2a4996c7e92}$$

Transforming this hexadecimal number into the binary digit first and then chopping off 5 bits each from both ends of the binary sequence, we shall have the remainder as follows:
$$00110000110111100101010010010011001011011000111111110100$$
Decimating this binary sequence in its size by picking 1 bit every three bits generates the required 18-bit authentication data:
$$AUTHR = 101111000000001100$$

8.11.5 Another Technique for 18-bit AUTHR Computation

Figure 8.18 illustrates another method to be considered for computation of 18-bit authentication data. There are two cases taken into consideration: (1) the 176-bit padded message, which consists of the 152-bit message plus the 24-bit padding; (2) the 192-bit padded message appended by the 40-bit padding.

AUTHR computation is demonstrated by presenting the following examples.

Example 8.8   Let the input parameters be set as follows:
M_1 (48 bits, key input) = 16c27a415f39
M_2 (64 bits) = 17b439a12f51c5a8
M_3 (64 bits) = 51cb36af4300000

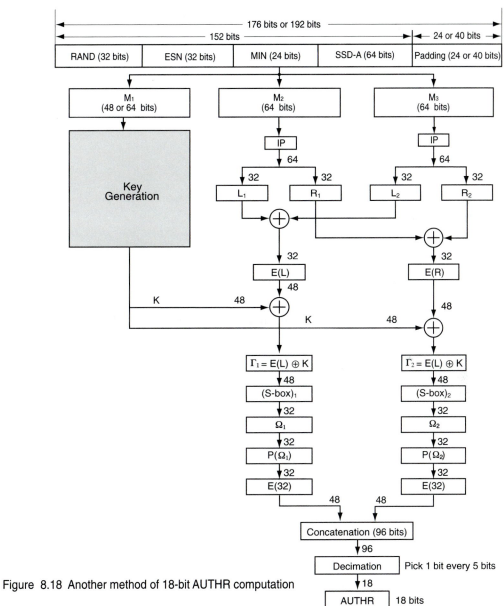

Figure 8.18 Another method of 18-bit AUTHR computation

As shown in Section 8.11.1 Key Generation Technique (I), the 48-bit key computed by means of column/row permutation was found to be

$$K = 01959d7857e4$$

Using Table 8.6, the initial permuted data of M_2 is computed as

$$(IP)_1 = 6027537dca9e9411$$

where $L_1 = 6027537d$ and $R_1 = ca9e9411$.

While the initial permuted data of M_3 becomes

$$(IP)_2 = 13050c1b0a0c0a1e$$

where $L_2 = 13050c1b$ and $R_2 = 0a0c0a1e$

EX-ORing L_1 with L_2 yields

$$L = L_1 \oplus L_2 = 73225f66$$

EX-ORing R_1 with R_2 yields

$$R = R_1 \oplus R_2 = c0929e0f$$

Using Table 8.7, the bit expansion of the 32-bit L is computed into 48 bits as

$$E(L) = 3a69042feb0c$$

Modulo-2 addition of E(L) with the 48-bit K is the input to (S-box)$_1$ as follows:

$$\Gamma_1 = E(L) \oplus K$$
$$= (3a69042feb0c) \oplus (01959d7857e4)$$
$$= 3bfc9957bce8$$

Using Table 8.8, the 32-bit output Ω_1 of the (S-box)$_1$ is computed as

$$\Omega_1 = 8911f059$$

Permutation of Ω_1 according to Table 8.9 produces

$$P(\Omega_1) = af9c4880$$

If the 32-bit $P(\Omega_1)$ is expanded into 48-bit data according to Table 8.7, we have

$$E[P(\Omega_1)] = 55fcf8251401$$

Next, using Table 8.7, the bit expansion of R into the 48-bit data is computed as

$$E(R) = e014a54fc05f$$

EX-ORing E(R) with the 48-bit key K produces the input Γ_2 to (S-box)$_2$:

$$\Gamma_2 = E(R) \oplus K$$
$$= (e014a54fc05f) \oplus (01959d7857e4)$$
$$= e181383797bb$$

Using Table 8.8, the 32-bit output Ω_2 of (S-box)$_2$ is computed as

$$\Omega_2 = 3c95d615$$

Following Table 8.9, permutation of Ω_2 yields

$$P(\Omega_2) = a72c1b3a$$

Using Table 8.7, the 32-bit $P(\Omega_2)$ is expanded into 48-bit data as shown below:

$$E[P(\Omega_2)] = 50e9580f69f5$$

Finally, concatenation of $E[P(\Omega_1)]$ and $E[P(\Omega_2)]$ is

$$55fcf825140150e9580f69f5 \text{ (96 bits)}$$

Transforming this hexadecimal number into the binary digits and then chopping off 12 bits from each end of the binary sequence, results in the following:

1100 1111 1000 0010 0101 0001 0100 0000 0001

0101 0000 1110 1001 0101 1000 0000 1111 0110

Decimating this binary sequence by picking 1 bit every four bits generates the 18-bit authentication data as shown below:

AUTHR = 010011001100110010

Example 8.9   Consider the case of the 192-bit padded message appended by the 40-bit padding. The input parameters are then set as follows:

M_1 (key input, 64 bits) = 7a138b2524af17c3

M_2 (input data, 64 bits) = 17b439a12f51c5a8

M_3 (input data, 64 bits) = 51cb360000000000

Utilizing Section 8.11.2 Key Generation Technique (II) (see Fig. 8.13), the 48-bit key data was computed to be

K = 058c4517a7a2

which is the result of PC–2.

The initial permuted sequence of M_2 resulting from Table 8.6 is:

$(IP)_1$ = 6027537dca9e9411

where L_1 = 6027537d and R_1 = ca9e9411

Likewise, the initial permuted sequence of M_3 is

$(IP)_2$ = 0305040302040206

where L_2 = 03050403 and R_2 = 02040206

EX-ORing L_1 with L_2 and R_1 with R_2, we have, respectively

$L = L_1 \oplus L_2$ = 6322577e

and $R = R_1 \oplus R_2$ = c89a9617

Using Table 8.7, the expansion of L and R to 48 bits each, is achieved as follows:

E(L) = 3069042aebfc

and   E(R) = e514f54ac0af

Modulo-2 addition of E(L) and the 48-bit key K is computed as

$\Gamma_1 = E(L) \oplus K$

= 35e5413d4c5e

This is the input data to $(S\text{-box})_1$.

Utilizing Table 8.8, the 32-bit output Ω_1 of $(S\text{-box})_1$ is obtained as

Ω_1 = da5d1397

Permutation of Ω_1 following Table 8.9 generates

$P(\Omega_1)$ = e6abb863

This 32-bit $P(\Omega_1)$ is expanded into the 48-bit data as shown below:

$E[P(\Omega_1)]$ = f0d557df0307

On the other hand, EX-ORing E(R) with the 48-bit key K obtains the input to (S-box)$_2$:

$\Gamma_2 = E(R) \oplus K$

$= \text{e098b05d670d}$

Using Table 8.8, the 32-bit output Ω_2 from (S-box)$_2$ is

$\Omega_2 = \text{3f6fa467}$

Permutation of Ω_2 by using Table 8.9 yields

$P(\Omega_2) = \text{c15b5efe}$

If the 32-bit $P(\Omega_2)$ is expanded according to Table 8.7, the 48-bit data sequence can be computed as

$E[P(\Omega_2)] = \text{602af6afd7fd}$

Finally, concatenation of $E[P(\Omega_1)]$ and $E[P(\Omega_2)]$ produces

$E[P(\Omega_1)] \parallel E[P(\Omega_2)] = \text{f0d557df0307602af6afd7fd}$

Transforming this hexadecimal number into the binary sequence and then chopping off 12 bits from each end of the binary data, the remaining sequence is

0101 0101 0111 1101 1111 0000 0011 0000 0111

0110 0000 0010 1010 1111 0110 1010 1111 1101

Decimating this binary sequence by picking 1 bit every four bits gives the 18-bit authentication data, as shown below:

AUTHR = 111110101000010011

Example 8.10   Refer to Fig. 8.18. If the expansion steps of $E[P(\Omega_1)]$ and $E[P(\Omega_2)]$ are omitted, $P(\Omega_1)$ is directly concatenated by $P(\Omega_2)$ like $P(\Omega_1) \parallel P(\Omega_2)$.

Case 1. For the 176-bit padded message:

$P(\Omega_1) \parallel P(\Omega_2) = \text{af9c4880a72c1b3a}$

After transforming this hexadecimal data into the binary sequence and then chopping off 5 bits from each end, we have

*111 1001 1100 0100 1000 1000 0000

1010 0111 0010 1100 0001 1011 001*

Decimating this binary sequence by choosing 1 bit every three bits, results in the 18-bit authentication data, as shown below:

AUTHR = 101000000110000111

Case 2. For the 192-bit padded message:

$P(\Omega_1) \parallel P(\Omega_2) = \text{e6abb863c15b5efe}$

From the transformed binary sequence, chop off 5 bits each from the left and right ends of the binary sequence. Then it produces

*110 1010 1011 1011 1000 0110 0011

1100 0001 0101 1011 0101 1110 111*

Decimating this chopped binary sequence by picking 1 bit every three bits, gives the 18-bit authentication data as

$$AUTHR = 010110101000111101$$

8.11.6 Modified CBC Scheme

As seen so far, there are many ways to consider the computation of authentication data. Modified CBC mode is also one of them. Block diagram for this scheme is illustrated in Fig. 8.19. The conceptual idea of this modified CBC scheme is somewhat similar to the DM scheme discussed in Section 8.11.4.

The 192-bit padded message to be hashed is divided into three blocks, M_1, M_2, and M_3, where M_i, $1 \leq i \leq 3$, is 64 bits long each.

Set H_0 as an initializing vector (IV). Then the encryption steps are shown as follows:

$$H_1 = E_{H_0 \oplus M_1}(H_0) \oplus H_0$$
$$H_2 = E_{H_1 \oplus M_2}(H_1) \oplus H_1$$
$$H_3 = E_{H_2 \oplus M_3}(H_2) \oplus H_2$$

where $H_i \oplus M_j$ denotes the enciphering key.

Since H_3 consists of 64 bits, chopping off 5 bits each from the left and right of H_3 results in 54 bits as the reduced sequence of H_3.

Decimating the chopped sequence by picking 1 bit every three bits forms the 18-bit authentication data as expected.

Computation for the 18-bit AUTHR is exactly similar to that of the DM scheme. Therefore, it is left as an exercise for the reader.

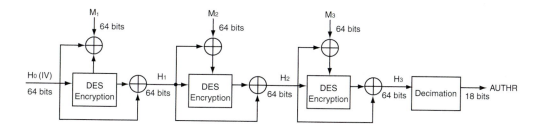

Figure 8.19 Modified CBC scheme for AUTHR computation

8.12 SSD Generation

SSD is a 128-bit shared secret data to be stored in semi-permanent memory in the mobile station. As shown in Fig. 8.4, SSD is divided into two distinct subsets: SSD-A and SSD-B. SSD-A is used to support the authentication procedure; and SSD-B is used for supporting voice privacy and message confidentiality.

SSD is updated using the SSD generation procedure as illustrated in Fig. 8.20. The A-key assigned to the mobile station is 64 bits long and is known only to the mobile station and to its associated Home Location Register/Authentication Center (HLR/AC). SSD updates are carried out only in the mobile station and its associated HLR/AC, not in the serving base station.

The SSD update procedure is performed as follows:

• The base station sends an SSD update order on either the paging channel or the forward traffic channel, with the RANDSSD field set to the same 56-bit random number used in the HLR/AC computations, to the mobile station via the SSD update message.

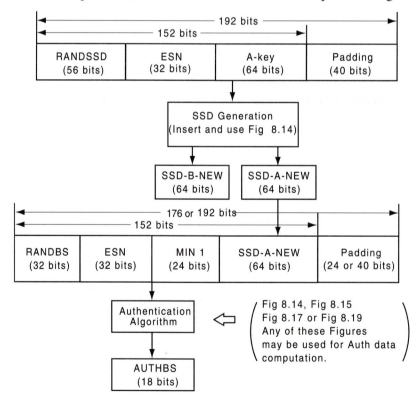

Figure 8.20 SSD generation and AUTHBS computation

- Upon receipt of the SSD update message, the mobile station sets the input parameters and executes the SSD-generation procedure. The mobile station computes SSD-A-NEW equal to the 64 most significant bits of the SSD-generated output and SSD-B-NEW to the 64 least significant bits of the SSD-generated output.

Example 8.11   Upon receipt of the SSD update message, the mobile station sets the following parameters (see Fig. 8.20):

RANDSSD (56 bits), ESN (32 bits), A-key (64 bits), and Padding (40 bits)

The 152-bit message length is expanded into the 192-bit padded message by appending the 40-bit padding to the message.

Referring to Fig. 8.14, we set

M_1 (64-bit input key): 7a138b2524af17c3

from which K (48-bit DES key): 058c4517a7a2

M_2 (64 bits): 17b439a12f51c5a8

M_3 (64 bits): 51cb360000000000

The permuted output $P(\Omega_i)$, $i = 1, 2, 3, 4$, of (S-box)$_i$ are, respectively,

$P(\Omega_1)$ = a34d397f

$P(\Omega_2)$ = 91193e8e

$P(\Omega_3)$ = c3cc8226

$P(\Omega_4)$ = 02234098

Final output data from the S-boxes will be a set of four blocks that are concatenated to form the 128-bit $P(\Omega)$:

$P(\Omega) = P(\Omega_1) \parallel P(\Omega_2) \parallel P(\Omega_3) \parallel P(\Omega_4)$

= (a34d397f91193e8ec3cc822602234098)

from which we obtain

SSD-B-NEW = a34d397f91193e8e

SSD-A-NEW = c3cc822602234098

Computation of the 18-bit message AUTHBS has been obtained by means of various authentication algorithms already presented in Section 8.11. The AUTHBS computation in Fig. 8.20 should be done in conjunction with SSD-A-NEW.

8.12.1 SSD-A and SSD-B Computation by MD5 Algorithm

The MD5 algorithm is an improved version of MD4, which is similar in design. MD5 takes a message of arbitrary length and produces a 128-bit message digest as output. MD5 is more conservative in design, but is slightly slower than MD4. MD5 processes the input message in 512-bit blocks which can be divided into sixteen 32-bit subblocks. The message digest is a set of four 32-bit blocks, which concatenate to form a single 128-bit hash value. Since SSD update is accomplished by a 128-bit pattern stored in the mobile station, it can be partitioned into two 64-bit distinct subsets, i.e., SSD-A NEW and SSD-B-NEW.

The following steps are performed to compute the message digest of the input message.

1. Append Padding Bits

The message is padded so that its length (in bits) is congruent to 448, modulo 512. That is, the padded message is just 64 bits short of being a multiple of 512.

Padding is formed by appending a single '1' bit to the message, and then '0' bits are appended as many as needed such that the length (in bits) of the padded message becomes congruent to 448 (= 512 – 64), modulo 512.

2. Initialize MD Register

Four 32-bit registers (A, B, C, D) are used for computation of the message digest. These registers are initialized by the following values in hexadecimal notation:

A :	01	23	45	67
B :	89	ab	cd	ef
C :	fe	dc	ba	98
D :	76	54	32	10

The four variables are copied into different variables : A as AA, B as BB, C as CC, and D as DD.

3. Define Four Auxiliary Functions (F, G, H, I)

F, G, H, and I are basic MD5 functions.

There are four nonlinear functions, one for each round.

$$F (X, Y, Z) = X \bullet Y + X' \bullet Z$$
$$G (X, Y, Z) = X \bullet Z + Y \bullet Z'$$
$$H (X, Y, Z) = Y \oplus Y \oplus Z$$
$$I (X, Y, Z) = Y \oplus (X + Z')$$

where the symbols denote: $X \bullet Y$ denotes the bitwise AND of X and Y, X+Y denotes the bitwise OR of X and Y, X' denotes the bitwise complement of X, i.e., NOT(X), and $X \oplus Y$ denotes the bitwise EX-OR of X and Y.

These four nonlinear functions are designed in such a way that if the bits of X, Y, and Z are independent and unbiased, then at each bit position the function F acts as a conditional: if X then Y else Z. The functions G, H, and I are similar to the function F, in that they act in 'bitwise parallel' to their product from the bits of X, Y, and Z. Notice that the function H is the bitwise EX-OR function of its inputs.

The truth table for computation of four nonlinear functions (F, G, H, I) is shown in Table 8. 14.

Table 8.14   Truth Table of Four Nonlinear Functions

X Y Z	F G H I
000	0000
001	1010
010	0110
011	1001
100	0011
101	0101
110	1100
111	1110

4. FF, GG, HH, and II Transformations for Rounds 1, 2, 3, and 4.

If M[k], $0 \le k \le 15$, denotes the kth subblock of the message, and <<< s represents a left shift s bits, the four operations are defined as follows:

FF(a, b, c, d, M[k], s, i) : a = b + ((a + F(b, c, d) + M[k] + T[i]) <<< s)
GG(a, b, c, d, M[k], s, i) : a = b + ((a + G(b, c, d) + M[k] + T[i]) <<< s)
HH(a, b, c, d, M[k], s, i) : a = b + ((a + H(b, c, d) + M[k] + T[i]) <<< s)
II(a, b, c, d, M[k], s, i) : a = b + ((a + I(b, c, d) + M[k] + T[i]) <<< s)

Computation uses a 64-element table T[i], i = 1, 2, $\cdots$, 64, which is constructed from the sine function. T[i] denotes the ith element of the table, which is equal to the integer part of 4294967296 times abs (sin(i)), where i is in radians :

$$T[i] = \text{integer part of } [2^{32} * \mid \sin(i) \mid]$$

where $0 \le \mid \sin(i) \mid \le 1$ and $0 \le 2^{32} * \mid \sin(i) \mid \le 2^{32}$.

Computation of T[i] for $1 \le i \le 64$ is tabulated as shown in Table 8.15.

5. Computation of Four Rounds (64 steps)

Each round consists of 16 operations. Each operation performs a nonlinear function on three of A, B, C, and D.

Let us show FF, GG, HH, and II transformations for rounds 1, 2, 3, and 4 in what follows.

Round 1

Let FF[a, d, c, d, M[k], s, i] denote the operation

$$a = b + ((a + F(b, c, d) + M[k] + T[i]) <<< s).$$

Then the following 16 operations are computed:

FF[a, b, c, d, M[0], 7, 1], FF[d, a, b, c, M[1], 12, 2], FF[c, d, a, b, M[2], 17, 3],
FF[b, c, d, a, M[3], 22, 4], FF[a, b, c, d, M[4], 7, 5], FF[d, a, b, c, M[5], 12, 6],
FF[c, d, a, b, M[6], 17, 7], FF[b, c, d, a, M[7], 22, 8], FF[a, b, c, d, M[8], 7, 9],
FF[d, a, b, c, M[9], 12, 10], FF[c, d, a, b, M[10], 17, 11], FF[b, c, d, a, M[11], 22, 12],
FF[a, b, c, d, M[12], 7, 13], FF[d, a, b, c, M[13], 12, 14], FF[c, d, a, b, M[14], 17, 15],
FF[b, c, d, a, M[15], 22, 16]

Table 8.15   Computation of T[*i*] for $1 \leq i \leq 64$

T[1] =	d76aa478	T[17]=	f61e2562	T[33]=	fffa3942	T[49]=	f4292244
T[2] =	e8c7b756	T[18]=	c040b340	T[34]=	8771f681	T[50]=	432aff97
T[3] =	242070db	T[19]=	265e5a51	T[35]=	69d96122	T[51]=	ab9423a7
T[4] =	c1bdcee	T[20]=	e9b6c7aa	T[36]=	fde5380c	T[52]=	fc93a039
T[5] =	f57c0faf	T[21]=	d62f105d	T[37]=	a4beea44	T[53]=	655b59c3
T[6] =	4787c62a	T[22]=	02441453	T[38]=	4bdecfa9	T[54]=	8f0ccc92
T[7] =	a8304613	T[23]=	d8a1e681	T[39]=	f6bb4b60	T[55]=	ffeff47d
T[8] =	fd469501	T[24]=	e7d3fbc8	T[40]=	bebfbc70	T[56]=	85845dd1
T[9] =	698098d8	T[25]=	21e1cde6	T[41]=	289b7ec6	T[57]=	6fa87e4f
T[10]=	8b44f7af	T[26]=	c33707d6	T[42]=	eaa127fa	T[58]=	fe2ce6e0
T[11]=	ffff5bb1	T[27]=	f4d50d87	T[43]=	d4ef3085	T[59]=	a3014314
T[12]=	895cd7be	T[28]=	455a14ed	T[44]=	04881d05	T[60]=	4e0811a1
T[13]=	6b901122	T[29]=	a9e3e905	T[45]=	d9d4d039	T[61]=	f7537e82
T[14]=	fd987193	T[30]=	fcefa3f8	T[46]=	e6db99e5	T[62]=	bd3af235
T[15]=	a679438e	T[31]=	676f02d9	T[47]=	1fa27cf8	T[63]=	2ad7d2bb
T[16]=	49b40821	T[32]=	8d2a4c8a	T[48]=	c4ac5665	T[64]=	eb86d391

Round 2

Let GG[a, d, c, d, M[*k*], s, *i*] denote the operation
$$a = b + ((a + G(b, c, d) + M[k] + T[i]) <<< s).$$
Then the following 16 operations are computed:
GG[a, b, c, d, M[1], 5, 17], GG[d, a, b, c, M[6], 9, 18], GG[c, d, a, b, M[11], 14, 19],
GG[b, c, d, a, M[0], 20, 20], GG[a, b, c, d, M[5], 5, 21], GG[d, a, b, c, M[10], 9, 22],
GG[c, d, a, b, M[15], 14, 23], GG[b, c, d, a, M[4], 20, 24], GG[a, b, c, d, M[9], 5, 25],
GG[d, a, b, c, M[14], 9, 26], GG[c, d, a, b, M[3], 14, 27], GG[b, c, d, a, M[8], 20, 28],
GG[a, b, c, d, M[13], 5, 29], GG[d, a, b, c, M[2], 9, 30], GG[c, d, a, b, M[7], 14, 31],
GG[b, c, d, a, M[12], 20, 32]

Round 3

Let HH[a, d, c, d, M[*k*], s, *i*] denote the operation
$$a = b + ((a + H(b, c, d) + M[k] + T[i]) <<< s).$$
Then the following 16 operations are computed:
HH[a, b, c, d, M[5], 4, 33], HH[d, a, b, c, M[8], 11, 34], HH[c, d, a, b, M[11], 16, 35],
HH[b, c, d, a, M[14], 23, 36], HH[a, b, c, d, M[1], 4, 37], HH[d, a, b, c, M[4], 11, 38],
HH[c, d, a, b, M[7], 16, 39], HH[b, c, d, a, M[10], 23, 40], HH[a, b, c, d, M[13], 4, 41],
HH[d, a, b, c, M[0], 11, 42], HH[c, d, a, b, M[3], 16, 43], HH[b, c, d, a, M[6], 23, 44],
HH[a, b, c, d, M[9], 4, 45], HH[d, a, b, c, M[12], 11, 46], HH[c, d, a, b, M[15], 16, 47],
HH[b, c, d, a, M[2], 23, 48]

Round 4

Let II[a, d, c, d, M[k], s, i] denote the operation

$$a = b + ((a + I(b, c, d) + M[k] + T[i]) <<< s).$$

Then the following 16 operations are computed:

II[a, b, c, d, M[0], 6, 49], II[d, a, b, c, M[7], 10, 50], II[c, d, a, b, M[14], 15, 51],
II[b, c, d, a, M[5], 21, 52], II[a, b, c, d, M[12], 6, 53], II[d, a, b, c, M[3], 10, 54],
II[c, d, a, b, M[10], 15, 53], II[b, c, d, a, M[1], 21, 56], II[a, b, c, d, M[8], 6, 57],
II[d, a, b, c, M[15], 10, 58], II[c, d, a, b, M[6], 15, 59], II[b, c, d, a, M[13], 21, 60],
II[a, b, c, d, M[4], 6, 61], II[d, a, b, c, M[11], 10, 62], II[c, d, a, b, M[2], 15, 63],
II[b, c, d, a, M[9], 21, 64]

After all of the above steps, A, B, C, and D are added to their respective increments AA, BB, CC, DD, as follows:

$$A = A + AA, B = B + BB$$
$$C = C + CC, D = D + DD$$

and the algorithm continues with the resulting block of data. The final output is the concatenation of A, B, C, and D.

Example 8.12   Set the initial buffer contents as follows:

A = 67452301 B = efcdab89
C = 98badcfe D = 10325476

The 512-bit padded message is produced from the 152-bit original message by appending the 360-bit padding as shown below:

Padded message (512 bits) = Original message (152 bits) + Padding (360 bits)

Namely,

7a138b25 24af17c3 17b439a1 2f51c5a8
51cb3600 00000000 00000000 00000000
00000000 00000000 00000000 00000000
00000000 00000000 00000000 00000000

1. Round 1 Computation (M[0], T[1], s = 7)

$$a = b + ((a + F(b, c, d) + M[k] + T[i]) <<< s).$$

where U <<< s denotes the 32-bit value obtained by circularly shifting U left by s bit positions.
Using Table 8.12, F(b, c, d) is computed as shown below:

b : 1110 1111 1100 1101 1010 1011 1000 1001
c : 1001 1000 1011 1010 1101 1100 1111 1110
d : 0001 0000 0011 0010 0101 0100 0111 0110

F(b, c, d) : 1001 1000 1011 1010 1101 1100 1111 1110
9   8   b   a   d   c   f   e

Compute U = (a + F(b, c, d) + M[0] + T[1]) <<< s , s = 7

a :	67452301
F(b, c, d) :	98badcfe
M[0] :	7a138b25
T[1] :	d76aa478

U: 517e2f9c

U = 517e2f9c <<< 7
=0101 0001 0111 1110 0010 1111 1001 1100

Since U <<< 7 denotes the 32-bit value obtained by circularly shifting U left by 7 bit positions, the shifted U value yields

U':	1011	1111	0001	0111	1100	1110	0010	1000
	b	f	1	7	c	e	2	8

From a = b + U', we have

b :	efcdab89
U':	bf17ce28

a :   aee57961

Hence, FF[a, b, c, d, M(0), 7, 1] can be computed as
aee57961, efcdab89, 98badcfe, 10325476

All FF transformations for Round 1 are similarly computed and consist of the results from the following 16 operations.

[1]   aee579b1, efcdab89, 98badcfe, 10325476
[2]   aee579b1, efcdab89, 98badcfe, 3bfb6779
[3]   aee579b1, efcdab89, 1e52ee63, 3bfb6779
[4]   aee579b1, 2279e391, 1e52ee63, 3bfb6779
[5]   224cc819, 2279e391, 1e52ee63, 3bfb6779
[6]   224cc819, 2279e391, 1e52ee63, 38e34369
[7]   224cc819, 2279e391, 249e0639, 38e34369
[8]   224cc819, 5526aa0a, 249e0639, 38e34369
[9]   82e240d3, 5526aa0a, 249e0639, 38e34369
[10] 82e240d3, 5526aa0a, 249e0639, 96033c04
[11] 82e240d3, 5526aa0a, 5ded0b8b, 96033c04
[12] 82e240d3, e40a2d3d, 5ded0b8b, 96033c04
[13] 21bfac5f, e40a2d3d, 5ded0b8b, 96033c04
[14] 21bfac5f, e40a2d3d, 5ded0b8b, 7f92ed5d
[15] 21bfac5f, e40a2d3d, 76c0395e, 7f92ed5d
[16] 21bfac5f, 25a998d6, 76c0395e, 7f92ed5d

2. Round 2 Computation (M[1], T[17], s = 5)

$$a = b + ((a + G(b, c, d) + M[1] + T[17]) <<< s)$$

Let V = (a + G(b, c, d) + M[1] + T[17]) <<< s , s = 5
where a = 21bfac5f, b = 25a998d6, c = 76c0395e, d = 7f92ed5d,
M[1] = 24af17c3, and T[17] = f61e2562.
Using Table 8.12, G(b, c, d) is computed as follows:

$$b : 0010\ 0101\ 1010\ 1001\ 1001\ 1000\ 1101\ 0110$$
$$c : 0111\ 0110\ 1100\ 0000\ 0011\ 1001\ 0101\ 1110$$
$$d : 0111\ 1111\ 1001\ 0010\ 1110\ 1101\ 0101\ 1101$$

G(b,c,d): 0010 0101 1100 0000 1001 1000 0101 0110

$$2 \quad 5 \quad c \quad 0 \quad 9 \quad 8 \quad 5 \quad 6$$

Compute V = a + G(b, c, d) + M[1] + T[17] <<< 5

$$
\begin{array}{rl}
a : & 21bfac5f \\
G(b, c, d) : & 25c09856 \\
M[1] : & 24af17c3 \\
T[17] : & f61e2562 \\
\hline
V : & 624d81da
\end{array}
$$

Thus, V = 624d81da <<< 5
= 0110 0010 0100 1101 1000 0001 1101 1010

Shifting V circularly to the left by 5 places yields

V'=	0100	1001	1011	0000	0011	1011	0100	1100
	4	9	b	0	3	b	4	c

From a = b + V', computation results in the following:

$$
\begin{array}{rl}
b : & 25a998d6 \\
V' : & 49b03b4c \\
\hline
a : & 6f59d422
\end{array}
$$

Thus, through the following 16 operations, GG transformation for round 2 can be accomplished as shown in the following:

[1]   6f59d422, 25a998d6, 76c0395e, 7f92ed5d
[2]   6f59d422, 25a998d6, 76c0395e, eaed1370
[3]   6f59d422, 25a998d6, e4f4d575, eaed1370
[4]   6f59d422, 164bbc01, e4f4d575, eaed1370
[5]   92a2cc8a, 164bbc01, e4f4d575, eaed1370
[6]   92a2cc8a, 164bbc01, e4f4d575, 4cca5389
[7]   92a2cc8a, 164bbc01, 8feaa429, 4cca5389
[8]   92a2cc8a, dd0c9181, 8feaa429, 4cca5389
[9]   52b693b1, dd0c9181, 8feaa429, 4cca5389
[10]   52b693b1, dd0c9181, 8feaa429, a2909476
[11]   52b693b1, dd0c9181, a4f2a227, a2909476
[12]   52b693b1, 724ed3b9, a4f2a227, a2909476
[13]   d1f4b09d, 724ed3b9, a4f2a227, a2909476
[14]   d1f4b09d, 724ed3b9, a4f2a227, 34bc09b1
[15]   d1f4b09d, 724ed3b9, a4f2a227, 34bc09b1
[16]   d1f4b09d, 3e23b74d, a4f2a227, 34bc09b1

3. Round 3 Computation (M[5], T[33], s = 4)

$$a = b + ((a + H(b, c, d) + M[5] + T[33]) <<< s)$$

where a = d1f4b09d, b = 3e23b74d, c = a6697938, d = 34bc09b1,
M[5] = 00000000, T[33] = fffa3942, and s = 4.
Let W = (a + H(b, c, d) + M[5] + T[33]) <<< 4.
Using Table 8.12, H(b, c, d) is computed as follows:

```
        b: 0011 1110 0010 0011 1011 0111 0100 1101
        c: 1010 0110 0110 1001 0111 1001 0011 1000
        d: 0011 0100 1011 1100 0000 1001 1011 0001
H(b,c,d): 1010 1100 1111 0110 1100 0111 1100 0100
             a    c    f    6    c    7    c    4
```

Compute W = a + H(b, c, d) + M[5] + T[33] <<< 4

```
         a :  d1f4b09d
G(b, c, d) :  acf6c7c4
      M[1] :  00000000
     T[17] :  fffa3942
         ─────────────
         W :  7ee5b1a3
```

Thus, W = 7ee5b1a3 <<< 4
       = 0111 1110 1110 0101 1011 0001 1010 0011

Shifting W circularly to the left by 4 places produces

W'=	1110	1110	0101	1011	0001	1010	0011	0111
	e	e	5	b	1	a	3	7

From a = b + W', computation results in:

```
    b :   3e23b74d
   W' :   ee5b1a37
   ────────────────
    a :   2c7ed184
```

Thus, HH[a, b, c, d, M[5], 4, 33] is computed as

$$2c7ed184, 3e23b74d, a6697938, 34bc09b1$$

All HH transformations for round 3 are similarly computed and consist of the results from the 16 operations as shown below:

[1]   2c7ed184, 3e23b74d, a6697938, 34bc09b1
[2]   2c7ed184, 3e23b74d, a6697938, 3d7fed07
[3]   2c7ed184, 3e23b74d, d3a8c930, 3d7fed07
[4]   2c7ed184, 5a2822a2, d3a8c930, 3d7fed07
[5]   8e5c4ac4, 5a2822a2, d3a8c930, 3d7fed07
[6]   8e5c4ac4, 5a2822a2, d3a8c930, 81f53790
[7]   8e5c4ac4, 5a2822a2, 6823d52c, 81f53790
[8]   8e5c4ac4, f9a0e2ae, 6823d52c, 81f53790
[9]   19151af2, f9a0e2ae, 6823d52c, 81f53790
[10] 19151af2, f9a0e2ae, 6823d52c, 19d6166c
[11] 19151af2, f9a0e2ae, d35f8234, 19d6166c
[12] 19151af2, 42486547, d35f8234, 19d6166c
[13] fd0fb1ee, 42486547, d35f8234, 19d6166c
[14] fd0fb1ee, 42486547, d35f8234, 4d472554
[15] fd0fb1ee, 42486547, 3e710a56, 4d472554
[16] fd0fb1ee, 5b477ba0, 3e710a56, 4d472554

4. Round 4 Computation (M[0], T[49], s = 6)
$$a = b + ((a + I(b, c, d) + M[0] + T[49]) <<< s), s = 6$$
where a= fd0fb1ee, b = 5b477ba0, c = 3e710a56, d = 4d472554,
$$M[0] = 7a138b25, T[49] = f4292244$$
Let Z = (a + I(b, c, d) + M[0] + T[49]) <<< 6 .

Using Table 8.12, I(b, c, d) is computed as follows:

$$
\begin{array}{ll}
\text{b:} & 0101\ 1011\ 0100\ 0111\ 0111\ 1011\ 1010\ 0000 \\
\text{c:} & 0011\ 1110\ 0111\ 0001\ 0000\ 1010\ 0101\ 0110 \\
\underline{\text{d:}} & \underline{0100\ 1101\ 0100\ 0111\ 0010\ 0101\ 0101\ 0100} \\
\text{I(b,c,d):} & 1100\ 0101\ 1000\ 1110\ 1111\ 0001\ 1111\ 1101 \\
& \quad\ \text{c}\quad\ 5\quad\ 8\quad\ \text{e}\quad\ \text{f}\quad\ 1\quad\ \text{f}\quad\ \text{d}
\end{array}
$$

Next, compute Z = (a + I(b, c, d) + M[0] + T[49] <<< 6

$$
\begin{array}{rl}
\text{a :} & \text{fd0fb1ee} \\
\text{I(b, c, d) :} & \text{c58ef1fd} \\
\text{M[0] :} & \text{7a138b25} \\
\underline{\text{T[49] :}} & \underline{\text{f4292244}} \\
\\
\text{Z :} & \text{30db5154}
\end{array}
$$

Thus, Z = 30db5154 <<< 6
     = 0011 0000 1101 1011 0101 0001 0101 0100
Shifting Z circularly to the left by 6 places yields

Z'=	0011	0110	1101	0100	0101	0101	0000	1100
	3	6	d	4	5	5	0	c

Using a = b + Z', the value of a is computed as follows:

$$
\begin{array}{rl}
\text{b :} & \text{5b477ba0} \\
\underline{\text{Z' :}} & \underline{\text{36d4550c}} \\
\\
\text{a :} & \text{921bd0ac}
\end{array}
$$

Thus, II[a, b, c, d, M[0]. 6, 33] is computed as
                921bd0ac, 5b477ba0, 3e710a56, 4d472554

All II transformations for round 4 are similarly computed and consist of the following 16 results:

[1] 921bd0ac, 5b477ba0, 3e710a56, 4d472554
[2] 921bd0ac, 5b477ba0, 3e710a56, bce7b111
[3] 921bd0ac, 5b477ba0, 86dfbd85, bce7b111
[4] 921bd0ac, dad51bea, 86dfbd85, bce7b111
[5] f958527f, dad51bea, 86dfbd85, bce7b111
[6] f958527f, dad51bea, 86dfbd85, c78bd4fa
[7] f958527f, dad51bea, 73cd27db, c78bd4fa
[8] f958527f, 07b3d57c, 73cd27db, c78bd4fa
[9] 569e32a9, 07b3d57c, 73cd27db, c78bd4fa
[10] 569e32a9, 07b3d57c, 73cd27db, 6e4ce124
[11] 569e32a9, 07b3d57c, 5c8bc0f4, 6e4ce124
[12] 569e32a9, da88ed56, 5c8bc0f4, 6e4ce124
[13] 95d282df, da88ed56, 5c8bc0f4, 6e4ce124
[14] 95d282df, da88ed56, 5c8bc0f4, ae6a0d43
[15] 95d282df, da88ed56, 63ac7a25, ae6a0d43
[16] 95d282df, cd5df50b, 63ac7a25, ae6a0d43

A four 32-bit buffer (A, B, C, D) is used to compute the message digest. Each of A, B, C, D is a 32-bit register. These registers are initialized to the following values:

$$aa = 67452301 \quad bb = efcdab89$$
$$cc = 98badcfe \quad dd = 10325476$$

The last operation [16] of II transformations are

$$a = 95d282df \quad b = cd5df50b$$
$$c = 63ac7a25 \quad d = ae6a0d43$$

SSD-update data, i.e., SSD-A-NEW and SSD-B-NEW, are generated as follows:

a :	95d282df	b :	cd5df50b
aa :	67452301	bb :	efcdab89
A :	fd17a5e0	B :	bd2ba094

SSD-A-NEW = fd17a5e0bd2ba094

Similarly,

c :	63ac7a25	d :	ae6a0d43
cc :	98badcfe	dd :	10325476
C :	fc675723	D :	be9c61b9

SSD-B-NEW = fc675723be9c61b9

8.13 MESSAGE ENCRYPTION AND SECURITY

To protect sensitive subscriber information, it is necessary to encipher certain fields of selected traffic channel signaling messages. However, messages should not be encrypted if authentication is not performed. Every reverse traffic channel message contains an encryption field which identifies the message encryption mode active at the time the message was created.

8.13.1 Enciphering Key Generation by Nonlinear Combiners

The key-bit sequence generated from an m-stage LFSR with taps g_1 through g_m and m memory cells is not appropriate for use as a cryptographic key because only $2m$ bits of either plaintext or ciphertext will be sufficient to break the key-bit stream by determining both taps and the initial contents of the shift register.

An m-stage LFSR with period $p=2^m-1$ is said to be a maximum-length shift register if it generates a PN sequence.

PN sequences generated from LFSRs combined by some nonlinear function have been proposed by several cryptologists for crypto-applications. Many proposed key-stream generators consist of a number of maximum-length shift registers combined by a nonlinear function. A nonlinear combination technique with LFSR sequence exhibits certain statistical properties which withstand any cryptographic attack.

Consider an m-stage LFSR generating a binary sequence which is applicable to a nonlinear filtering function f, as illustrated in Fig. 8.21.

A nonlinear keystream should be secure against cryptanalysis. For achieving high unpredictability of the generated key sequence, we will develop a nonlinear theory of binary sequences which reflects directly the sum of k variable products in a Boolean function. Let $f(x)$ be an arbitrary nonlinear function of variable x_i, $1 \leq i \leq n$, such that

$$f(x) = a_1x_1 + a_2x_2 + \cdots + a_nx_n + a_{12}x_1x_2 + \cdots + a_{n-1,\,n}x_{n-1}x_n$$
$$+ a_{123}x_1x_2x_3 + \cdots + a_{n-2,\,n-1,\,n}x_{n-2}x_{n-1}x_n$$
$$\cdots\cdots\cdots\cdots\cdots\cdots\cdots\cdots\cdots$$
$$+ a_{12\cdots n}x_1x_2\cdots x_n$$

where a product of variables is called a kth order product. Therefore, the order of $f(x)$ will be the maximum among the orders of its product terms.

Example 8.13 Figure 8.21 illustrates the 32-stage LFSR combined with a nonlinear filtering function $f(x)$ which is applicable to CDMA mobile communication systems. Assume that the generator polynomial of maximum-length LFSR is $g(x) = 1 + x + x^2 + x^{22} + x^{32}$.

A 128-bit SSD, which is stored in the mobile station's semi-permanent memory, is partitioned into two distinct subsets: The 64-bit SSD-A is used for supporting the authentication procedure; and the 64-bit SSD-B is for supporting the message confidentiality for CDMA.

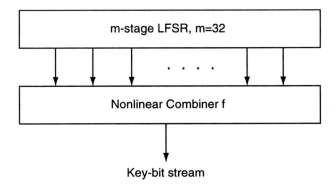

Figure 8.21 Key-bit generator filtered by the nonlinear combiner

The 64-bit SSD-B is divided into two halves (i.e., 32 bits each). The upper 32 bits are assigned to the 32-stage LFSR as the initial values; the lower 32 bits are assigned to the nonlinear function as shown in Fig. 8.21. The lower 32 bits are again partitioned evenly into four sets of 8 bits.

Assume that the 64-bit SSD-B is 30a7f415 83c7519a (in hexadecimal notation). Then, the upper 32-bit initial value is 30a7f415 (hexadecimal) = 00110000101001111111010000010101 (binary); while the lower 32 bits for the nonlinear function are 83c7519a (hexadecimal) = 10000011110001110101000110011010 (binary).

The lower 32 bits are divided into four 8-bit identical lengths in order to utilize them as the coefficients of nonlinear function $f(x)$, as follows:

First nonlinear combiner's coefficient: $(a_1, a_2, a_3, \cdots)$

83 (hexadecimal) = 10000011 (binary) $\rightarrow x_1 + x_7 + x_8$

Second nonlinear combiner's coefficient: $(a_{12}, a_{13}, a_{14}, \cdots)$

c7 (hexadecimal) = 11000111 (binary) $\rightarrow x_1 x_2 + x_1 x_3 + x_1 x_7 + x_1 x_8 + x_1 x_9$

Third nonlinear combiner's coefficient: $(a_{123}, a_{124}, a_{125}, a_{126}, \cdots)$

51 (hexadecimal) = 01010001 (binary) $\rightarrow x_1 x_2 x_4 + x_1 x_2 x_6 + x_1 x_2 x_{10}$

Fourth nonlinear combiner's coefficient: $(a_{1234}, a_{1235}, a_{1236}, \cdots)$

9a (hexadecimal) = 10011010 (binary) $\rightarrow x_1 x_2 x_3 x_4 + x_1 x_2 x_3 x_7 + x_1 x_2 x_3 x_8 + x_1 x_2 x_3 x_{10}$

Thus, the nonlinear function $f(x)$ is obtained as

$$f(x) = x_1 + x_7 + x_8 + x_1 x_2 + x_1 x_3 + x_1 x_7 + x_1 x_8 + x_1 x_9 + x_1 x_2 x_4 + x_1 x_2 x_6 + x_1 x_2 x_{10}$$
$$+ x_1 x_2 x_3 x_4 + x_1 x_2 x_3 x_7 + x_1 x_2 x_3 x_8 + x_1 x_2 x_3 x_{10}$$

In conjunction with the upper 32-bit initial values, the corresponding LFSR contents are given as

$$x_1 = 0, x_2 = 0, x_3 = 1, x_4 = 1, x_5 = 0,$$
$$x_6 = 0, x_7 = 0, x_8 = 0, x_9 = 1, x_{10} = 0.$$

Using these LFSR contents, the nonlinear combiner's coefficients are determined as follows:

1st coefficient: $x_1 + x_7 + x_8 = 0 \oplus 0 \oplus 0 = 0$

2nd coefficient: $x_1 x_2 + x_1 x_3 + x_1 x_7 + x_1 x_8 + x_1 x_9 = 0 \oplus 0 \oplus 0 \oplus 0 \oplus 0 = 0$

3rd coefficient: $x_1 x_2 x_4 + x_1 x_2 x_6 + x_1 x_2 x_{10} = 0 \oplus 0 \oplus 0 = 0$

4th coefficient: $x_1 x_2 x_3 x_4 + x_1 x_2 x_3 x_7 + x_1 x_2 x_3 x_8 + x_1 x_2 x_3 x_{10} = 0 \oplus 0 \oplus 0 \oplus 0 = 0$

Thus, the combiner's output data is $0 \oplus 0 \oplus 0 \oplus 0 = 0$.

Applying the 32-bit initial value to LFSR in Fig. 8.21 and shifting LFSR bitwise to the right, the PN sequence being outputted from the nonlinear combiner is computed as shown in Table 8.16

Table 8.16   Computation of Nonlinear Combiner's Output

Shift No.	LFSR Contents	Combiner's output (bits)
0	00110000101001111111010000010101	0
1	11111000010100111111100000001010	1
2	01111100001010011111110000000101	0
3	11011110000101001111110000000010	0
4	01101111000010100111111000000001	0
5	11010111100001010011110100000000	1
. . .	. . . .	. . .
14	11100000011010111100000010011110	1
15	01110000001101011110000001001111	0
. . .	. . . .	. . .
28	10101011110110011000001000100110	1
29	01010101111011001100000100010011	1
30	11001010111101100110001010001001	0
. . .	. . . .	. . .
49	11101110100011011110001101111101	0
50	10010111010001101111001110111110	1
. . .	. . . .	. . .

The nonlinear combiner's output Z in Table 8.16, which is used as a key sequence, is found as shown below:

Combiner's output sequence Z: 0100010111101010110010101 10011

0111000100101010011010100100111100110101 0011100100110000000101001100101

This PN sequence Z is used for the enciphering key.

Example 8.14 The LFSR content corresponding to the shift number 14 in Table 8.16 is 11100000011010111100000010011110.

Using the contents at the shift No. 14, the ith stage content of LFSR will be $x_1 = 1$, $x_2 = 1$, $x_3 = 1$, $x_4 = 0$, $x_5 = 0$, $x_6 = 0$, $x_7 = 0$, $x_8 = 0$, and $x_{10} = 0$. Hence, the nonlinear combiner's coefficients are computed as follows:

1st coefficient: $x_1 + x_7 + x_8 = 1 \oplus 0 \oplus 0 = 1$

2nd coefficient: $x_1 x_2 + x_1 x_3 + x_1 x_7 + x_1 x_8 + x_1 x_9 = 1 \oplus 1 \oplus 0 \oplus 0 \oplus 0 = 0$

3rd coefficient: $x_1 x_2 x_4 + x_1 x_2 x_6 + x_1 x_2 x_{10} = 0 \oplus 0 \oplus 1 = 1$

4th coefficient: $x_1 x_2 x_3 x_4 + x_1 x_2 x_3 x_7 + x_1 x_2 x_3 x_8 + x_1 x_2 x_3 x_{10} = 0 \oplus 0 \oplus 0 \oplus 1 = 1$

Thus, the nonlinear combiner's output bit corresponding to the shift No. 14 is $1 \oplus 0 \oplus 1 \oplus 1 = 1$.

8.13.2 Encryption and Message Security

Data transmitted on the reverse traffic channel is grouped into 20 ms frames. Data frames may be transmitted on the reverse traffic channel at variable data rates of 9600, 4800, 2400, and 1200 bps. The reverse traffic channel is used for the transmission of user and signaling information to the base station during a call.

Each data frame is 20 ms in duration and consists of either information-CRC-tail bits or information and tail bits depending on transmission rates. For example, for 4800 bps, the frame bits (96 bits) consist of the 80-bit information, 8-bit CRC, and 8-bit encoder tail.

The message encryption can be considered in two ways: 'external encryption' and 'internal encryption', as shown in Fig. 8.22.

The message information m is first enciphered with the key stream K generated by Fig. 8.21 in Section 8.13.1 and then encoded with the (3,1,8) convolutional encoder. We call this kind of encryption scheme '*external encryption*'. The second scheme to be considered is illustrated as shown in Fig. 8.22(b). Encoding precedes encryption and deciphering precedes decoding. We call this kind of encryption scheme '*internal encryption*'.

Example 8.14 Consider the 'external encryption' case in this example. As shown in Example 3.12, the 80-bit information data for the 4800-bps frame (20ms) was assumed as:

          1010110011 1001101111 0010010100 0100110011 0011001000

          1110010110 1001100110 0110011110

The PN key sequence produced from the nonlinear key generator was computed in Section 8.13.1 as shown below:

          0100010111 1010101100 1010110011 0111000100 1010100110

          1010010011 1100110101 0011100100 1100000010 1001100101

Using these information and key sequences, the encryption data sequence then becomes:

          1110100100 0011000011 1000100111 0011110111 1001101110

          0100000101 0101010011 0101111010

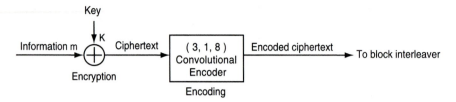

(a) External encryption

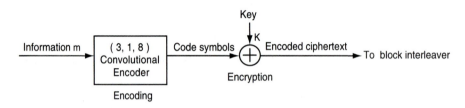

(b) Internal encryption

Figure 8.22 Message encryption scheme

Next, using the generator polynomial, $g(x) = 1 + x + x^3 + x^4 + x^7 + x^8$, for computing CRC at 4800 bps, CRC computation is accomplished using $g(x)$ and the all-one initial contents of the register as shown below:

$$CRC = 11000001$$

Thus, concatenation of the 80-bit encryption data with the 8-bit CRC and the 8-bit all-zero encoder tail yields

  1110100100 0011000011 1000100111 0011110111 100110110
  0100000101 0101010011 0101111010 1100000100 000000

This encrypted ciphertext is subject to inputting to the (3, 1, 8) convolutional encoder. The encoder output will be

c_0 : 1100101110 1110011111 0100110001 1101100101 1111010000 0110000011
  1001101101 1100000010 1001101100 101111
c_1 : 1000111100 0000000111 0100000000 0101111000 1100010000 1111111100
  0000111010 0101101000 1101101011 110011
c_2 : 1010000001 1010000100 0111001010 1111101111 1111101111 1110110111
  1000010001 1110111111 1011110100 001001

As a result, the final encoded ciphertext symbols are obtained as

1111000010001100101101101000001 10110010100000010010011111010110

0001110010011001000000000001100 10011100111111100100111000011100

1111111011010011100010010010001 01111111101001101101010011101101

1010000001001100111101000101011 10111100101001100101100110100101

11101000111111100111010101010010 1100101011001010111

Example 8.15 Consider this time the enciphering problem of all 96-bit concatenated data per frame, i.e., information data at the transmission rate of 4800 bps were assumed to be

1010110011 1001101111 0010010100 0100110011

0011001000 1110010110 1001100110 0110011110

The CRC bits were computed as

10111001

Thus, the 96-bit frame data for the 4800 bps is simply computed by adding the encoder tail bits (extra all-zero bits) as follows:

1010110011 1001101111 0010010100 0100110011 0011001000

1110010110 1001100110 0110011110 1011100100 000000

This is the data sequence to be enciphered with the following key sequence which is generated from the nonlinear combiner discussed in Section 8.13.1. The key sequence is

0100010111 1010101100 1010110011 0111000100 1010100110

1010010011 1100110101 0011100100 1100000010 100110

EX-ORing the information frame data with the key sequence generates the following ciphertext data sequence:

1110100100 0011000011 1000100111 0011110111 100110110

0100000101 0101010011 0101111010 0101110110 100110

This is the encrypted frame data sequence to be used as the (3,1,8) convolutional encoder input.

The mobile station convolutionally encodes the encrypted data sequence transmitted on the reverse traffic channel prior to block interleaving. The encoder generates three code symbols (c_0, c_1, c_2) for each bit of encoder input. Three code symbols are computed as shown below:

c_0 : 1100101110 1110011111 0100110001 1101100101 1111010000 0110000011

c_1 : 1000111100 0000000111 0100000000 0101111000 1100010000 1111111100

c_2 : 1010000001 1010000100 0111000010 0111001010 1111101111 1110110111

Therefore, the final encrypted code symbols are

1111000010001100101101101000001 10110010100000010010011111010110

0001110010011001000000000001100 10011100111111100100111000011100

1111111011010011100010010010001 01111111101001101101010011101101

1010000001001100111101000101011 10111100101001100101100110100101

00000110011000110101001001101100 0110100111111100010

So far, the case for the external message encryption has been discussed; the internal encryption for the message information follows. Data frames transmitted on the reverse traffic channel are grouped into 20 ms blocks. Each frame at the transmission rate of 4800 bps is 96 bits long which is composed of the 80-bit information, 8-bit CRC, and 8-bit encoder tail.

The internal encryption scheme is depicted as shown in Fig. 8.22(b). The 96-bit message frame is first encoded by the $(3,1,8)$ convolutional encoder. As a result, the code symbols output from the convolutional encoder are 3 times longer than the message frame bits at the encoder input. These code symbols are encrypted with the key stream generated from the key-bit generator filtered by the nonlinear combiner discussed in Section 8.13.1.

The internal encryption method is demonstrated by means of the following example.

Example 8.16   Based on Fig. 8.22(b), the internal encryption problem at the 4.8 kbps rate is considered in this example.

Assumed information sequence:

  1010110011 1001101111 0010010100 0100110011
  0011001000 1110010110 1001100110 0110011110

Computed CRC bits: 10111001

Encoder tail bits: 00000000

Thus, the information data concatenated with CRC bits and tail bits is expressed by

  1010110011 1001101111 0010010100 0100110011 0011001000
  1110010110 1001100110 0110011110 1011100100 000000

This data sequence is input to the $(3, 1, 8)$ convolutional encoder as shown in Fig. 3.4. These code symbols, corresponding to each input bit to the encoder, are computed as

 $c0$:1001010001 1000100001 0111110000 1001100111 1001100011 1111110111
  1011100111 0011001010 1100010011 011011

 $c1$:1110010000 1100100001 0001000000 0001101100 1010101100 0101110010
  1101010001 0101010001 0100101001 010111

 $c2$:1101010001 1001111110 0110100011 0100111110 1100110100 0010000111
  0101000101 1001100011 0001111010 110101

Concatenation of all three (c_0, c_1, c_2) symbols from the modulo-2 adders results in a single output sequence which is input to the symbol repeater and block interleaver. Since the code rate is 1/3, the code symbol output is 288 bits as shown below:

  1110110101 0100011100 0000000101 1110100000 0111100100
  1001001110 0001011011 1010110000 0000001001 1000010001
  1011100101 1111101100 1110010101 0011100101 0011100100
  1001101011 1011011000 0101111101 1100111001 1110001000
  0101100111 0010101001 1100101010 0000101011 1001100000
  0101110101 1000101110 0011111000 11110111

Next, the key stream (288 bits only) generated from the key generator filtered by the non-linear combiner is obtained as

0100010111 1010101100 1010110011 0111000100 1010100110
1010010011 1100110101 0011100100 1100000010 1001100101
0001010110 0001110010 1011111101 0001111101 0101000110
0100001110 0100010010 1110100101 1110010001 0101000010
1010110100 1100001101 0011111110 0001111110 0010101111
1000101000 0100101000 0110000001 1110000

Finally, the encrypted data sequence is computed by EX-ORing the code-symbol output with the key stream as follows:

1010100010 1110110000 1010110110 1001100100 1101000010
0011011101 1101101110 1001010100 1100001011 0001110100
1010110011 1110011110 0101101000 0010011000 0110100010
1101100101 1111001010 1011011000 0010101000 1011001010
1111010011 1110100100 1111010100 0001010101 1011001111
1101011101 1100000110 0101111001 00010111

Reverse Wideband CDMA Channel

$\mathbf{T}$he wideband CDMA PCS standard is specified by the following features:

- Unified services for high and low tier applications.
- Toll quality speech services using the adaptive differential pulse code modulation (ADPCM) coding scheme.
- System capacity of 128 channels per cell is provided by 5 MHz bandwidth, 256 channels by 10 MHz, or 384 channels by 15 MHz with interference canceller and voice activity detection. An interference canceller system (ICS) cancels intercell interference and increases available channels when it is installed.
- High speed voiceband data services up to 64 kbps using wideband modulation. But the data rate also can be extended to 144 kbps to support ISDN.

The basic W-CDMA system features 5 MHz channel spacing. Transmission bit errors are mitigated by a (2,1,8) convolutional encoder. An orthogonal code is assigned uniquely to each channel for discrimination (or channelization) in the forward link. Direct sequence spreading is employed and the chip rate is 4.096 Mcps. Each channel is modulated by QPSK after BPSK spreading modulation method. However, the channel spacings can be expanded to 10 MHz or 15 MHz with corresponding higher spreading rates.

The reverse W-CDMA channel is the communication link from the personal (mobile) station to the base station. The reverse link contains an access channel and a traffic channel. Each channel includes a reverse pilot channel. The reverse traffic channel also includes a signaling channel. A personal station transmits a reverse pilot channel, which is synchronized to the pilot channel from the base station.

The reverse CDMA channel is composed of access and reverse traffic channels. The access channel consists of two channels, i.e., (1) reverse pilot channel and (2) reverse access channel. The reverse traffic channel consists of three channels whose types are (1) reverse pilot channel, (2) reverse information channel, and (3) reverse signaling channel.

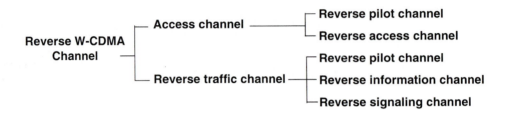

These channels share the same CDMA frequency assignment using direct-sequence CDMA techniques. Each of the access and reverse traffic channels is identified by a distant user long code sequence.

Chapters 9 and 10 are confined to describe the W-CDMA PCS standard which covers system capacity of 128 channels per cell provided by 5 MHz bandwidth and high speed voiceband data services up to 64 kbps using wideband modulation. The data transmission rates are 64, 32 and 16 kbps. The 64 kbps data is transmitted to support G4 FAX and video; the 32 kbps data is transmitted to support 32 kbps ADPCM; and the 16 kbps data is transmitted to support 16 kbps LD-CELP. The code excited linear prediction (CELP) is a speech coding algorithm. CELP coders use random excitation, long-term pitch prediction filters, and short-term format prediction filters.

9.1 REVERSE TRAFFIC CHANNEL

The reverse traffic channel supports variable data rates operating at 64, 32, and 16 kbps for a reverse information channel and rates of 4, 4, and 2 kbps for a reverse signaling channel. The frame of reverse information and signaling channels is 5 ms in duration. The reverse traffic channel is used for the simultaneous transmission of user and signaling information to the base station during a call.

The reverse CDMA channel structure for the reverse traffic channel is shown in Fig. 9.1. All data transmitted on the reverse CDMA channel is convolutionally encoded, interleaved, and modulated by direct-sequence spreading prior to transmission.

The modulation parameters for the reverse information channel and the reverse signaling channel are shown in Tables 9.1 and 9.2.

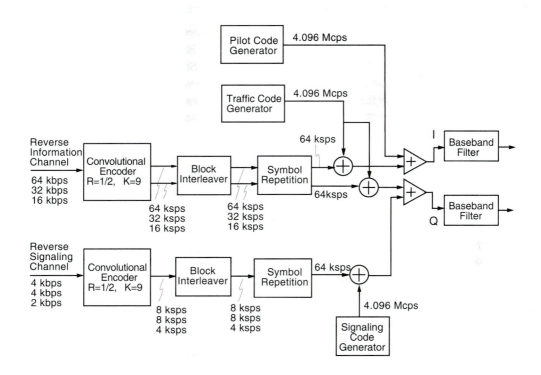

Figure 9.1 Reverse W-CDMA channel structure for reverse traffic channel
(After JTC (AIR) /95)

Table 9.1   Reverse Information Channel Modulation Parameters

Parameter	Data Rate (bps)			Units
	64000	**32000**	**16000**	
PN chip rate	4.096	4.096	4.096	Mcps
Code rate	1/2	1/2	1/2	bits/symbols
Symbol repetition	1	2	4	
Symbol rate	64000	64000	64000	sps
PN chips/symbol	64	64	64	PNchips/symbol
PN chips/bit	64	128	256	PN chips/bit

Table 9.2   Reverse Signaling Channel Modulation Parameters

Parameter	Data Rate (bps)		Units
	4000	**2000**	
PN chip rate	4.096	4.096	Mcps
Code rate	1/2	1/2	bits/symbols
Symbol repetition	8	16	
Symbol rate	64000	64000	sps
PN chips/symbol	64	64	PN chips/symbol
PN chips/bit	512	1024	PN chips/bit

9.2 REVERSE INFORMATION CHANNEL (RIC)

Reverse information channel 5 ms frames sent at the 64, 32, and 16 kbps transmission rate consist of 320, 160, and 80 bits, respectively.

9.2.1 RIC Convolutional Encoding

The personal station convolutionally encodes the data transmitted on the reverse traffic channel prior to interleaving. The convolutional code rate is R=1/2 and has a constraint length K=9.

The generator polynomials are, respectively:

$$g_0 (x) = 1 + x + x^2 + x^3 + x^5 + x^7 + x^8$$
$$g_1 (x) = 1 + x^2 + x^3 + x^4 + x^8$$

The (2,1,8) convolutional code generates two code symbols c_0 and c_1 for each data bit input to the encoder. Both in the information channel and in the signaling channel, these code symbols are output such that the first code symbol c_0 encoded with the generator polynomial g_0 is output first; the second code symbol c_1 encoded with the generator polynomial g_1 is output last. Suppose the state of the convolutional encoder, upon initialization, is the all-zero state.

Convolutional encoding involves the modulo-2 addition of selected taps of a serially time-delayed data sequence, as shown in Fig. 9.2.

Example 9.1   Consider the data rate of 16 kbps for the reverse information channel of the reverse traffic channel. Since the reverse information channel frame at 16 kbps is 80 bits, let us arbitrarily assume that the data sequence transmitted on this channel is

1100011100 0010100100 1100110111 0101011001
1100110010 1001110101 1001111100 0110011110

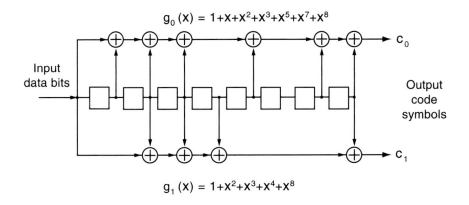

$$g_0(x) = 1+x+x^2+x^3+x^5+x^7+x^8$$

$$g_1(x) = 1+x^2+x^3+x^4+x^8$$

Figure 9.2 The (2, 1, 8) convolutional encoder

Under the all-zero initial contents of the encoder, the (2,1,8) convolutional encoder shown in Fig. 9.2 generates two code symbols c_0 and c_1 for each input data bit d as shown below:

d	c_0	c_1
1	1	1
1	0	1
0	0	1
0	0	0
0	1	0
1	0	0
.	.	.
.	.	.
.	.	.

Thus, the encoder output symbol sequences are respectively computed as :

c_0 : 1000101011 0111011111 1100101011 0010000101
    0100010001 1110010001 1011011011 1010011011

c_1 : 1110001010 0111100011 1000111100 0010101100
    0100001100 0111001011 0110011101 0100100010

9.2.2   RIC Block Interleaving

The personal station interleaves all code symbols on the reverse traffic channel prior to modulation and transmission. An interleaving span of 5, 10 or 20 ms is used. The interleaver span of 10 or 20 ms is optional and only applies to the reverse information channel of the reverse traffic channel. The coded symbols on the reverse information channel are interleaved as specified in the following. The interleaver is an array with 20 rows and 16 columns (i.e., 320 cells) for an interleaving span of 5 ms. For the reverse information channel of the reverse traffic channel, code symbols are written into the interleaver by columns, filling the complete 20 x 16 matrix. For the signaling channel of the reverse traffic channel, code symbols are written into the interleaver by columns, filling the complete 8 x 5 matrix (see Table 9.5)

For the interleaver span of 5 ms, Table 9.3 shows the ordering of the write operations of code symbols into the interleaving input array for the reverse information I channel of the reverse traffic channel. Table 9.4 shows the write operation for the reverse Q channel of the reverse traffic channel.

Table 9.3   Reverse Information and Access Channel (I-Channel) Interleaver Input
(Write Operation at 16 kbps)

Interleaver Span of 5 ms							
1	11	21	31	41	51	61	71
2	12	22	32	42	52	62	72
3	13	23	33	43	53	63	73
4	14	24	34	44	54	64	74
5	15	25	35	45	55	65	75
6	16	26	36	46	56	66	76
7	17	27	37	47	57	67	77
8	18	28	38	48	58	68	78
9	19	29	39	49	59	69	79
10	20	30	40	50	60	70	80

Table 9.4 Reverse Information and Access Channel (Q-channel) Interleaver Input (Write Operation at 16 kbps)

Interleaver Span of 5 ms							
36	46	56	66	76	6	16	26
37	47	57	67	77	7	17	27
38	48	58	68	78	8	18	28
39	49	59	69	79	9	19	29
40	50	60	70	80	10	20	30
41	51	61	71	1	11	21	31
42	52	62	72	2	12	22	32
43	53	63	73	3	13	23	33
44	54	64	74	4	14	24	34
45	55	65	75	5	15	25	35

Tables 9.5 and 9.6 show the ordering of write operations of code symbols into the interleave array for the signaling channel for both 4 kbps (8 x 5) and 2 kbps (4 x 5), respectively. The reverse information channel and the signaling channel code symbols are output from the interleaver by rows.

Table 9.5 Interleaver Output for Signaling Channel (Write Operation at 4 kbps)

1	9	17	25	33
2	10	18	26	34
3	11	19	27	35
4	12	20	28	36
5	13	21	29	37
6	14	22	30	38
7	15	23	31	39
8	16	24	32	40

Table 9.6 Interleaver Output for Signaling Channel (Write Operation at 2 kbps)

1	5	9	13	17
2	6	10	14	18
3	7	11	15	19
4	8	12	16	20

Example 9.2 Consider again the case of the 16 kbps data rate for the 80-bit reverse information channel of the reverse traffic channel. If the I-channel symbol sequence c_0 of the (2,1,8) convolutional encoder, using Table 9.3., is written into the interleaver by columns such that

Input symbols c_0s
(write in column by column)
⇩
10100111
01101100
01010111
01000010
10100000   ⇨   Output symbols c_0s
01001111   (Read out row by row)
11100011
01010000
11100011
11111111

then the I-channel code symbols c_0s will be output from the interleaver by rows as follows:

10100111   01101100 01010111 01000010   10100000
01001111   11100011 01010000 11100011   11111111

Next, for the Q-channel symbol sequence c_1, code symbols c_1s are arranged in the following manner in order to easily utilize Table 9.4:

Bit position	Code symbols
1–10	1110001010
11–20	0111100011
21–30	1000111100
31–40	0010101100
41–50	0100001100
51–60	0111101011
61–70	0110011101
71–80	0100100010

The interleaver output (Q-channel symbol sequence c_1) can be computed from the ordering of the write operation of code symbols c_1s in accordance with Table 9.4. Therefore, the Q-channel interleaved symbols c_1s are obtained using Table 9.4 by reading row by row as follows:

36, 46, 56, 66, 76, 6, 16, 26   →   00010001
.   11110101
.   11010001
.   00101110
.   00110010

41, 51, 61, 71, 1, 11, 21, 31   →   00001010
.   11111100
.   01101101
.   01000100

45, 55, 65, 75, 5, 15, 25, 35   →   00010111

Thus, the Q-channel interleaved symbol sequence (interleaver output) is expressed as

00010001 11110101 11010001 00101110 00110010
00001010 11111100 01101101 01000100 00010111

9.2.3   RIC Symbol Repetition

After the interleaver, the symbol repetition on the reverse traffic channels varies with the data rate. Each symbol at 64 kbps is not repeated. Each symbol at 32 kbps is repeated one time (each symbol occurs 2 consecutive times). Each symbol at 16 kbps is repeated 3 times (each symbol occurs 4 consecutive times) prior to data scrambling.

Example 9.3   Show the symbol repetition at the data rate of 16 kbps. The I-channel interleaved symbol sequence is

| 10100111 | 01101100 | 01010111 | 01000010 | 10100000 |
| 01001111 | 11100011 | 01010000 | 11100011 | 11111111 |

Since each symbol at 16 kbps is repeated 3 times, the symbol repetition output becomes:

1111	0000	1111	0000	0000	1111	1111	1111
0000	1111	1111	0000	1111	1111	0000	0000
0000	1111	0000	1111	0000	1111	1111	1111
0000	1111	0000	0000	0000	0000	1111	0000
1111	0000	1111	0000	0000	0000	0000	0000
0000	1111	0000	0000	1111	1111	1111	1111
1111	1111	1111	0000	0000	0000	1111	1111
0000	1111	0000	1111	0000	0000	0000	0000
1111	1111	1111	0000	0000	0000	1111	1111
1111	1111	1111	1111	1111	1111	1111	1111

The Q-channel interleaved symbol sequence is

00010001   11110101   11010001   00101110   00110010

00001010   11111100   01101101   01000100   00010111

The symbol repetition output at the 16 kbps rate is as follows:

0000	0000	0000	1111	0000	0000	0000	1111
1111	1111	1111	1111	0000	1111	0000	1111
1111	1111	0000	1111	0000	0000	0000	1111
0000	0000	1111	0000	1111	1111	1111	0000
0000	0000	1111	1111	0000	0000	1111	0000
0000	0000	0000	0000	1111	0000	1111	0000
1111	1111	1111	1111	1111	1111	0000	0000
0000	1111	1111	0000	1111	1111	0000	1111
0000	1111	0000	0000	0000	1111	0000	0000
0000	0000	0000	1111	0000	1111	1111	1111

9.2.4   RIC Direct-Sequence Spreading

Prior to transmission, the reverse information channel and signaling channel of the reverse traffic channel are direct sequence spread by the pilot code sequence, traffic code I sequence, traffic code Q sequence, and the signaling code sequence. These code sequences are generated by a code sequence with a period of 81920 chips, generated from the modulo-2 addition of the long code generator and the appropriate Walsh function, as shown in Fig. 9.3. As illustrated in Fig. 9.3(b), the Walsh function with index zero is assigned to the pilot code. The Walsh function with index one is assigned to the traffic I and Q and access codes. The Walsh function with index two is assigned to the signaling code. Each code sequence results in a period of 81920 chips with 20 ms. PN code generation for the long code is devised as shown in Fig. 9.3(a).

The long code seed is sent in the message of the paging channel from the base station. This long code is a shortened linear code with period of $2^{23}-1$ chips and satisfies the linear recursion specified by the following generator polynomial:

$$p(x)=1 + x + x^2 + x^{22} + x^{32}$$

The Walsh function at a fixed chip rate of 4.096 Mcps provides orthogonal channelization among all code channels (reverse pilot channel, reverse information channel and reverse signaling channel) on the reverse traffic channel or access channel (see Fig. 9.3(b)). Three of the 64 Walsh functions, as defined in Table 9.7, are used.

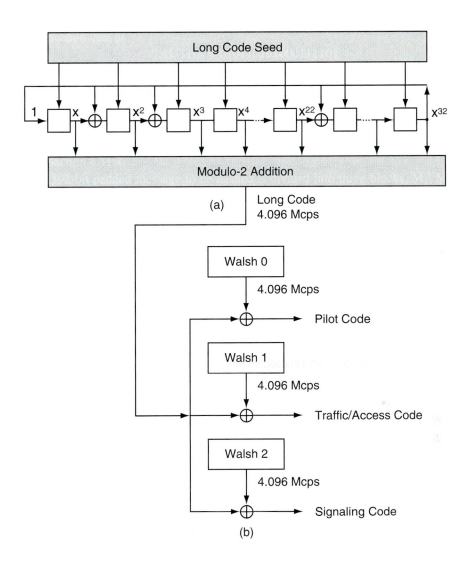

Figure 9.3 Orthogonal channelization among all code channels
(a) PN code generator for long code;
(b) PN sequence generator for long code

Table 9.7   Walsh Functions for Reverse Link

Walsh chip within a Walsh function

Walsh index	0123	4567	11 8901	1111 2345	1111 6789	2222 0123	2222 4567	2233 8901	3333 2345	3333 6789	4444 0123	4444 4567	4455 8901	5555 2345	5555 6789	6666 0123
0	0000	0000	0000	0000	0000	0000	0000	0000	0000	0000	0000	0000	0000	0000	0000	0000
1	0101	0101	0101	0101	0101	0101	0101	0101	0101	0101	0101	0101	0101	0101	0101	0101
2	0011	0011	0011	0011	0011	0011	0011	0011	0011	0011	0011	0011	0011	0011	0011	0011

The Walsh function time alignment will be such that the first Walsh chip, indicated by 0 by 0 in the column heading of Table 9.7, begins at the first chip of each frame with length 5 ms.

After direct sequence spreading, the binary data (0s and 1s) is mapped in phase. The '0' data is mapped into $+\pi$ phase and the '1' data is mapped into 0 phase.

Example 9.4   Referring to Fig. 9.3(a), the tap polynomial of the PN code generator is $g(x) = 1 + x + x^2 + x^{22} + x^{32}$, whose vector form is 1110000000 0000000000 0010000000 001; while the initial seed vector is assumed to be 1001001100 1010011011 0101111000 11. Using these two vectors, the long code sequence is generated as shown in Table 9. 8.

Table 9.8   Long Code Computation Using PN Code Generator

Shift No.	Register Contents	Σ '1's	Long Code Bit
0	10010011001010011011010111100011	17	1
1	10101001100101001101100011110001	16	0
2	10110100110010100110111001111000	17	1
3	01011010011001010011011100111100	17	1
4	00101101001100101001101110011110	17	1
5	00010110100110010100110111001111	17	1
6	11101011010011001010010011100111	18	0
7	10010101101001100101000001110011	15	1
8	10101010110100110010101000111001	16	0
9	10110101011010011001011100011100	17	1
10	01011010101101001100101110001110	17	1

Thus, the generated long code sequence is as shown below:

1011110101110100 0010111010110010 1001111011101011 0111001111000001
0010100011011010 0110110111011100 1010110100101100 0100101000000101
1100001010110101 0101101101001111 1001111011101101 0011101110100000
0011000010001111 1000000110000000 1111001110010010 0110000000011010
1101110001001000 0011001001001000 1000011111011001 0110100110100001
1111111101110001 1100000000010101 1111100100101011 0000101010101011
0100010100000000 0111010101001101 1000100001100011 0101001000110100
0011010000100101 1011011111100001 0111010010100101 0010111011101101
1111010000000110 1100100100110011 1000000011110000 0100001110001010
0110000001001001 11011000

Example 9.5   As illustrated in Fig. 9.3(b), the pilot code, traffic/access code, and signaling code are generated from the modulo-2 addition of the long code sequence and the appropriate Walsh function. Using three of the 64 Walsh functions (see Table 9.7), these code sequences are generated by a code sequence with a period of 81920 chips.

1. Pilot Code Generation
The Walsh function with index zero, W0, is assigned to the pilot code.

$$\text{Pilot code} = \text{Long code} \ \oplus \ \text{Walsh } 0$$

Long code: 1011110101110100 0010111010110010 1001111011101011 011100111100001
  $\oplus$
Walsh 0:   0000000000000000 0000000000000000 0000000000000000 000000000000000
Pilot code: 1011110101110100 0010111010110010 1001111011101011 011100111100001

This sample computation demonstrates only the first row of the long code sequence.

Thus, the pilot code is generated from the modulo-2 addition of the long code sequence and the Walsh function with index zero as follows:

```
1011110101110100 0010111010110010 1001111011101011 0111001111000001
0010100011011010 0110110111011100 1010110100101100 0100101000000101
1100001010110101 0101101101001111 1001111011101101 0011101110100000
0011000010001111 1000000110000000 1111001110010010 0110000000011010
1101110001001000 0011001001001000 1000011111011001 0110100110100001
1111111101110001 1100000000010101 1111100100101011 0000101010101011
0100010100000000 0111010101001101 1000100001100011 0101001000110100
0011010000100101 1011011111100001 0111010010100101 0010111011101101
```

We see that the pilot code is nothing but a reproduction of the long code.

2. Traffic/Access Code Generation

The Walsh function with index one, W1, is assigned to traffic/access codes.

$$\text{Traffic/access codes} = \text{Long code} \oplus \text{Walsh 1}$$

Long code: 1011110101110100 0010111010110010 1001111011101011 0111001111000001
$\oplus$

Walsh 1: 0101010101010101 0101010101010101 0101010101010101 0101010101010101

Traffic/
access 1110100000100001 0111101111100111 1100101110111110 0010011010010100
codes:

The traffic/access codes are generated from the modulo-2 addition of the long code sequence and the Walsh function with index one as shown below:

```
1110100000100001    0111101111100111    1100101110111110    0010011010010100
0111110110001111    0011100010001001    1111100001111001    0001111101010000
1001011111100000    0000111000011010    1100101110111000    0110111011110101
0110010111011010    1101010011010101    1010011011000111    0011010101001111
1000100100011101    0110011100011101    1101001010001100    0011110011110100
1010101000100100    1001010101000000    1010110001111110    0101111111111110
0001000001010101    0010000000011000    1101110100110110    0000011101100001
0110000101110000    1110001010110100    0010000111110000    0111101110111000
```

3. Signaling code generation

The Walsh function with index two is assigned to the signaling code.

Signaling code = Long code $\oplus$ Walsh 2

Long code: 1011110101110100 0010111010110010 1001111011101011 0111001111000001

      $\oplus$

Walsh 2:   0011001100110011 0011001100110011 0011001100110011 0011001100110011

Signaling code: 1000111001000111 0001110110000001 1010110111011000 0100000011110010

The signaling code is generated from the modulo-2 addition of the long code sequence and the Walsh function with index two as shown below:

1000111001000111	0001110110000001	1010110111011000	0100000011110010
0001101111101001	0101111011101111	1001111000011111	0111100100110110
1111000110000110	0110100001111100	1010110111011110	0000100010010011
0000001110111100	1011001010110011	1100000010100001	0101001100101001
1110111101111011	0000000101111011	1011010011101010	0101101010010010
1100110001000010	1111001100100110	1100101000011000	0011100110011000
0111011000110011	0100011001111110	1011101101010000	0110000100000111
0000011100010110	1000010011010010	0100011110010110	0001110111011110

Prior to transmission, the reverse information channel and the signaling channel are direct sequence spread by the traffic code I sequence, traffic code Q sequence, and the signaling code sequence, as shown in Fig. 9.1.

Data transmitted on the reverse W-CDMA channel is grouped into 5 ms frames. The reverse information channel frames sent at the 64, 32, and 16 kbps transmission rates consist of 320, 160, and 80 bits, respectively.

Example 9.6 The symbol repeater output at the 16 kbps rate consists of 320 bits during the 5 ms frame. The traffic code sequence at 4.096 Mcps consists of 64 x 320 bits during the 5 ms frame. Hence, the corresponding bit ratio between these two sequences is 1 to 64.

1. The reverse I-channel information is direct sequence spread by traffic code I sequence as follows:

The I-channel symbol repetition output (first row only) is 1111 0000 1111 0000 0000 1111 1111 1111.

The traffic code I sequence is

1110100000100001	0111101111100111	1100101110111110	0010011010010100
0111110110001111	0011100010001001	1111100001111001	0001111101010000
1001011111100000	0000111000011010	1100101110111000	0110111011110101
0110010111011010	1101010011010101	1010011011000111	0011010101001111
1000100100011101	0110011100011101	1101001010001100	0011110011110100
1010101000100100	1001010101000000	1010110001111110	0101111111111110
0001000001010101	0010000000011000	1101110100110110	0000011101100001
0110000101110000	1110001010110100	0010000111110000	0111101110111000

Since one symbol of the I-channel symbol repetition sequence corresponds to 64 bits of the traffic code I sequence, the modulo-2 addition of these two sequences can be found as follows:

$1 \rightarrow$ 0001011111011110 1000010000011000 0011010001000001 1101100101101011
$1 \rightarrow$ 1000001001110000 1100011101110110 0000011110000110 1110000010101111
$1 \rightarrow$ 0110100000011111 1111000111100101 0011010001000111 1001000100001010
$1 \rightarrow$ 1001101000100101 0010101100101010 0101100100111000 1100101010110000
$0 \rightarrow$ 1000100100011101 0110011100011101 1101001010001100 0011110011110100
$0 \rightarrow$ 1010101000100100 1001010101000000 1010110001111110 0101111111111110
$0 \rightarrow$ 0001000001010101 0010000000011000 1101110100110110 0000011101100001
$0 \rightarrow$ 0110000101110000 1110001010110100 0010000111110000 0111101110111000

2. On the other hand, since the Q-channel symbol repetition sequence (first row only) is represented by 0000 0000 0000 1111 0000 0000 0000 1111, the output sequence resulting from the modulo-2 addition of the Q-channel repetition sequence and the traffic code Q sequence is computed as shown below:

$0 \rightarrow$ 1110100000100001 0111101111100111 1100101110111110 0010011010010100
$0 \rightarrow$ 0111110110001111 0011100010001001 1111100001111001 0001111101010000
$0 \rightarrow$ 1001011111100000 0000111000011010 1100101110111000 0110111011110101
$0 \rightarrow$ 0110010111011010 1101010011010101 1010011011000111 0011010101001111
$0 \rightarrow$ 1000100100011101 0110011100011101 1101001010001100 0011110011110100
$0 \rightarrow$ 1010101000100100 1001010101000000 1010110001111110 0101111111111110
$0 \rightarrow$ 0001000001010101 0010000000011000 1101110100110110 0000011101100001
$0 \rightarrow$ 0110000101110000 1110001010110100 0010000111110000 0111101110111000

9.3 REVERSE SIGNALING CHANNEL (RSC)

The reverse traffic channel supports two data rates operating at 4 and 2 kbps for a reverse signaling channel (RSC). The frame of the signaling channel is 5 ms in duration. The reverse signaling channel of the reverse traffic channel is illustrated in Fig. 9.1. All data transmitted on the reverse signaling channel is also convolutionally encoded, interleaved, and modulated by direct-sequence spreading prior to transmission.

The reverse signaling channel frames sent at 4 kbps transmission rate consist of 20 bits. These 20 bits are composed of a 2-bit discontinuous transmission indicator and the 18-bit signaling information. The reverse signaling channel sent at the 2 kbps transmission rate consists of 10 bits.

The modulation parameters for the reverse signaling channel are shown in Table 9.2.

9.3.1 RSC Convolutional Encoding

The reverse signaling channel data is convolutionally encoded prior to transmission. The convolutional code rate is 1/2 and its encoder has a constraint length K of 9. The generator tap polynomials are

$$g_0(x) = 1 + x + x^2 + x^3 + x^5 + x^7 + x^8$$
$$g_1(x) = 1 + x + x^2 + x^3 + x^4 + x^8$$

The reverse signaling channel uses a (2,1,8) convolutional encoder as shown in Fig. 9.2. The convolutional encoder is composed of K–1 = 8 stages of one-bit storage element each and the modulo-2 adders of selected taps of a serially time-delayed data sequence. In the reverse signaling channel, two code symbols are generated for each data bit input to the encoder. The first code symbol c_0 is generated with the generator tap polynomial $g_0(x)$, while the second code symbol c_1 is generated with the generator tap polynomial $g_1(x)$. These two output symbols (c_0, c_1) are pairwise output and concatenated into a serial symbol sequence.

Example 9.7   Consider the data rate of 2 kbps for an reverse signaling channel of the reverse traffic channel. The reverse signaling channel frame at 2 kbps is assumed to be 1101000101 (10 bits). The state of the convolutional encoder, upon initialization, is the all-zero state. Two output symbols c_0 and c_1 for each input data bit are computed as shown in Table 9.9.

Table 9.9   Computation of Convolutional Encoder Output Symbols for the Reverse Signaling Channel Frame at 2 kbps.

Shift No.	Contents of Encoder Stage	Output Pair (c_0, c_1)
1	100000000	(1, 1)
2	110000000	(0, 1)
3	011000000	(0, 1)
4	101100000	(1, 1)
5	010110000	(0, 0)
6	001011000	(0, 0)
7	000101100	(0, 1)
8	100010110	(0, 0)
9	010001011	(0, 1)
10	101000101	(1, 1)

Thus, the encoder output symbols are

$c_0 = (1001000001)$

$c_1 = (1111001011)$

The code symbol sequence generated from the convolutional encoder is expressed as

$c = (1101011100\ 0001000111)$

Example 9.8   Consider the data rate of 4 kbps for a reverse signaling channel of the reverse traffic channel. The reverse signaling channel frame at 4 kbps is 20 bits and can be assumed as 11010001011011010011. Under the all-zero initialization state, the encoder output symbols c_0 and c_1 for each input data bit are computed as

$c_0 = (1001000001\ 0101011001)$

$c_1 = (1111001011\ 0111000110)$

Thus, the code symbol sequence generated from the convolutional encoder is expressed as

$c = (1101011100\ 0001000111\ 0011011100\ 1010010110)$

9.3.2   RSC Block Interleaving

All code symbols on the reverse signaling channel of the reverse traffic channel are interleaved prior to modulation and transmission. For the RTC signaling channel, code symbols are written into the interleaver by columns, filling the complete 8 x 5 matrix for 4 kbps (see Table 9.5); and the complete 4 x 5 matrix for 2 kbps (see Table 9.6). The reverse signaling channel code symbols are output from the interleaver by rows.

Example 9.9   Consider the data rate of 2 kbps for a reverse signaling channel of the reverse traffic channel. From Example 9.7, the convolutional encoder output sequence was computed as c = (11010111000001000111). If code symbols in c are written into the interleaver by columns, filling the complete 4 x 5 matrix for 2 kbps, as shown below, we have

Interleaver input by columns

⇩

10000

11011   ⇨ Interleaver output by rows

01001

11001

Then the interleaver output will be composed of row after row of the above 4 x 5 matrix. This is,

10000 11011 01001 11001

Example 9.10   Consider the 4 kbps data rate for the RTC signaling channel. As computed in Example 9.8, the convolutional encoder output is

c = (1101011100 0001000111 0011011100 1010010110)

If all code symbols of c are written into the interleaver by columns, filling the complete 8 x 5 matrix, it can be shown that:

Interleaver input by columns

⇩

10001

10110

00110

10111   ⇨ Interleaver output by rows

00000

11001

10111

10100

Thus, the interleaver output is composed of

10001 10110 00110 10111

00000 11001 10111 10100

9.3.3   RSC Symbol Repetition

After interleaving, the symbol repetition on the RTC reverse signaling channel varies with the data rate. Each symbol at 4 kbps rate is repeated 7 times (each symbol occurs 8 consecutive times). Each symbol at 2 kbps rate is repeated 15 times (each symbol occurs 16 consecutive times).

Example 9.11   At the data rate of 2 kbps, the interleaver output was computed as 1 0 0 0 0 11011 01001 11001 (see Example 9.9). Each symbol at 2 kbps is repeated 15 times as shown below:

111111111111111 000000000000000 000000000000000 000000000000000
000000000000000 111111111111111 111111111111111 000000000000000
111111111111111 111111111111111 000000000000000 111111111111111
000000000000000 000000000000000 111111111111111 111111111111111
111111111111111 000000000000000 000000000000000 111111111111111

Example 9.12   At the data rate of 4 kbps, the interleaver output was found as

10001 10110 00110 10111
00000 11001 10111 10100

Each symbol at 4 kbps data rate is repeated 7 times because the transmission rate at the interleaver output is 8 ksps, while the transmission rate at the symbol repeater output is 64 ksps, that is,

8 x 8 (each symbol occurring 8 consecutive times) = 64 ksps.

Therefore, the symbol repeater output is computed as follows:

11111111 00000000 00000000 00000000 11111111
11111111 00000000 11111111 11111111 00000000
00000000 00000000 11111111 11111111 00000000
11111111 00000000 11111111 11111111 11111111
00000000 00000000 00000000 00000000 00000000
11111111 11111111 00000000 00000000 11111111
11111111 00000000 11111111 11111111 11111111
11111111 00000000 11111111 00000000 00000000

9.3.4   RSC Direct Sequence Spreading

The reverse signaling channel of the reverse traffic channel is direct sequence spread by the signaling code sequence. Referring to Fig. 9.3, the signaling code sequence is generated from the modulo-2 addition of the long code and the Walsh function with index two. This operation is called orthogonal modulation because the modulation symbol is one of the 64 mutually orthogonal waveforms generated using Walsh functions. The signaling code sequence results in a period of 81920 chips (20 ms). PN code generation for the long code is derived from Fig. 9.3(a).

The signaling code sequence was already computed in Section 9.2.4-(3) using the following formula:

Signaling code = Long code $\oplus$ Walsh 2

The generated signaling code computed in Section 9.2.4-(3) is shown below:

1000111001000111	0001110110000001	1010110111011000	0100000011110010
0001101111101001	0101111011101111	1001111000011111	0111100100110110
1111000110000110	0110100001111100	1010110111011110	0000100010010011
0000001110111100	1011001010110011	1100000010100001	0101001100101001
1110111101111011	0000000101111011	1011010011101010	0101101010010010
1100110001000010	1111001100100110	1100101000011000	0011100110011000
0111011000110011	0100011001111110	1011101101010000	0110000100000111
0000011100010110	1000010011010010	0100011110010110	0001110111011110
1100011100110101	1111101000000000	1011001111000011	0111000010111001
0101001101111010	1110101111110010	0010001010001010	0101011100101101
0100000010000001	1110001111011100	0101001011110111	0111000010101111
1111000101011110	0110001010110001	0001110010111010	0110010111101000

The RSC symbol repeater output at the transmission rate of 64 ksps is 320 symbols/frame during the 5 ms duration; the signaling code at a fixed chip rate of 4.096 Mcps is 64 x 320 chips during the 5 ms duration. That is, PN chips/symbol is 4.096 x 10^6/64 x 10^3 = 64.

Example 9.13 Consider direct-sequence spreading of the RSC symbol repeater output by the signaling code sequence. Picking only the first 32 symbols of repeater output, i.e., 1111111111111111 0000000000000000, direct-sequence spreading of the 32-symbol repeater output is computed as follows:

1 → 0111000110111000 1110001001111110 0101001000100111 1011111100001101
1 → 1110010000010110 1010000100010000 0110000111100000 1000011011001001
1 → 0000111001111001 1001011110000011 0101001000100001 1111011101101100
1 → 1111110001000011 0100110101001100 0011111101011110 1010110011010110
1 → 0001000010000100 1111111010000100 0100101100010101 1010010101101101
1 → 0011001110111101 0000110011011001 0011010111100111 1100011001100111
1 → 1000100111001100 1011100110000001 0100010010101111 1001111011111000
1 → 1111100011101001 0111101100101101 1011100001101001 1110001000100001

    .   .
    .   .
    .   .

This analysis is based on the data rate of 2 kbps.

Example 9.14   Consider the case for the data rate of 4 kbps.

Direct-sequence spreading of the RSC symbol repeater output can be achieved by modulo-2 addition of the repeater output with the signaling code such that DS spreading = Repeater output $\oplus$ Signaling code.

Using the symbol repeater output, 11111111 00000000 00000000 $\cdots$, computed in Example 9.12 and the signaling code sequence, direct-sequence spreading is computed as follows:

```
1 →    0111000110111000 1110001001111110 0101001000100111 1011111100001101
1 →    1110010000010110 1010000100010000 0110000111100000 1000011011001001
1 →    0000111001111001 1001011110000011 0101001000100001 1111011101101100
1 →    1111110001000011 0100110101001100 0011111101011110 1010110011010110
1 →    0001000010000100 1111111010000100 0100101100010101 1010010101101101
1 →    0011001110111101 0000110011011001 0011010111100111 1100011001100111
1 →    1000100111001100 1011100110000001 0100010010101111 1001111011111000
1 →    1111100011101001 0111101100101101 1011100001101001 1110001000100001
0 →    1100011100110101 1111101000000000 1011001111000011 0111000010111001
0 →    0101001101111010 1110101111110010 0010001010001010 0101011100101101
0 →    0100000010000001 1110001111011100 0101001011110111 0111000010101111
0 →    1111000101011110 0110001010110001 0001110010111010 0110010111101000
```

$\cdot$

$\cdot$ $\cdot$

$\cdot$ $\cdot$

$\cdot$ $\cdot$

9.3.5   Baseband Filtering

Following the DS spreading operation, the reverse I and Q information channel and the reverse signaling channel are filtered as follows. After direct-sequence spreading, the I and Q impulses are applied to the inputs of the I and Q baseband filters, as shown in Fig. 9.1. The normalized frequency response, $20\log_{10}|S(f)|$, of the filter is contained within $\delta_1 = \pm 1.5$ dB in the passband $0 \leq f \leq f_p$ where $f_p = 4.94$ MHz and is less than or equal to $\delta_2 = -40$ dB in the stopband $f \geq f_s$ where $f_s = 4.94$ MHz.

9.4 REVERSE PILOT CHANNEL

The reverse pilot channel is used by the base station to acquire, track, and derive a phase reference for the reverse information channel, reverse signal channel, and reverse access channel. The reverse pilot channel consists of an unmodulated long code sequence, which is specified in Section 9.3.4. The power level of the reverse pilot channel is typically 6 dB below the power level of the reverse information channel.

9.5 ACCESS CHANNEL

Access channels consist of two channels. The types of channels are (1) reverse pilot channel and (2) reverse access channel. The access channel is used by the personal station to initiate communication with the base station and to respond to paging channel messages. An access channel is an encoded, interleaved, and modulated spread-spectrum signal.

The personal station transmits information on the access channel at a fixed data rate of 16 kbps. An access channel frame is 5 ms in duration and transmits the slot number designated by the paging channel message.

The reverse channel may contain up to 32 access channels numbered 0 through 31 per supported paging channel. At least one access channel exists on the reverse channel for each paging channel on the corresponding forward channel. Each access channel is associated with a single paging channel. The long code seed of the access channel is set by the paging channel from the base station.

The reverse pilot channel may be used by the base station to acquire, track, and derive a phase reference for the reverse access channel.

Each reverse access channel frame contains 80 bits (5 ms frame). The structure of the access channel is shown in Fig. 9.4. The modulation parameters for the access channel is shown in Table 9.10.

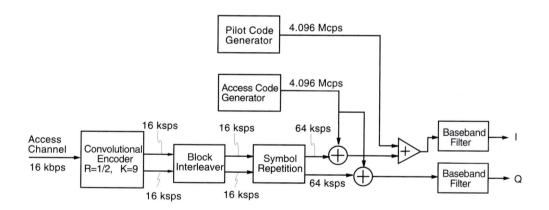

Figure 9.4 Reverse W-CDMA channel structure for access channel
(After JTC (AIR) /95)

Table 9.10   Access Channel Modulation Parameters

Parameter	Data Rate (bps) 16000	Units
PN chip rate	4.096	Mcps
Code rate	1/2	bits/symbols
Symbol repetition	4	
Symbol rate	64000	sps
PN chips/symbol	64	PN chips/symbol
PN chips/bit	256	PN chips/bit

9.5.1   Reverse Access Channel Convolutional Encoding

The reverse access channel information bits are convolutionally encoded prior to interleaving as specified in Section 9.2.1. The (2,1,8) convolutional code is used. The convolutional code rate is 1/2 and a constraint length K is 9. The generator tap polynomials are

$$g_0(x) = 1 + x + x^2 + x^3 + x^5 + x^7 + x^8$$
$$g_1(x) = 1 + x^2 + x^3 + x^4 + x^8$$

The convolutional encoder is the same as that shown in Fig. 9.2. When generating reverse access channel information bits, the encoder is initialized to the all-zero state at the end of each 5 ms frame.

Example 9.15   Suppose the information sequence on the reverse access channel at the 16 kbps data rate is

1001011100 0110100111 1101000010 0101110100
1110100100 1001111001 0100011010 1101000111

This 80 bits (5 ms frame) is input to the (2,1,8) convolutional encoder. With the all-zero initial state, the encoder output symbols are, respectively:

c_0 = (1110111010 1100110110 1110000000 0111101101
0000000010 1010110001 1101110101 1010111010)
c_1 = (1010100111 0110010001 0000010001 1110011001
0111010000 1100011101 1101001011 1000100001)

where the c_0 sequence is for the reverse access I channel and the c_1 sequence for the reverse access Q channel. These two coded sequences c_0 and c_1 are input to the block interleaver.

9.5.2 Reverse Access Channel Block Interleaving

The personal station interleaves all code symbols on the reverse access channel prior to modulation and transmission. The interleaver is an array with 10 rows and 8 columns (i.e., 80 cells) for an interleaving span of 5 ms. For the reverse access channel, coded symbols are written into the interleaver by columns, filling the complete 10 x 8 matrix.

For the interleaver span of 5 ms, Tables 9.3 and 9.4 show the ordering of the write operations of code symbols into the interleaving input array for the reverse access channel. The reverse access channel code symbols are output from the interleaver by rows.

Example 9.16 Consider the 16 kbps data rate for the 80-bit reverse access channel. Using Table 9.3, the I-channel symbol sequence c_0 of the $(2,1,8)$ convolutional encoder is written into the interleaver by columns as follows:

Access Channel (I-channel) Interleaver Input (write in column by column)

1	11	21	31	41	51	61	71	11100111	
2	12	22	32	42	52	62	72	11110010	
3	13	23	33	43	53	63	73	10110101	
4	14	24	34	44	54	64	74	00010010	**Access Channel**
5	15	25	35	45	55	65	75 ⇨	11010111	**(I-channel)**
6	16	26	36	46	56	66	76	11000111	**Interleaver**
7	17	27	37	47	57	67	77	10010001	**output**
8	18	28	38	48	58	68	78	01010010	**(read out row by**
9	19	29	39	49	59	69	79	11001001	**row)**
10	20	30	40	50	60	70	80	00010110	

Thus, the I-channel interleaver output can be computed as

11100111 11110010 10110101 00010010 11010111
11000111 10010001 01010010 11001001 00010110

Example 9.17 Using Table 9.4 (write operation at 16 kbps), the Q-channel symbol sequence c_1 of the $(2,1,8)$ convolutional encoder is written into the block interleaver by columns as shown below:

Access Channel (Q-channel) Interleaver
Input (write in column by column)

36	46	56	66	76	6	16	26	11100011
37	47	57	67	77	7	17	27	10110000
38	48	58	68	78	8	18	28	00100100
39	49	59	69	79	9	19	29	00010100
40	50	60	70	80	10	20	30	⇨ 10111111
41	51	61	71	1	11	21	31	01111001
42	52	62	72	2	12	22	32	11100101
43	53	63	73	3	13	23	33	10001101
44	54	64	74	4	14	24	34	10100000
45	55	65	75	5	15	25	35	00011000

Access Channel (Q-channel) Interleaver Output (read out row by row)

Thus, the Q-channel interleaver output can be computed as

11100011 10110000 00100100 00010100 10111111
01111001 11100101 10001101 10100000 00011000

9.5.3 Reverse Access Channel Symbol Repetition

For the access channel which has a fixed data rate of 16 kbps, each interleaved symbol is repeated three times (each symbol occurs 4 consecutive times) prior to data scrambling.

Example 9.18 Referring to the I-channel interleaver output obtained in Example 9.16, the repeater output is found as shown below:

1111 1111 1111 0000 0000 1111 1111 1111
1111 1111 1111 1111 0000 0000 1111 0000
1111 0000 1111 1111 0000 1111 0000 1111
0000 0000 0000 1111 0000 0000 1111 0000
1111 1111 0000 1111 0000 1111 1111 1111
1111 1111 0000 0000 0000 1111 1111 1111
1111 0000 0000 1111 0000 0000 0000 1111
0000 1111 0000 1111 0000 0000 1111 0000
1111 1111 0000 0000 1111 0000 0000 1111
0000 0000 0000 1111 0000 1111 1111 0000

Example 9.19   Using the Q-channel interleaver output computed in Example 9.17, the repeater output of the Q-channel is found as shown below:

1111 1111 1111 0000 0000 0000 1111 1111
1111 0000 1111 1111 0000 0000 0000 0000
0000 0000 1111 0000 0000 1111 0000 0000
0000 0000 0000 1111 0000 1111 0000 0000
1111 0000 1111 1111 1111 1111 1111 1111
0000 1111 1111 1111 1111 0000 0000 1111
1111 1111 1111 0000 0000 1111 0000 1111
1111 0000 0000 0000 1111 1111 0000 1111
1111 0000 1111 0000 0000 0000 0000 0000
0000 0000 0000 1111 1111 0000 0000 0000

9.5.4   Reverse Access Channel Direct-Sequence Spreading

The access channel is direct sequence spread by the pilot code sequence and access code sequence (or traffic code I sequence, traffic code Q sequence) prior to transmission. These code sequences are generated by a code sequence with a period of 81920 chips, generated from the modulo–2 addition of the long code and the appropriate Walsh function. As illustrated in Fig. 9.3, the Walsh function with index zero is assigned to the pilot code. The Walsh function with index one is assigned to the access code. Each code sequence results in a period of 81920 chips (20 ms).

The Walsh function at a fixed chip rate of 4.096 Mcps provides orthogonal channelization among all code channels (reverse pilot channel, reverse information channel, and reverse signaling channel) on the reverse traffic channel or access channel.

To achieve direct-sequence spreading, each repeater symbol on either the access I channel or the access Q channel at a fixed transmission rate of 64 ksps should be modulo-2 added with the access code sequence at a fixed chip rate of 4.096 Mcps, as shown in Fig. 9 . 4 .

Referring to Fig. 9.3(a), the long code is generated by the generator tap polynomial $p(x)= 1 + x + x^2 + x^{22} + x^{32}$, along with the long code seed, which is sent in the message of the paging channel from the base station. This long code is a shortened linear code with period $2^{32}-1$.

Example 9.20 Suppose the long code seed vector is 100100110010100011 011010111100011. Since the generating tap vector is 1110000000000000 0000001000000000, the long code from PN code generator is computed as shown in Table 9 . 1 1 .

Table 9-1   Long Code Generation by PN Code Generator

Shift No.	LFSR Contents	Σ '1's	Long Code Bit
0	100100110010100110110101111100011	17	1
1	101010011001010011011000111110001	16	0
2	101101001100101001101110011111000	17	1
3	010110100110010100110111001111100	17	1
. . .	. . .	. . .	. . .
17	000101011011010101101001111100111	18	0
18	111010101101101010110110111110011	21	1
19	100101010110110101011001011111001	18	0
. . .	. . .	. . .	. . .
43	111100101100110010011001001101111	18	0
44	100110010110011001001110100110011	17	1
45	101011001011001100100101010011011	16	0
. . .	. . .	. . .	. . .
71	001011000010011011101101011011101101	18	0
72	111101100001001110110101011011010110	19	1
73	011110110000100111011010101011011	19	1
. . .	. . .	. . .	. . .
97	100011100110000001001011011101011	16	0
98	101001110011000000100111101111011	17	1
99	101100111001100000010001110111101	16	0

Thus, the rightmost column in Table 9.11 represents the long code sequence generated by PN code generator as shown below:

1011110101110100 0010111010110010 1001111011101011 0111001111000001
0010100011011010 0110110111011100 1010110100101100 0100101000000101
1100001010110101 0101101101001111 1001111011101101 0011101110100000
0011000010001111 1000000110000000 1111001110010010 0110000000011010
1101110001001000 0011001001001000 1000011111011001 0110100110100001
1111111101110001 1100000000010101 1111100100101011 0000101010101011
0100010100000000 0111010101001101 1000100001100011 0101001000110100
0011010000100101 1011011111100001 0111010010100101 0010111011101101
1111010000000110 1100100100110011 1000000011110000 0100001110001010
0110000001001001 11011000

Example 9.21 The pilot code sequence is generated from the modulo-2 addition of the long code and the Walsh function with index zero (W0), i.e., Pilot code = Long code $\oplus$ W0. Since the Walsh function with index zero consists of the all-zero 64 chips, the pilot code sequence is identical to the long code sequence computed in Example 9.20.

Example 9.22 Consider the generation of the access code sequence. The access code (A.C.) is generated from the modulo-2 addition of the long code (L.C.) and the Walsh function with index one (W1) such that A.C. = L.C. $\oplus$ W1. This code sequence results in a period of 81920 chips (20 ms).

Consider only the first 64 chips of the long code and W1. Then, the access code is generated using the format of A.C. = L.C. $\oplus$ W1 as shown below:

```
L.C.:   1011110101110100 0010111010110010 1001111011101011 0111001111000001
  ⊕
W1:     0101010101010101 0101010101010101 0101010101010101 0101010101010101
A.C.:   1110100000100001 0111101111100111 1100101110111110 0010011010010100
```

Thus, the generated access code sequence is

```
1110100000100001 0111101111100111 1100101110111110 0010011010010100
0111110110001111 0011100010001001 1111100001111001 0001111101010000
1001011111100000 0000111000011010 1100101110111000 0110111011110101
0110010111011010 1101010011010101 1010011011000111 0011010101001111
1000100100011101 0110011100011101 1101001010001100 0011110011110100
1010101000100100 1001010101000000 1010110001111110 0101111111111110
0001000001010101 0010000000011000 1101110100110110 0000011101100001
0110000101110000 1110001010110100 0010000111110000 0111101110111000
```

Prior to transmission, the access I channel and the access Q channel of the reverse access channel are direct sequence spread by the access code sequence, which is generated from the modulo-2 addition of the long code and the Walsh function with index one (W1).

Example 9.23 Using the I-channel repeater output computed in Example 9.18, direct-sequence spreading can be achieved from the modulo-2 addition of the access code and the I-channel repeater output.

The first row of the I-channel repeater output is

1111 1111 1111 0000 0000 1111 1111 1111

Considering the first two rows of the access code obtained in Example 9.22, it gives

```
1110100000100001 0111101111100111 1100101110111110 0010011010010100
0111110110001111 0011100010001001 1111100001111001 0001111101010000
```

The DS spreading output corresponding to the first two symbols, 11, of the repeater output are computed as

1 → 0001011111011110 1000010000011000 0011010001000001 1101100101101011
1 → 1000001001110000 1100011101110110 0000011110000110 1110000010101111

This example shows for demonstrative purposes how to compute direct-sequence spreading.

Example 9.24   Using the Q-channel repeater output obtained in Example 9.19, DS spreading can be accomplished by the modulo-2 addition of the access code and the Q-channel repeater output, as follows:
    The first row of the Q-channel repeater output is

1111 1111 1111 0000 0000 0000 1111 1111

Comparing the first row of both the I-channel and Q-channel repeater outputs, we find that the sixth pair, 1111 and 0000, are different from each other. The DS spreading output corresponding to the first two bits of the sixth block, 00, of the Q-channel repeater output can be computed as

0 → 0110010001101111 1010010100101010 1101110100010011 1001010110011011
0 → 1101111100100111 1000101110001011 0101010110001001 0111101001000100

9.5.5   Reverse Access Channel Baseband Filtering

After direct-sequence spreading, the impulses of the access I and Q channels are applied to the inputs of the I and Q baseband filters as specified in Section 9.3.5.

Forward Wideband CDMA Channel

$\mathbf{T}$he forward W-CDMA channel consists of the following code channels: (1) one pilot channel, (2) one sync channel, (3) up to eight paging channels, and (4) a number of forward traffic channels. The forward traffic channel consists of a forward information channel and a forward signaling channel.

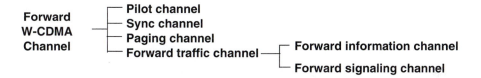

Each of these code channels is orthogonally spread by the appropriate Walsh code and is then spread by a PN sequence at a fixed chip rate of 4.096 Mcps.

10.1 PILOT CHANNEL

A pilot channel is transmitted at all times by the base station when it is active. The pilot channel is an unmodulated spread spectrum signal that is used for synchronization by a remote personal station operating within the coverage area of the base station. The pilot channel has the structure shown in Fig. 10.1.

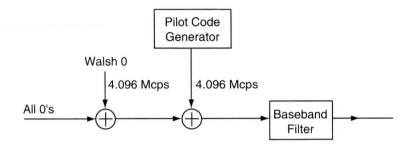

Figure 10.1 Pilot channel structure

Each base station uses a time offset of the pilot PN sequence to identify a forward channel. Time offsets may be reused within a PCS system. Distinct pilot channels are identified by an offset index (0 through 319 inclusive). This offset index specifies the offset value from the zero offset pilot PN sequence.

Three hundred twenty unique values are possible for the pilot PN sequence offset. The offset (in chips) for a given pilot PN sequence from the zero shift PN sequence equals the index value multiplied by 256. For example, if the pilot PN sequence offset index is 15, the pilot PN sequence offset will be (15)(256)=3480 PN chips. The same pilot PN sequence offset is used on all frequency assignments for a given base station.

10.1.1 Pilot Channel Orthogonal and DS Spreading

The pilot channel is orthogonally spread with Walsh code index zero prior to transmission.

Each code channel transmitted on the forward CDMA channel is spread with a Walsh code at a fixed chip rate of 4.096 Mcps to provide orthogonal channelization among all code channels on a given forward W-CDMA channel. A code channel that is spread using Walsh code n is assigned to code index number n. The Walsh code with index zero is assigned to the pilot channel.

Example 10.1   Consider the 80-bit all 0s pilot channel input per frame at the 16 kbps data rate as follows:

0000000000 0000000000 0000000000 0000000000
0000000000 0000000000 0000000000 0000000000

The Walsh code with index zero consists of 64 zeros:

00000000 00000000 00000000 00000000
00000000 00000000 00000000 00000000

Since each 0-bit input is spread using Walsh code 0 (64 zero chips), the total symbols orthogonally spread with the Walsh function with index zero are (80)(64) = 5120 symbols (all 0s)

Example 10.2   The tap polynomial of the PN code generator is given as $g(x)=1+x+x^2+x^{22}+x^{32}$ whose vector form is 11100000000000000000001000000001 (see Fig. 10.2). The chip rate for the pilot PN sequence is 4.096 Mcps. The long code seed is assumed to be 10010011001010011011010111100011. Applying these two vectors to Fig. 10.2, the long code is generated as shown below:

```
10111101 01110100 00101110 10110010 10011110 11101011 01110011 11000001
00101000 11011010 01101101 11011100 10101101 00101100 01001010 00000101
11000010 10110101 01011011 01001111 10011110 11101101 00111011 10100000
00110000 10001111 10000001 10000000 11110011 10010010 01100000 00011010
11011100 01001000 00110010 01001000 10000111 11011001 01101001 10100001
11111111 01110001 11000000 00010101 11111001 00101011 00001010 10101011
01000101 00000000 01110101 01001101 10001000 01100011 01010010 00110100
00110100 00100101 10110111 11100001 01110100 10100101 00101110 11101101
11110100 00000110 11001001 00110011 10000000 11110000 01000011 10001010
01100000 01001001 11011000
```

Direct-sequence spreading can thus be achieved from the modulo-2 addition of the all-zero symbol sequence orthogonally spread with Walsh code index 0 and the long code sequence computed above. However, the output symbols by DS spreading are nothing but a reproduction of the long code sequence itself.

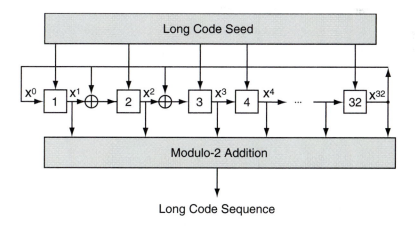

Figure 10.2 Long code sequence generated from PN code generator

10.1.2 Pilot Channel Filtering

Following the spreading operation, the pilot channel is filtered as follows: The impulses by DS spreading are applied to the input of the baseband filter as shown in Fig. 10.1 The baseband filter has a frequency response $S(f)$ that satisfies the limits given in Fig. 10.3. Specifically, the normalized frequency response of the filter is contained within $\pm\delta_1$ in the passband $0 \leq f \leq f_P$ and is less than or equal to $-\delta_2$ in the stopband $f \geq f_S$. The numerical values for these parameters are $\delta_1 = 1.5$ dB, $\delta_2 = 40$ dB, $f_P = 1.96$ MHz, and $f_S = 2.47$ MHz.

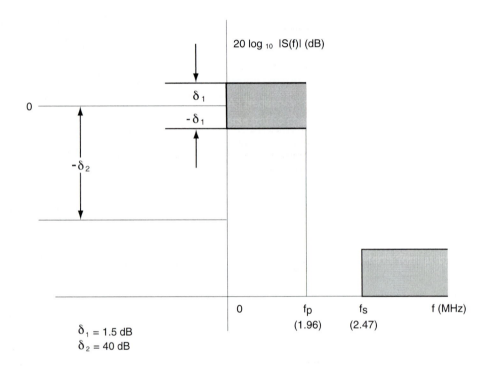

Figure 10.3 Baseband filter frequency response limits

10.2 SYNC CHANNEL

The sync channel is convolutionally encoded, interleaved, spread, and modulated to become a spread spectrum signal that is used by personal stations operating within the coverage area of the base station to acquire initial frame synchronization.

The bit rate for the sync channel is 16 kbps. The I and Q channel pilot PN sequences for the sync channel use the same pilot PN sequence offset as the pilot channel for a given station.

Once the personal station achieves the pilot PN sequence synchronization by acquiring the pilot channel, the synchronization for the sync channel is immediately known. This is because the sync channel is spread with the same pilot PN sequence, and because the frame timing on the sync channel is aligned with the pilot PN sequence.

The start of the interleaver block and frame of the sync channel aligns with the start of the pilot PN sequence being used to spread the forward channel.

10.2.1 Sync Channel Structure and Modulation Parameters

The length of the sync channel superframe is 20 ms. Messages transmitted on the sync channel begin only at the start of a sync channel superframe. The overall structure of the sync channel is shown in Fig. 10.4.

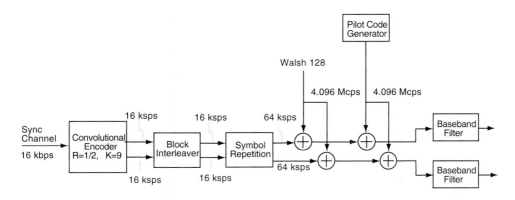

Figure 10.4 Sync channel structure (After JTC (AIR) /95)

The modulation parameters for the sync channel are as shown in Table 10.1.

Table 10.1   Sync Channel Modulation Parameters

Parameter	Data Rate (bps) 16000	Units
PN chip rate	4.096	Mcps
Code rate	1/2	bits/code symbols
Symbol repetition	4	
Symbol rate	64000	sps
PN chips/symbol	64	PN chips/symbol
PN chips/bit	256	PN chips/bit

10.2.2 Sync Channel Convolutional Encoding

The sync channel data is convolutionally encoded prior to transmission. The $(2, 1, 8)$ convolutional code has its code rate $R=1/2$ with a constraint length $K = 9$. The generator tap polynomials are, respectively:

$$g_0\,(x) = 1 + x + x^2 + x^3 + x^5 + x^7 + x^8$$
$$g_1\,(x) = 1 + x^2 + x^3 + x^4 + x^8$$

The $(2, 1, 8)$ convolutional encoder generates two code symbols (c_0, c_1) for each data bit input to the encoder. The first code symbol c_0 is output encoded with the generator tap polynomial $g_0(x)$, and the last code symbol c_1 is output encoded with the generator tap polynomial $g_1(x)$. The state of the convolutional encoder, upon initialization, is the all-zero state.

Convolutional encoding involves the modulo-2 addition of selected taps of a serially time-delayed data sequence. The length of the data sequence delay is equal to $K{-}1 = 8$. Figure 10.5 shows the $(2, 1, 8)$ convolutional encoder that is also used for the sync channel.

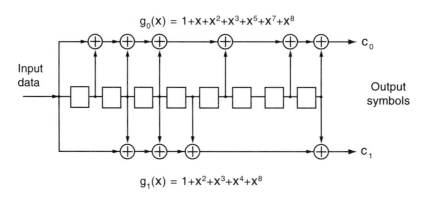

Figure 10.5 The $(2, 1, 8)$ convolutional encoder

Example 10.3   The base station transmits the data sequence on the sync channel at a 16 kbps bit rate. The sync channel frame is 5 ms in duration and consists of 80 bits.

Let us assume the following data sequence as the encoder input:

1001100101 0000110101 1000101101 1101001000

0110110010 1001110101 1001111000 0110011010

Applying this input data to Fig. 10.5, the convolutional output symbols are computed without difficulty as follows:

c_0: 1110010111 1110111001 0110111110 1110110100
1101011110 0110010001 1011011100 0000101100

c_1: 1010010100 1101101001 0011010111 1110110000
1011010010 1111001011 0110011000 1000110111

10.2.3 Sync Channel Interleaving

The coded symbols c_0 and c_1 on the sync channel are interleaved as follows. The base station interleaves all coded symbols on the sync channel prior to modulation and transmission. A block interleaving span of 5 ms is used.

The sync channel interleaver is an array with 8 rows and 10 columns (80 cells). Tables 10.2 and 10.3 show the ordering of write operations of symbols into the I and Q interleaving array at 16 kbps for both the I and Q sync channels. The sync channel code symbols are output from the interleaver by rows.

Table 10.2 Sync and Paging I Channel Interleaver Input
(Write Operation at 16 kbps, 5 ms Span)

1	9	17	25	33	41	49	57	65	73
2	10	18	26	34	42	50	58	66	74
3	11	19	27	35	43	51	59	67	75
4	12	20	28	36	44	52	60	68	76
5	13	21	29	37	45	53	61	69	77
6	14	22	30	38	46	54	62	70	78
7	15	23	31	39	47	55	63	71	79
8	16	24	32	40	48	56	64	72	80

Table 10.3 Sync and Paging Q Channel Interleaver Input
(Write Operation at 16 kbps, 5 ms Span)

37	45	53	61	69	77	5	13	21	29
38	46	54	62	70	78	6	14	22	30
39	47	55	63	71	79	7	15	23	31
40	48	56	64	72	80	8	16	24	32
41	49	57	65	73	1	9	17	25	33
42	50	58	66	74	2	10	18	26	34
43	51	59	67	75	3	11	19	27	35
44	52	60	68	76	4	12	20	28	36

Example 10.4   Writing the coded symbols c_0 by columns in the I-channel interleaver according to Table 10.2 yields:

```
1111111000
1101010010
1101100011                    I-channel interleaver output
0111111110      ⇨                  (Read row by row)
0101001101
1010110001
0111010100
1101011100
```

Hence, the I-channel interleaver output is

1111111000 1101010010 1101100011 0111111110
0101001101 1010110001 0111010100 1101011100

Example 10.5   Writing the coded symbols c_1 by columns in the Q-channel interleaver in accordance with Table 10.3 yields:

```
0010000001
0111011101
0001110111
0000011011      ⇨           Q-channel interleaver output
1110010101                      (Read row by row)
0001000010
1111111001
1110101111
```

Thus, the Q-channel interleaver output becomes

0010000001 0111011101 0001110111 0000011011
1110010101 0001000010 1111111001 1110101111

10.2.4 Sync Channel Symbol Repetition

After the interleaver, each output symbol at 16 kbps is repeated three times (each symbol occurs 4 consecutive times) prior to data scrambling.

Example 10.6   Refer to the I-channel interleaver output computed in Example 10.4. Each symbol in the interleaver output is repeated three times. The symbol repetition results in:

1111 1111 1111 1111 1111 1111 1111 0000 0000 0000
1111 1111 0000 1111 0000 1111 0000 0000 1111 0000
1111 1111 0000 1111 1111 0000 0000 0000 1111 1111
0000 1111 1111 1111 1111 1111 1111 1111 1111 0000
0000 1111 0000 1111 0000 0000 1111 1111 0000 1111
1111 0000 1111 0000 1111 1111 0000 0000 0000 1111

Example 10.7 The Q-channel interleaver output was computed in Example 10.5. Each symbol in the Q-channel interleaver output is repeated three times (each symbol occurs 4 consecutive times). The symbol repetition for the Q-channel results in:

0000 0000 1111 0000 0000 0000 0000 0000 0000 1111
0000 1111 1111 1111 0000 1111 1111 1111 0000 1111
0000 0000 0000 1111 1111 1111 0000 1111 1111 1111
0000 0000 0000 0000 0000 1111 1111 0000 1111 1111
1111 1111 1111 0000 0000 1111 0000 1111 0000 1111
0000 0000 0000 1111 0000 0000 0000 0000 1111 0000

10.2.5 Sync Channel Walsh Code Spreading

The sync channel transmitted on the forward CDMA channel is spread with Walsh 128 at a fixed chip rate of 4.096 Mcps. For the sync channel, the repeater output symbol rate of orthogonal spreading with Walsh 128 is 64 ksps. Walsh code spreading is accomplished by performing the modulo-2 addition of the repeater output symbol with Walsh 128 at a fixed chip rate of 4.096 Mcps.

In general, a code channel that is spread using Walsh code n is assigned to code index number n ($0 \leq n \leq 64$ for 64 kbps, $0 \leq n \leq 127$ for 32 kbps, and $0 \leq n \leq 255$ for 16 kbps). One of the time-orthogonal Walsh codes, defined in Tables 10.4, 10.5 and 10.6, is used for 64 kbps, 32 kbps and 16 kbps, respectively. $\overline{\text{Walsh code 64}}$ in Tables 10.5 and 10.6 represents the complement of Walsh code 64.

Table 10.4   Walsh Code for 64 kbps

Walsh Chip within a Walsh Code

```
          0                          63
        ┌────────────────────────────┐
  Code  │                            │
  index │          Walsh 64          │
        └────────────────────────────┘
   63
```

Table 10.5   Walsh Code for 32 kbps

Walsh chip within a Walsh code

	0	63 64	127
63	Walsh 64	Walsh 64	
Code index 64			
127	Walsh 64	Walsh 64	

Table 10.6   Walsh Code for 16 kbps

Walsh chip within a Walsh code

	0	63 64	127 128	191 192	255
63	Walsh 64	Walsh 64	Walsh 64	Walsh 64	
64					
127	Walsh 64	Walsh 64	Walsh 64	Walsh 64	
Code index 128					
191	Walsh 64	Walsh 64	Walsh 64	Walsh 64	
192					
255	Walsh 64	Walsh 64	Walsh 64	Walsh 64	

Example 10.8   Suppose Walsh code 128 is assigned to code index number $n = 128$ for 16 kbps, as shown below:

Walsh chip within a Walsh code

	0			255
128 Code index $128 \leq n \leq 191$	Walsh 64	Walsh 64	Walsh 64	Walsh 64
191				

00000000 00000000 00000000 00000000 00000000 00000000 00000000 00000000
00000000 00000000 00000000 00000000 00000000 00000000 00000000 00000000
11111111 11111111 11111111 11111111 11111111 11111111 11111111 11111111
11111111 11111111 11111111 11111111 11111111 11111111 11111111 11111111

Example 10.9 Consider Walsh code 64 corresponding to code index number $n = 64$ for 32 kbps.

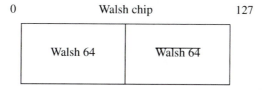

01100110 01100110 01100110 01100110
01100110 01100110 01100110 01100110
10011001 10011001 10011001 10011001
10011001 10011001 10011001 10011001

Since Walsh code chip of Walsh 64 is as shown in Table 10.7, Walsh code chip of $\overline{\text{Walsh}}$ $\overline{64}$ can be as shown in Table 10.8.

Table 10.7 Walsh code chip of Walsh 64

Walsh Chip within a Walsh Code

		0000 0123	0000 4567	0011 8901	1111 2345	1111 6789	2222 0123	2222 4567	2233 8901	3333 2345	3333 6789	4444 0123	4444 4567	4455 8901	5555 2345	5555 6789	6666 0123
W a l s h	0	0000	0000	0000	0000	0000	0000	0000	0000	0000	0000	0000	0000	0000	0000	0000	0000
	1	0101	0101	0101	0101	0101	0101	0101	0101	0101	0101	0101	0101	0101	0101	0101	0101
	2	0011	0011	0011	0011	0011	0011	0011	0011	0011	0011	0011	0011	0011	0011	0011	0011
	3	0110	0110	0110	0110	0110	0110	0110	0110	0110	0110	0110	0110	0110	0110	0110	0110
C o d e	4	0000	1111	0000	1111	0000	1111	0000	1111	0000	1111	0000	1111	0000	1111	0000	1111
	5	0101	1010	0101	1010	0101	1010	0101	1010	0101	1010	0101	1010	0101	1010	0101	1010
	6	0011	1100	0011	1100	0011	1100	0011	1100	0011	1100	0011	1100	0011	1100	0011	1100
I n	7	0110	1001	0110	1001	0110	1001	0110	1001	0110	1001	0110	1001	0110	1001	0110	1001
d e x	8	0000	0000	1111	1111	0000	0000	1111	1111	0000	0000	1111	1111	0000	0000	1111	1111
	9	0101	0101	1010	1010	0101	0101	1010	1010	0101	0101	1010	1010	0101	0101	1010	1010
	10	0011	0011	1100	1100	0011	0011	1100	1100	0011	0011	1100	1100	0011	0011	1100	1100
	11	0110	0110	1001	1001	0110	0110	1001	1001	0110	0110	1001	1001	0110	0110	1001	1001

Table 10.7   Walsh code chip of Walsh 64 *(continued)*

Walsh Code Index

Idx																
12	0000	1111	1111	0000	0000	1111	1111	0000	0000	1111	1111	0000	0000	1111	1111	0000
13	0101	1010	1010	0101	0101	1010	1010	0101	0101	1010	1010	0101	0101	1010	1010	0101
14	0011	1100	1100	0011	0011	1100	1100	0011	0011	1100	1100	0011	0011	1100	1100	0011
15	0110	1001	1001	0110	0110	1001	1001	0110	0110	1001	1001	0110	0110	1001	1001	0110
16	0000	0000	0000	0000	1111	1111	1111	1111	0000	0000	0000	0000	1111	1111	1111	1111
17	0101	0101	0101	0101	1010	1010	1010	1010	0101	0101	0101	0101	1010	1010	1010	1010
18	0011	0011	0011	0011	1100	1100	1100	1100	0011	0011	0011	0011	1100	1100	1100	1100
19	0110	0110	0110	0110	1001	1001	1001	1001	0110	0110	0110	0110	1001	1001	1001	1001
20	0000	1111	0000	1111	1111	0000	1111	0000	0000	1111	0000	1111	1111	0000	1111	0000
21	0101	1010	0101	1010	1010	0101	1010	0101	0101	1010	0101	1010	1010	0101	1010	0101
22	0011	1100	0011	1100	1100	0011	1100	0011	0011	1100	0011	1100	1100	0011	1100	0011
23	0110	1001	0110	1001	1001	0110	1001	0110	0110	1001	0110	1001	1001	0110	1001	0110
24	0000	0000	1111	1111	1111	1111	0000	0000	0000	0000	1111	1111	1111	1111	0000	0000
25	0101	0101	1010	1010	1010	1010	0101	0101	0101	0101	1010	1010	1010	1010	0101	0101
26	0011	0011	1100	1100	1100	1100	0011	0011	0011	0011	1100	1100	1100	1100	0011	0011
27	0110	0110	1001	1001	1001	1001	0110	0110	0110	0110	1001	1001	1001	1001	0110	0110
28	0000	1111	1111	0000	1111	0000	0000	1111	0000	1111	1111	0000	1111	0000	0000	1111
29	0101	1010	1010	0101	1010	0101	0101	1010	0101	1010	1010	0101	1010	0101	0101	1010
30	0011	1100	1100	0011	1100	0011	0011	1100	0011	1100	1100	0011	1100	0011	0011	1100
31	0110	1001	1001	0110	1001	0110	0110	1001	0110	1001	1001	0110	1001	0110	0110	1001
32	0000	0000	0000	0000	0000	0000	0000	0000	1111	1111	1111	1111	1111	1111	1111	1111
33	0101	0101	0101	0101	0101	0101	0101	0101	1010	1010	1010	1010	1010	1010	1010	1010
34	0011	0011	0011	0011	0011	0011	0011	0011	1100	1100	1100	1100	1100	1100	1100	1100
35	0110	0110	0110	0110	0110	0110	0110	0110	1001	1001	1001	1001	1001	1001	1001	1001
36	0000	1111	0000	1111	0000	1111	0000	1111	1111	0000	1111	0000	1111	0000	1111	0000
37	0101	1010	0101	1010	0101	1010	0101	1010	1010	0101	1010	0101	1010	0101	1010	0101
38	0011	1100	0011	1100	0011	1100	0011	1100	1100	0011	1100	0011	1100	0011	1100	0011
39	0110	1001	0110	1001	0110	1001	0110	1001	1001	0110	1001	0110	1001	0110	1001	0110
40	0000	0000	1111	1111	0000	0000	1111	1111	1111	1111	0000	0000	1111	1111	0000	0000
41	0101	0101	1010	1010	0101	0101	1010	1010	1010	1010	0101	0101	1010	1010	0101	0101
42	0011	0011	1100	1100	0011	0011	1100	1100	1100	1100	0011	0011	1100	1100	0011	0011
43	0110	0110	1001	1001	0110	0110	1001	1001	1001	1001	0110	0110	1001	1001	0110	0110
44	0000	1111	1111	0000	0000	1111	1111	0000	1111	0000	0000	1111	1111	0000	0000	1111
45	0101	1010	1010	0101	0101	1010	1010	0101	1010	0101	0101	1010	1010	0101	0101	1010
46	0011	1100	1100	0011	0011	1100	1100	0011	1100	0011	0011	1100	1100	0011	0011	1100
47	0110	1001	1001	0110	0110	1001	1001	0110	1001	0110	0110	1001	1001	0110	0110	1001
48	0000	0000	0000	0000	1111	1111	1111	1111	1111	1111	1111	1111	0000	0000	0000	0000
49	0101	0101	0101	0101	1010	1010	1010	1010	1010	1010	1010	1010	0101	0101	0101	0101
50	0011	0011	0011	0011	1100	1100	1100	1100	1100	1100	1100	1100	0011	0011	0011	0011
51	0110	0110	0110	0110	1001	1001	1001	1001	1001	1001	1001	1001	0110	0110	0110	0110

Table 10.7 Walsh code chip of Walsh 64 *(continued)*

Walsh Code Index

Idx																
52	0000	1111	0000	1111	1111	0000	1111	0000	1111	0000	1111	0000	0000	1111	0000	1111
53	0101	1010	0101	1010	1010	0101	1010	0101	1010	0101	1010	0101	0101	1010	0101	1010
54	0011	1100	0011	1100	1100	0011	1100	0011	1100	0011	1100	0011	0011	1100	0011	1100
55	0110	1001	0110	1001	1001	0110	1001	0110	1001	0110	1001	0110	0110	1001	0110	1001
56	0000	0000	1111	1111	1111	1111	0000	1111	1111	1111	0000	0000	0000	0000	1111	1111
57	0101	0101	1010	1010	1010	1010	0101	1010	1010	1010	0101	0101	0101	0101	1010	1010
58	0011	0011	1100	1100	1100	1100	0011	1100	1100	1100	0011	0011	0011	0011	1100	1100
59	0110	0110	1001	1001	1001	1001	0110	1001	1001	1001	0110	0110	0110	0110	1001	1001
60	0000	1111	1111	0000	1111	0000	0000	1111	1111	0000	0000	1111	0000	1111	1111	0000
61	0101	1010	1010	0101	1010	0101	0101	1010	1010	0101	0101	1010	0101	1010	1010	0101
62	0011	1100	1100	0011	1100	0011	0011	1100	1100	0011	0011	1100	0011	1100	1100	0011
63	0110	1001	1001	0110	1001	0110	0110	1001	1001	0110	0110	1001	0110	1001	1001	0110

Table 10.8 Walsh code chip of Walsh 64

Walsh Chip within a Walsh Code

Walsh Code Index

Idx	0000 0123	0000 4567	0011 8901	1111 2345	1111 6789	2222 0123	2222 4567	2233 8901	3333 2345	3333 6789	4444 0123	4444 4567	4455 8901	5555 2345	5555 6789	6666 0123
0	1111	1111	1111	1111	1111	1111	1111	1111	1111	1111	1111	1111	1111	1111	1111	1111
1	1010	1010	1010	1010	1010	1010	1010	1010	1010	1010	1010	1010	1010	1010	1010	1010
2	1100	1100	1100	1100	1100	1100	1100	1100	1100	1100	1100	1100	1100	1100	1100	1100
3	1001	1001	1001	1001	1001	1001	1001	1001	1001	1001	1001	1001	1001	1001	1001	1001
4	1111	0000	1111	0000	1111	0000	1111	0000	1111	0000	1111	0000	1111	0000	1111	0000
5	1010	0101	1010	0101	1010	0101	1010	0101	1010	0101	1010	0101	1010	0101	1010	0101
6	1100	0011	1100	0011	1100	0011	1100	0011	1100	0011	1100	0011	1100	0011	1100	0011
7	1001	0110	1001	0110	1001	0110	1001	0110	1001	0110	1001	0110	1001	0110	1001	0110
8	1111	1111	0000	0000	1111	1111	0000	0000	1111	1111	0000	0000	1111	1111	0000	0000
9	1010	1010	0101	0101	1010	1010	0101	0101	1010	1010	0101	0101	1010	1010	0101	0101
10	1100	1100	0011	0011	1100	1100	0011	0011	1100	1100	0011	0011	1100	1100	0011	0011
11	1001	1001	0110	0110	1001	1001	0110	0110	1001	1001	0110	0110	1001	1001	0110	0110
12	1111	0000	0000	1111	1111	0000	0000	1111	1111	0000	0000	1111	1111	0000	0000	1111
13	1010	0101	0101	1010	1010	0101	0101	1010	1010	0101	0101	1010	1010	0101	0101	1010
14	1100	0011	0011	1100	1100	0011	0011	1100	1100	0011	0011	1100	1100	0011	0011	1100
15	1001	0110	0110	1001	1001	0110	0110	1001	1001	0110	0110	1001	1001	0110	0110	1001
16	1111	1111	1111	1111	0000	0000	0000	0000	1111	1111	1111	1111	0000	0000	0000	0000
17	1010	1010	1010	1010	0101	0101	0101	0101	1010	1010	1010	1010	0101	0101	0101	0101
18	1100	1100	1100	1100	0011	0011	0011	0011	1100	1100	1100	1100	0011	0011	0011	0011
19	1001	1001	1001	1001	0110	0110	0110	0110	1001	1001	1001	1001	0110	0110	0110	0110
20	1111	0000	1111	0000	0000	1111	0000	1111	1111	0000	1111	0000	0000	1111	0000	1111
21	1010	0101	1010	0101	0101	1010	0101	1010	1010	0101	1010	0101	0101	1010	0101	1010
22	1100	0011	1100	0011	0011	1100	0011	1100	1100	0011	1100	0011	0011	1100	0011	1100
23	1001	0110	1001	0110	0110	1001	0110	1001	1001	0110	1001	0110	0110	1001	0110	1001

Table 10.8   Walsh code chip of Walsh 64 *(continued)*

Index																
24	1111	1111	0000	0000	0000	0000	1111	1111	1111	1111	0000	0000	0000	0000	1111	1111
25	1010	1010	0101	0101	0101	0101	1010	1010	1010	1010	0101	0101	0101	0101	1010	1010
26	1100	1100	0011	0011	0011	0011	1100	1100	1100	1100	0011	0011	0011	0011	1100	1100
27	1001	1001	0110	0110	0110	0110	1001	1001	1001	1001	0110	0110	0110	0110	1001	1001
28	1111	0000	0000	1111	0000	1111	1111	0000	1111	0000	0000	1111	0000	1111	1111	0000
29	1010	0101	0101	1010	0101	1010	1010	0101	1010	0101	0101	1010	0101	1010	1010	0101
30	1100	0011	0011	1100	0011	1100	1100	0011	1100	0011	0011	1100	0011	1100	1100	0011
31	1001	0110	0110	1001	0110	1001	1001	0110	1001	0110	0110	1001	0110	1001	1001	0110
32	1111	1111	1111	1111	1111	1111	1111	1111	0000	0000	0000	0000	0000	0000	0000	0000
33	1010	1010	1010	1010	1010	1010	1010	1010	0101	0101	0101	0101	0101	0101	0101	0101
34	1100	1100	1100	1100	1100	1100	1100	1100	0011	0011	0011	0011	0011	0011	0011	0011
35	1001	1001	1001	1001	1001	1001	1001	1001	0110	0110	0110	0110	0110	0110	0110	0110
36	1111	0000	1111	0000	1111	0000	1111	0000	0000	1111	0000	1111	0000	1111	0000	1111
37	1010	0101	1010	0101	1010	0101	1010	0101	0101	1010	0101	1010	0101	1010	0101	1010
38	1100	0011	1100	0011	1100	0011	1100	0011	0011	1100	0011	1100	0011	1100	0011	1100
39	1001	0110	1001	0110	1001	0110	1001	0110	0110	1001	0110	1001	0110	1001	0110	1001
40	1111	1111	0000	0000	1111	1111	0000	0000	0000	0000	1111	1111	0000	0000	1111	1111
41	1010	1010	0101	0101	1010	1010	0101	0101	0101	0101	1010	1010	0101	0101	1010	1010
42	1100	1100	0011	0011	1100	1100	0011	0011	0011	0011	1100	1100	0011	0011	1100	1100
43	1001	1001	0110	0110	1001	1001	0110	0110	0110	0110	1001	1001	0110	0110	1001	1001
44	1111	0000	0000	1111	1111	0000	0000	1111	0000	1111	1111	0000	0000	1111	1111	0000
45	1010	0101	0101	1010	1010	0101	0101	1010	0101	1010	1010	0101	0101	1010	1010	0101
46	1100	0011	0011	1100	1100	0011	0011	1100	0011	1100	1100	0011	0011	1100	1100	0011
47	1001	0110	0110	1001	1001	0110	0110	1001	0110	1001	1001	0110	0110	1001	1001	0110
48	1111	1111	1111	1111	0000	0000	0000	0000	0000	0000	0000	0000	1111	1111	1111	1111
49	1010	1010	1010	1010	0101	0101	0101	0101	0101	0101	0101	0101	1010	1010	1010	1010
50	1100	1100	1100	1100	0011	0011	0011	0011	0011	0011	0011	0011	1100	1100	1100	1100
51	1001	1001	1001	1001	0110	0110	0110	0110	0110	0110	0110	0110	1001	1001	1001	1001
52	1111	0000	1111	0000	0000	1111	0000	1111	0000	1111	0000	1111	1111	0000	1111	0000
53	1010	0101	1010	0101	0101	1010	0101	1010	0101	1010	0101	1010	1010	0101	1010	0101
54	1100	0011	1100	0011	0011	1100	0011	1100	0011	1100	0011	1100	1100	0011	1100	0011
55	1001	0110	1001	0110	0110	1001	0110	1001	0110	1001	0110	1001	1001	0110	1001	0110
56	1111	1111	0000	0000	0000	0000	1111	0000	0000	0000	1111	1111	1111	1111	0000	0000
57	1010	1010	0101	0101	0101	0101	1010	0101	0101	0101	1010	1010	1010	1010	0101	0101
58	1100	1100	0011	0011	0011	0011	1100	0011	0011	0011	1100	1100	1100	1100	0011	0011
59	1001	1001	0110	0110	0110	0110	1001	0110	0110	0110	1001	1001	1001	1001	0110	0110
60	1111	0000	0000	1111	0000	1111	1111	0000	0000	1111	1111	0000	1111	0000	0000	1111
61	1010	0101	0101	1010	0101	1010	1010	0101	0101	1010	1010	0101	1010	0101	0101	1010
62	1100	0011	0011	1100	0011	1100	1100	0011	0011	1100	1100	0011	1100	0011	0011	1100
63	1001	0110	0110	1001	0110	1001	1001	0110	0110	1001	1001	0110	1001	0110	0110	1001

(Left margin label: Walsh Code Index)

Consider Walsh code spreading which can be accomplished by performing the modulo-2 addition of the repeater output symbol with Walsh n, $0 \le n \le 255$ for 16 kbps in the following.

Example 10.10   Consider Walsh code spreading with Walsh 128.

In accordance with Table 10.6, Walsh chip within Walsh code 128 was shown in Example 10.8. Walsh code spreading is computed from the following modulo-2 addition:

Spreading = I-channel repetition symbol $\oplus$ Walsh chip within Walsh 128

The I-channel symbol repetition was already obtained in Example 10.6. Walsh code 128 is listed in Example 10.8. Therefore, the orthogonal spreading by Walsh code 128 is easily computed as shown below:

**Repetition
Symbol   Walsh code spreading**

$1 \rightarrow$ 11111111 11111111 11111111 11111111 11111111 11111111 11111111 11111111

$1 \rightarrow$ 11111111 11111111 11111111 11111111 11111111 11111111 11111111 11111111

$1 \rightarrow$ 00000000 00000000 00000000 00000000 00000000 00000000 00000000 00000000

$1 \rightarrow$ 00000000 00000000 00000000 00000000 00000000 00000000 00000000 00000000

Walsh code spreading was shown against four repetition symbols only.

Example 10.11   Now consider Walsh code spreading with the Q-channel repetition symbols. The Q-channel repetition symbols were computed in Example 10.7. Hence, Walsh code spreading is accomplished by performing the modulo-2 addition of Q-channel repetition symbols with Walsh code 128 as follows:

**Repetition
Symbol   Walsh code spreading**

$0 \rightarrow$ 00000000 00000000 00000000 00000000 00000000 00000000 00000000 00000000

$0 \rightarrow$ 00000000 00000000 00000000 00000000 00000000 00000000 00000000 00000000

$0 \rightarrow$ 11111111 11111111 11111111 11111111 11111111 11111111 11111111 11111111

$0 \rightarrow$ 11111111 11111111 11111111 11111111 11111111 11111111 11111111 11111111

Walsh code spreading was shown against Q-channel four repetition symbols only.

10.2.6 Sync Channel Quadrature Spreading

Following Walsh code spreading, the sync channel is spread in quadrature as shown in Fig. 10.4. The chip rate for the pilot PN sequence is 4.096 Mcps. A pilot sequence is generated by a code sequence with a period of 81920 chips generated from the long code generator. This long code with period of $2^{32}-1$ chips is specified by the tap polynomial, $g(x) = 1 + x + x^2 + x^{22} + x^{32}$. A pilot code sequence generated from the long code generator is shown in Fig. 10.2.

Example 10.12 Compute the pilot code sequence. The generator tap vector is (11100000000000000000001000000001) and the long code seed vector is (10010011001010011011010111100011) as assumed previously. Using these two vectors, the pilot code sequence can be generated from the operation of Fig. 10.2, as shown below:

1011110101110100 0010111010110010 1001111011101011 0111001111000001
0010100011011010 0110110111011100 1010110100101100 0100101000000101
1100001010110101 0101101101001111 1001111011101101 0011101110100000
0011000010001111 1000000110000000 1111001110010010 0110000000011010
1101110001001000 0011001001001000 1000011111011001 0110100110100001
1111111111110001 1100000000010101 1111100100101011 0000101010101011
0100010100000000 0111010101001101 1000100001100011 0101001000110100
0011010000100101 1011011111100001 0111010010100101 0010111011101101
1111010000000110 1100100100110011 1000000011110000 0100001110001010
0110000001001001 11011000

Since the pilot code was generated, quadrature spreading is easily accomplished by performing the modulo-2 addition of the Walsh code spreading with the pilot code sequence.

Example 10.13 Perform the modulo-2 addition of sync I-channel Walsh spreading with the pilot code. Using the pilot code obtained in Example 10.12 and the sync I-channel Walsh spreading, quadrature spreading can be achieved by performing the modulo-2 addition of these two sequences as follows:

Quadrature spreading = I-channel Walsh spreading $\oplus$ Pilot code sequence

01000010 10001011 11010001 01001101 01100001 00010100 10001100 00111110
11010111 00100101 10010010 00100011 01010010 11010011 10110101 11111010
11000010 10110101 01011011 01001111 10011110 11101101 00111011 10100000
00110000 10001111 10000001 10000000 11110011 10010010 01100000 00011010
00100011 10110111 11001101 10110111 01111000 00100110 10010110 01011110
00000000 10001110 00111111 11101010 00000110 11010100 11110101 01010100
01000101 00000000 01110101 01001101 10001000 01100011 01010010 00110100
00110100 00100101 10110111 11100001 01110100 10100101 00101110 11101101

Example 10.14 In a similar fashion, quadrature spreading for the sync Q channel is accomplished by performing the modulo-2 addition of Q-channel Walsh spreading computed in Example 10.11 with the pilot code obtained in Example 10.12, as shown in the following:

Quadrature spreading = Q-channel Walsh spreading $\oplus$ Pilot code sequence

10111101 01110100 00101110 10110010 10011110 11101011 01110011 11000001
00101000 11011010 01101101 11011100 10101101 00101100 01001010 00000101
00111101 01001010 10100100 10110000 01100001 00010010 11000100 01011111
11001111 01110000 01111110 01111111 00001100 01101101 10011111 11100101
11011100 01001000 00110010 01001000 10000111 11011001 01101001 10100001
11111111 01110001 11000000 00010101 11111001 00101011 00001010 10101011
10111010 11111111 10001010 10110010 01110111 10011100 10101101 11001011
11001011 11011010 01001000 00011110 10001011 01011010 11010001 00010010

10.2.7 Sync Channel Baseband Filtering

Following the quadrature spreading operation, the I and Q impulses are applied to the inputs of the I and Q baseband filters as specified in Section 10.1.2 (see Fig. 10.3). After baseband filtering, the binary I and Q data are mapped into phase according to Table 10.9, as shown in Fig. 10.6.

Table 10.9   Forward CDMA Channel I and Q Mapping

I	Q	Phase
0	0	1/4
1	0	3/4
1	1	−3/4
0	1	−1/4

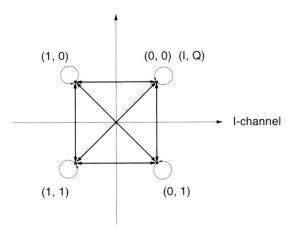

Figure 10.6   Forward CDMA channel signal constellation and phase mapping

10.3 PAGING CHANNEL

The paging channel is encoded, interleaved, spread, and modulated to become a spread spectrum signal that is monitored by personal stations operating within the coverage area of the base station. The base station uses multiple paging channels to transmit system information and personal station specific messages.

The paging channels transmit information at a fixed data rate of 16 kbps. The paging channel frame is 5 ms in duration. The first paging channel frame occurs at the start of base station transmission time. The paging channel is divided into paging channel slots that are each 20 ms in duration.

The pilot PN sequences for the paging channel use the same pilot PN sequence offset as the pilot channel for a given base station. The paging channel structure is shown in Fig. 10.7. The modulation parameters for the paging channel are shown in Table 10.10.

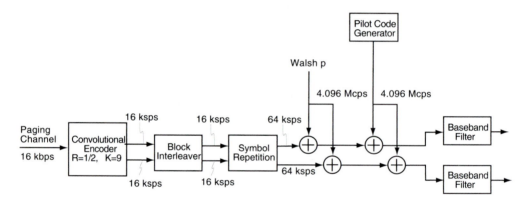

Figure 10.7 Paging channel structure

Table 10.10   Paging Channel Modulation Parameters

Parameters	Data Rate (bps) 16000	Units
PN chip rate	4.096	Mcps
Code rate	1/2	bits/symbols
Symbol repetition	4	
Symbol rate	64000	sps
PN chips/symbol	64	PN chips/symbol
PN chips/bit	256	PN chips/bit

10.3.1 Paging Channel Convolutional Encoding

The paging channel data is convolutionally encoded prior to transmission as specified in Section 10.2.2. The (2, 1, 8) convolutional encoder is used for generation of two code symbols (c_0, c_1) for each data bit to the encoder.

The convolutional encoder involves the modulo-2 addition in accordance with the following tap polynomials:

$$g_0(x) = 1 + x + x^2 + x^3 + x^5 + x^7 + x^8$$
$$g_1(x) = 1 + x^2 + x^3 + x^4 + x^8$$

The (2, 1, 8) convolutional encoder with these tap polynomials is shown in Fig. 10.5.

Example 10.15   The base station transmits the data sequence on the paging channel at a 16 kbps bit rate. The paging channel frame consists of 80 bits over the 5 ms duration.

Suppose the data sequence to the encoder is as follows:

1000110101 0001110101 1100101101 1101000001

0110110010 1001111101 1001011010 0110011010

The coded symbols, c_0 and c_1, of the encoder are computed according to each data bit in this input sequence. The coded output of the (2,1,8) convolutional encoder is as shown below:

c_0 :   1111110011 1000000011 1101010101 1110111010
        0110010010 0110011110 1110000010 1011110100

c_1 :   1011011111 1001110101 0010100110 1110111010
        0100110110 1111000000 1110010100 0100111111

10.3.2 Paging Channel Interleaving

The base station interleaves all coded symbols on the paging channel prior to modulation and transmission. A block interleaving span of 5 ms is used for the paging channel. The paging channel interleaver is an array with 8 rows and 10 columns as shown in Tables 10.2 and 10.3. These tables show the ordering of write operations of coded symbols into the I and Q interleaving array at 16000 bps for both the I and Q channels. The paging channel code symbols are output from the interleaver by rows.

Example 10.16   Fill out all coded symbols of c_0 according to Table 10.2. Then, an array with 8 rows and 10 columns is arranged as shown in the following:

```
1100101101
1101010101
1110110101
1011101001          ⇨        Read row by row
1010101110
1011010101
0001100110
0011000100
```

Thus, reading this array by rows, the I-channel interleaver output is obtained as shown below:

1100101101 1101010101 1110110101 1011101001
1010101110 1011010101 0001100110 0011000100

which is the I-channel input to the symbol repeater.

Example 10.17 Applying c_1 computed in Example 10.15 to Table 10.3, the Q-channel output symbols of the block interleaver is obtained by writing c_1 into the 8 x 10 array of Table 10.3. That is,

```
1111010001
0111011100
1001011111
0100111101          ⇨        Read row by row
0100011011
1001001100
0100111001
0101110111
```

By reading this array row by row, the Q-channel interleaver output yields:

1111010001 0111011100 1001011111 0100111101
0100011011 1001001100 0100111001 0101110111

This symbol sequence is the Q-channel input to the symbol repeater.

10.3.3 Paging Channel Symbol Repetition

For the paging channel, each interleaved symbol from the convolutionally encoded symbols is repeated three times (each symbol occurs 4 consecutive times) prior to data scrambling.

Example 10.18 Using the I-channel interleaver output computed in Example 10.16, the symbol repetition resulting from repeating each symbol three times is shown below:

```
1111 1111 0000 0000 1111 0000 1111 1111 0000 1111
1111 1111 0000 1111 0000 1111 0000 1111 0000 1111
1111 1111 1111 0000 1111 1111 0000 1111 0000 1111
1111 0000 1111 1111 1111 0000 1111 0000 0000 1111
1111 0000 1111 0000 1111 0000 1111 1111 1111 0000
1111 0000 1111 1111 0000 1111 0000 1111 0000 1111
0000 0000 0000 1111 1111 0000 0000 1111 1111 0000
0000 0000 1111 1111 0000 0000 0000 1111 0000 0000
```

Example 10.19 Using the Q-channel interleaver output obtained in Example 10.17, the symbol repetition for the Q-channel repeater results from repeating each symbol three times as shown in the following:

1111 1111 1111 1111 0000 1111 0000 0000 0000 1111
0000 1111 1111 1111 0000 1111 1111 1111 0000 0000
1111 0000 0000 1111 0000 1111 1111 1111 1111 1111
0000 1111 0000 0000 1111 1111 1111 1111 0000 1111
0000 1111 0000 0000 0000 1111 1111 0000 1111 1111
1111 0000 0000 1111 0000 0000 1111 1111 0000 0000
0000 1111 0000 0000 1111 1111 1111 0000 0000 1111
0000 1111 0000 1111 1111 1111 0000 1111 1111 1111

10.3.4 Paging Channel Orthogonal Spreading

Prior to transmission, the paging channel is orthogonally spread by a Walsh function, with index equal to the paging channel number. Each code channel transmitted on the forward W-CDMA channel is spread with a Walsh code at a fixed chip rate of 4.096 Mcps to provide orthogonal channelization among all code channels on a given forward CDMA channel. The paging channel that is spread using Walsh code n is assigned to code index number n, $0 \leq n \leq 255$ for 16 kbps. One of the 256 time-orthogonal Walsh codes, defined in Table 10.6, is used for 16 kbps.

The paging channels are assigned code indices one through seven (inclusive), in sequence. There can be at most 256 channels for 16 kbps. If one of 128 Walsh codes is used for a 32 kbps channel, the same code cannot be simultaneously used on any 16 kbps channel.

Orthogonal spreading is accomplished by performing the modulo-2 addition of the repeater output with Walsh code index n, $1 \leq n \leq 7$.

Example 10.20 If Walsh code index 1 is chosen, then from Table 10.6 Walsh chip within a Walsh code is

01010101 01010101 01010101 01010101 01010101 01010101 01010101 01010101
01010101 01010101 01010101 01010101 01010101 01010101 01010101 01010101
01010101 01010101 01010101 01010101 01010101 01010101 01010101 01010101
01010101 01010101 01010101 01010101 01010101 01010101 01010101 01010101

Performing the modulo-2 addition of each symbol (symbol positions from 8 to 12) of the I-channel repeater output with Walsh code index 1 yields as follows:

Symbol No.
(8~12)   **Result of modulo-2 addition (I-channel)**

1 →   10101010 10101010 10101010 10101010 10101010 10101010 10101010 10101010

1 →   10101010 10101010 10101010 10101010 10101010 10101010 10101010 10101010

1 →   10101010 10101010 10101010 10101010 10101010 10101010 10101010 10101010

1 →   10101010 10101010 10101010 10101010 10101010 10101010 10101010 10101010

0 →   01010101 01010101 01010101 01010101 01010101 01010101 01010101 01010101

0 →   01010101 01010101 01010101 01010101 01010101 01010101 01010101 01010101

0 →   01010101 01010101 01010101 01010101 01010101 01010101 01010101 01010101

0 →   01010101 01010101 01010101 01010101 01010101 01010101 01010101 01010101

Thus, orthogonal spreading on the paging I channel, resulted from the modulo-2 addition, is expressed by the above sequence.

Example 10.21   Choosing Walsh code index 1 of Table 10.6, the modulo-2 addition of each symbol (between 8 to 12 position) of the Q-channel repeater output with Walsh code index 1 results in:

Symbol No.
(8~12)   **Result of modulo-2 addition (Q-channel)**

1 →   10101010 10101010 10101010 10101010 10101010 10101010 10101010 10101010

1 →   10101010 10101010 10101010 10101010 10101010 10101010 10101010 10101010

1 →   10101010 10101010 10101010 10101010 10101010 10101010 10101010 10101010

1 →   10101010 10101010 10101010 10101010 10101010 10101010 10101010 10101010

1 →   10101010 10101010 10101010 10101010 10101010 10101010 10101010 10101010

1 →   10101010 10101010 10101010 10101010 10101010 10101010 10101010 10101010

1 →   10101010 10101010 10101010 10101010 10101010 10101010 10101010 10101010

1 →   10101010 10101010 10101010 10101010 10101010 10101010 10101010 10101010

This sequence represents the orthogonal spreading based on the paging Q channel.

10.3.5 Paging Channel Quadrature Modulation

Following the orthogonal spreading, the paging channel is spread in quadrature. The chip rate for the pilot PN sequence is 4.096 Mcps. A pilot sequence is generated by a code sequence with a period of 81920 chips generated from the long code generator (see Fig. 10.2)

Example 10.22 Referring to Fig. 10.2, a pilot code sequence is generated from the long code generator specified by the generator tap polynomial $g(x) = 1 + x + x^2 + x^{22} + x^{32}$.

Generator tap vector: 11100000000000000000001000000001
Initial seed vector: 10010011001010011011010111100011

Using these two vectors, the pilot code sequence is generated from the mechanism of Fig. 10.2, as shown below:

10111101 01110100 00101110 10110010 10011110 11101011 01110011 11000001
00101000 11011010 01101101 11011100 10101101 00101100 01001010 00000101
11000010 10110101 01011011 01001111 10011110 11101101 00111011 10100000
00110000 10001111 10000001 10000000 11110011 10010010 01100000 00011010
11011100 01001000 00110010 01001000 10000111 11011001 01101001 10100001
11111111 01110001 11000000 00010101 11111001 00101011 00001010 10101011
01000101 00000000 01110101 01001101 10001000 01100011 01010010 00110100
00110100 00100101 10110111 11100001 01110100 10100101 00101110 11101101
11110100 00000110 11001001 00110011 10000000 11110000 01000011 10001010
01100000 01001001 1101100

Thus, quadrature spreading is achieved by performing the modulo-2 addition of the symbol sequence resulting from orthogonal spreading with the pilot code sequence.

Example 10.23 Using the I-channel orthogonal modulation sequence obtained in Example 10.20 and the pilot code computed in Example 10.22, the quadrature spreading sequence based on the paging I channel is computed as follows:

00010111 11011110 10000100 00011000 00110100 01000001 11011001 01101011
10000010 01110000 11000111 01110110 00000111 10000110 11100000 10101111
01101000 00011111 11110001 11100101 00110100 01000111 10010001 00001010
10011010 00100101 00101011 00101010 01011001 00111000 11001010 10110000
10001001 00011101 01100111 00011101 11010010 10001100 00111100 11110100
10101010 00100100 10010101 01000000 10101100 01111110 01011111 11111110
00010000 01010101 00100000 00011000 11011101 00110110 00000111 01100001
01100001 01110000 11100010 10110100 00100001 11110000 01111011 10111000

Example 10.24   Using the Q-channel orthogonal modulation sequence obtained in Example 10.21 and the pilot code computed in Example 10.22, quadrature spreading based on the paging Q channel is computed as shown below:

> 00010111 11011110 10000100 00011000 00110100 01000001 11011001 01101011
> 10000010 01110000 11000111 01110110 00000111 10000110 11100000 10101111
> 01101000 00011111 11110001 11100101 00110100 01000111 10010001 00001010
> 10011010 00100101 00101011 00101010 01011001 00111000 11001010 10110000
> 01110110 11100010 10011000 11100010 00101101 01110011 11000011 00001011
> 01010101 11011011 01101010 10111111 01010011 10000001 10100000 00000001
> 11101111 10101010 11011111 11100111 00100010 11001001 11111000 10011110
> 10011110 10001111 00011101 01001011 11011110 00001111 10000100 01000111

10.3.6 Paging Channel Filtering and Mapping

Following the quadrature spreading operation, the I and Q impulses are applied to the inputs of the I and Q baseband filters as shown in Fig. 10.7. The baseband filters have a frequency response $S(f)$ that satisfies the limits given in Fig. 10.3.

After baseband filtering, the binary I and Q data is mapped into phase according to Table 10.9, as shown in Fig. 10.6.

10.4 Forward Traffic Channel (FTC)

The forward traffic channel is used for the transmission of user and signaling information to a specific personal station during a call. The maximum number of forward traffic channels that can be simultaneously supported by a given forward channel is equal to 64 for 64 kbps, 128 for 32 kbps, and 256 for 16 kbps. The number of pilot channels, paging channels, and sync channels operating on the same forward channel must be subtracted from these numbers.

The base station transmits information on the forward traffic channel at the variable data rates of 64, 32, and 16 kbps. The output symbol from the punctured multiplexing is kept constant at 64 kbps. The forward traffic channel frame is 5 ms in duration. The pilot PN sequences for the forward traffic channel use the same pilot PN sequence offset as the pilot channel for a given base station. A base station may implement staggered forward traffic channel frames.

The forward information channel of the forward traffic channel frames sent at the 64, 32, and 16 kbps data rates consist of 320, 160, and 80 bits, respectively. The forward information channel of forward traffic channel frame offset is the same as the reverse traffic channel time offset. The forward signaling channel of forward channel frames sent at 4 and 2 kbps data rates consist of 20 and 10 bits, respectively.

10.4.1 Forward Traffic Channel Structure and Modulation Parameters

The forward traffic channel has the overall structure shown in Fig. 10.8. The forward traffic channel consists of a forward information channel and a forward signaling channel. The forward traffic channel is orthogonally spread by the appropriate Walsh code and is then spread by a PN sequence at a fixed chip rate of 4.096 Mcps.

The modulation parameters for the forward traffic channel are as shown in Tables 10.11 and 10.12.

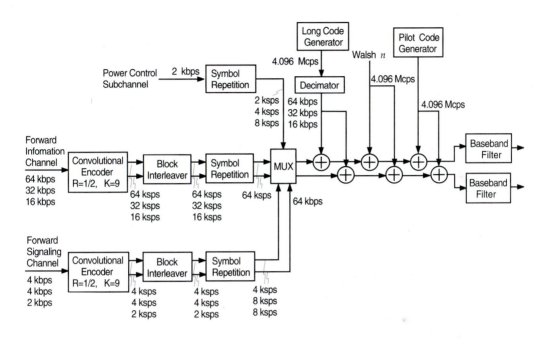

Figure 10.8 Forward traffic channel structure (After JTC(AIR)/95)

Table 10.11 Forward Information Channel for Forward Traffic Channel Modulation Parameters

Parameter	Data Rate (bps)			Units
	64000	32000	16000	
PN chip rate	4.096	4.096	4.096	Mcps
Code rate	1/2	1/2	1/2	bits/symbols
Symbol repetition	1	2	4	
Punctured rate	29/32	26/32	24/32	
Effective code rate	16/29	8/13	4/6	bits/symbols
Symbol rate	64000	64000	64000	sps
PN chips/symbol	64	64	64	PN chips/symbol
PN chips/bit	58	104	192	PN chips/bit

Table 10.12   Forward Signaling Channel for Forward Traffic Channel Modulation Parameters

Parameter	Data Rate (bps)		Units
	4000	2000	
PN chip rate	4.096	4.096	Mcps
Code rate	1/2	1/2	bits/symbols
Symbol repetition	1, 2	4	
Symbol rate	4000, 8000	8000	sps
PN chips/symbol	64	64	PN chips/symbol
PN chips/bit	64, 128	256	PN chips/bit

10.4.2   Forward Information Channel Convolutional Encoding

The forward information channel and forward signaling channel of the forward traffic channel are convolutionally encoded prior to transmission. The (2, 1, 8) convolutional code is used and its code rate R=1/2 with a constraint length K=9. The generator tap polynomials are

$$g_0(x) = 1 + x + x^2 + x^3 + x^5 + x^7 + x^8$$
$$g_1(x) = 1 + x^2 + x^3 + x^4 + x^8$$

The (2, 1, 8) convolutional encoder generates the two code symbol pair (c_0, c_1) for each data bit input to the encoder. The first code symbol c_0 is output encoded with $g_0(x)$, and the last code symbol c_1 is output encoded with $g_1(x)$. The state of the convolutional encoder, upon initialization, is the all-zero state.

Convolutional encoding involves the modulo-2 addition of selected taps of a serially time-delayed data sequence. The length of the data sequence delay is equal to K−1 = 8. Fig. 10.5 shows the (2, 1, 8) convolutional encoder that is used for the forward traffic channel.

The base station transmits information on the forward traffic channel at the variable data rates of 64, 32, and 16 kbps. The forward signaling channel of the forward traffic channel frames are sent at 4 and 2 kbps data rates.

Example 10.25   Consider the case that the base station transmits a data sequence on the forward information channel at the 16 kbps data rate. The FTC forward information channel at the 16 kbps rate consists of 80 bits and its frame is 5 ms in duration.

The input data sequence to the encoder is assumed as

1011000101 0110100011 1101010001 1001011100
1010011101 1101110001 1001011100 0100011011

The initial all-zero state of the $(2, 1, 8)$ convolutional encoder is $1\,0\,0\,0\,0\,0\,0\,0\,0$ where the binary value '1' represents the flag bit. Applying these input data and the all-zero initial state to Fig. 10.5, the convolutionally encoded symbols are computed as shown in the following:

c_0 : 1101011111 0100111101 0100101100 0111010110
   0100100100 1011011001 0001010110 1111101000

c_1 : 1000000010 0000110000 1100010101 1000100011
   1000110010 0101000011 1010100011 0100011110

10.4.3 FTC Information Channel Interleaving

The forward traffic channel has an interleaver for both information and signaling. The coded symbols c_0 and c_1 on both the information and signaling channels of the forward traffic channel are interleaved as specified in the following. The base station interleaves all coded symbols on the FTC information or signaling channel prior to multiplexing, modulation, and transmission. A block interleaving span of 5/10/20 ms may be used. The interleaver span of 10 and 20 ms is optional and only applies to the information channel of the forward traffic channel.

The information channel of the traffic channel interleaver has puncturing functions. The input data of the FTC information channel interleaver is punctured at the rate of 29/32, 13/16, and 6/8 for 64, 32, and 16 kbps, respectively. The FTC information channel interleaver is an array with 29 rows and 10 columns, 13 rows and 10 columns, and 6 rows and 10 columns at 64, 32, and 16 kbps, respectively.

The FTC signaling channel interleaver is an array with 5 rows and 4 columns (i.e., 20 cells) at the data rate of 4 kbps, and 5 rows and 2 columns (i.e., 10 cells) at the data rate of 2 kbps.

Tables 10.13 and 10.14 show the ordering of write operations of symbols into the I and Q interleaving array at 16 kbps for both the I and Q information channels. Forward traffic channel code symbols are output from the interleaver by rows.

Table 10.13 FTC Information I-channel Interleaver Input
     (Write Operation at 16000 bps; Interleaver Span of 5 ms)

1	9	17	25	33	41	49	57	65	73
2	10	18	26	34	42	50	58	66	74
3	11	19	27	35	43	51	59	67	75
5	13	21	29	37	45	53	61	69	77
6	14	22	30	38	46	54	62	70	78
7	15	23	31	39	47	55	63	71	79

Table 10.14   FTC Information Q-channel Interleaver Input
(Write Operation at 16000 bps; Interleaver Span of 5 ms)

37	45	53	61	69	77	5	13	21	29
39	47	55	63	71	79	7	15	23	31
40	48	56	64	72	80	8	16	24	32
41	49	57	65	73	1	9	17	25	33
43	51	59	67	75	2	11	19	27	35
44	52	60	68	76	4	12	20	28	36

   Notice that the input data of both information I and Q channels of the forward traffic channel interleaver are punctured at the rate of 6/8, i.e., 6 rows and 10 columns (from 80 cells to 60 cells) for 16000 bps.

Example 10.26   Using the coded symbol sequence c_0 obtained in Example 10.25, writing c_0 by columns into the information I-channel interleaver according to Table 10.13 yields:

**Interleaver input by columns
(I-channel)**

⇩

1111100101
1110110011
0001001001   ⇨   **Interleaver output
0000011011   by rows**
1010101000
1100100010

   Thus, the interleaver output sequence for the information I channel is computed as
1111100101 1110110011 0001001001
0000011011 1010101000 1100100010

Example 10.27   For the information Q channel of forward traffic channel, writing the coded symbol sequence c_1 by columns into the information Q-channel interleaver in accordance with Table 10.14 yields:

**Interleaver input by columns
(Q-channel)**

0101110010
1011010101
1000100100   ⇨   **Interleaver output
by rows**
1101011000
0000000001
0110100010

Thus, the interleaver output sequence for the information Q channel is obtained as

0101110010 1011010101 1000100100
1101011000 0000000001 0110100010

10.4.4 FTC Information Channel Symbol Repetition

After the interleaver, the symbol repetition rate on the forward traffic channels varies with the data rate. Each symbol at 64 kbps must not be repeated prior to multiplexing. Each symbol at 32 kbps is repeated one time (each symbol occurs 2 consecutive times) prior to multiplexing. Each symbol at 16 kbps is repeated 3 times (each symbol occurs 4 consecutive times) prior to multiplexing.

Example 10.28   Each symbol at 16 kbps is repeated 3 times. Using the symbol output from the information I-channel interleaver computed in Example 10.26, the symbol repetition on the FTC information I channel is

1111 1111 1111 1111 1111 0000 0000 1111 0000 1111
1111 1111 1111 0000 1111 1111 0000 0000 1111 1111
0000 0000 0000 1111 0000 0000 1111 0000 0000 1111
0000 0000 0000 0000 0000 1111 1111 0000 1111 1111
1111 0000 1111 0000 1111 0000 1111 0000 0000 0000
1111 1111 0000 0000 1111 0000 0000 0000 1111 0000

Example 10.29 Using the symbol output from the information Q-channel interleaver obtained in Example 10.27, the symbol repetition on the FTC information Q channel is
    0000 1111 0000 1111 1111 1111 0000 0000 1111 0000
1111 0000 1111 1111 0000 1111 0000 1111 0000 1111
1111 0000 0000 0000 1111 0000 0000 1111 0000 0000
1111 1111 0000 1111 0000 1111 1111 0000 0000 0000
0000 0000 0000 0000 0000 0000 0000 0000 0000 1111
0000 1111 1111 0000 1111 0000 0000 0000 1111 0000

10.4.5 Forward Signaling Channel Convolutional Encoding

The forward signaling channel of the forward traffic channel (FTC) is convolutionally encoded prior to transmission. The $(2, 1, 8)$ convolutional code with the generator tap polynomials

$$g_0(x) = 1 + x + x^2 + x^3 + x^5 + x^7 + x^8$$
$$g_1(x) = 1 + x^2 + x^3 + x^4 + x^8$$

also is used for the FTC forward signaling channel.

Referring to Fig. 10.5, the $(2, 1, 8)$ convolutional encoder generates the code symbol pair (c_0, c_1) for each data bit input to the encoder. The forward signaling I and Q channels of forward traffic channel frames are sent at 4 and 2 kbps data rates.

Example 10.30 The forward signaling I and Q channel frames sent at 2 kbps data rate consist of 10 bits. Suppose the input data to the $(2, 1, 8)$ convolutional encoder is 1001110001 and the initial contents of the register is all-zero. Then, the encoder coded symbols are computed as shown below:

Shift No.	Register contents	Code symbols (c_0, c_1)
0	100000000	(1, 1)
1	010000000	(1, 0)
2	001000000	(1, 1)
3	100100000	(0, 0)
4	110010000	(0, 0)
5	111001000	(0, 0)
6	011100100	(1, 0)
7	001110010	(1, 1)
8	000111001	(1, 1)
9	100011100	(0, 0)

Thus, the code symbol c_0 encoded with the generator polynomial
$g_0(x) = 1 + x + x^2 + x^3 + x^5 + x^7 + x^8$ are
$$c_0 = (1110001110)$$

The code symbol c_1 encoded with the generator polynomial
$g_1(x) = 1 + x^2 + x^3 + x^4 + x^8$ are
$$c_1 = (1010000110)$$

10.4.6 Forward Signaling Channel Interleaving

The forward traffic channel has an interleaver for signaling. A block interleaving span of 5 ms is used. The forward signaling channel of the traffic channel interleaver is an array with 5 rows and 2 columns (i.e., 10 cells) at the data rate of 2000 bps, as shown in Table 10.15.

Table 10.15   Forward Signaling Channel Interleaver Input
(I-channel and Q-channel Write Operation at 2000 bps)

$$
\begin{array}{cc}
1 & 6 \\
2 & 7 \\
3 & 8 \\
4 & 9 \\
5 & 10 \\
\end{array}
$$

Example 10.31   Using the coded symbol $c_0 = (1110001110)$ computed in Example 10.30, writing c_0 by columns into the forward signaling I-channel interleaver according to Table 10.15 results in:

Input by columns
⇩

$$
\begin{array}{cc}
1 & 0 \\
1 & 1 \\
1 & 1 \\
0 & 1 \\
0 & 0 \\
\end{array}
$$

⇨   **Output by rows**

Thus, the interleaver output of the forward signaling I channel is
1011110100

Example 10.32   Using the coded symbol $c_1 = (1010000110)$ computed in Example 10.30, the interleaver output of the forward signaling Q channel is computed by writing c_1 by columns according to Table 10.15 as follows:

Input by columns

⇩

1 0
0 0
1 1   ⇨   **Output by rows**
0 1
0 0

Therefore, the interleaver output of forward signaling Q channel is
1000110100

10.4.7 Forward Signaling Channel Symbol Repetition

After interleaving, the symbol repetition rate on the forward signaling channel varies with the data rate. Each symbol at 4000 bps is repeated 1 time (each symbol occurs 2 consecutive times) prior to multiplexing.

Example 10.33   Each symbol at 2000 bps is repeated 3 times. The symbol output from the forward signaling I and Q channel interleaver is repeated as follows:

1. Symbol repetition on FTC signaling I channel
   Using the interleaver output of signaling I channel computed in Example 10.31, the symbol repetition for the I channel is obtained as
   1111 0000 1111 1111 1111 1111 0000 1111 0000 0000

2. Symbol repetition on FTC signaling Q channel
   Using the interleaver output of signaling channel computed in Example 10.32, the symbol repetition for Q channel is found as
   1111 0000 0000 0000 1111 1111 0000 1111 0000 0000

10.4.8 FTC Power Control Subchannel

The base station generates the power control bit and inserts on every forward traffic channel. A power control subchannel is continuously transmitted on the forward traffic channel. The subchannel transmits at a rate of one bit ('0' or '1') every 500μ s (i.e., 2000 bps).

The mean power is referenced to the nominal CDMA channel bandwidth of 5.0 MHz. The power difference between the mean input power and the mean output power at the base station is defined and set to be a constant value to control the mean output power of the personal station.

A '0' bit indicates to the personal station to increase the output power level and a '1' bit indicates to the personal station to decrease the average output power level. The amount that the personal station increases and decreases its power for every power control bit should be specified.

The reverse traffic channel receiver at the base station estimates the received signal strength of the particular personal station it is assigned to over a 500μ s period, which is equivalent to 32 code symbols. The base station receiver uses the estimate to determine the value of the power control bit ('0' or '1'). The base station transmits the power control bit on the corresponding forward traffic channel using the power control subchannel as shown in Fig. 10.8.

10.4.9 Forward Traffic Channel Multiplexing

For forward traffic channels, the effective output symbol rate of multiplexing is 64 ksps. The forward information channel, forward signaling and power control subchannel of the forward traffic channel are multiplexed as specified in the following. The base station multiplexes the power control bits, signaling bits, and information bits after symbol repetition as shown in Tables 10.16, 10.17, and 10.18 with data rates of 64, 32, and 16 kbps, respectively. In these tables, p, s, and i represent power control bit, signaling symbol and information symbol, respectively. The MUX output is called as the multiplexed forward traffic channel.

Table 10.16 Multiplexing Matrix for Forward Traffic Channel (Write Operation at 64 kbps)

p1	s1	s2	i1	i2	i3	i4	i5	i6	i7	i8	i9	i10	i11	i12	i13
i14	i15	i16	i17	i18	i19	i20	i21	i22	i23	i24	i25	i26	i27	i28	i29
p2	s3	s4	i30	i31	i32	i33	i34	i35	i36	i37	i38	i39	i40	i41	i42
i43	i44	i45	i46	i47	i48	i49	i50	i51	i52	i53	i54	i55	i56	i57	i58
p3	s5	s6	i59	i60	i61	i62	i63	i64	i65	i66	i67	i68	i69	i70	i71
i72	i73	i74	i75	i76	i77	i78	i79	i80	i81	i82	i83	i84	i85	i86	i87
p4	s7	s8	i88	i89	i90	i91	i92	i93	i94	i95	i96	i97	i98	i99	i100
i101	i102	i103	i104	i105	i106	i107	i108	i109	i110	i111	i112	i113	i114	i115	i116
p5	s5	s10	i117	i118	i119	i120	i121	i122	i123	i124	i125	i126	i127	iI28	i129
i130	i131	i132	i133	i134	i135	iI36	i137	i138	i139	i140	i141	i142	i143	i144	iI45
p6	s11	s12	i146	i147	i148	i149	i150	i151	i152	iI53	i154	i155	i156	i157	i158
i159	i160	i161	i162	i163	i164	i165	i166	i167	i168	i169	iI70	i171	i172	i173	i174
p7	s13	s14	i175	i176	i177	i778	i179	i180	i181	i182	i183	i184	i185	i186	i187
i188	i189	i190	i191	i192	i193	i194	i195	i196	i197	i198	i199	i200	i201	i202	i203
p8	s15	s16	i204	i205	i206	i207	i208	i209	i210	i211	i212	i213	i214	i215	i216
i217	i218	i219	i220	i221	i222	i223	i224	i225	i226	i227	i228	i229	i230	i321	i232
p9	s17	s18	i233	i234	i235	i236	i237	i238	i239	i240	i241	i242	i243	i244	i245

Table 10.16 Multiplexing Matrix for Forward Traffic Channel
(Write Operation at 64 kbps) *(continued)*

i246	i247	i248	i249	i250	i251	i252	i253	i254	i255	i256	i257	i258	i259	i260	i261
p10	s19	s20	i262	i263	i264	i265	i266	i267	i268	i269	i270	i271	i272	i273	i274
i275	i276	i277	i278	i279	i280	i281	i282	i283	i284	i285	i286	i287	i288	i289	i290

(After JTC (AIR) / 95.02.02 – 037R1)

Table 10.17 Multiplexing Matrix for Forward Traffic Channel
(Write Operation at 32 kbps)

p1	p1	s1	s1	s2	s2	i1	i1	i2	i2	i3	i3	i4	i4	i5	i5
i6	i6	i7	i7	i8	i8	i9	i9	i10	i10	i11	i11	i12	i12	i13	i13
p2	p2	s3	s3	s4	s4	i14	i14	i15	i15	i16	i16	i17	i17	i18	i18
i19	i19	i20	i20	i21	i21	i22	i22	i23	i23	i24	i24	i25	i25	i26	i26
p3	p3	s5	s5	s6	s6	i27	i27	i28	i28	i29	i29	i30	i30	i31	i31
i32	i32	i33	i33	i34	i34	i35	i35	i36	i36	i37	i37	i38	i38	i39	i39
p4	p4	s7	s7	s8	s8	i40	i40	i41	i41	i42	i42	i43	i43	i44	i44
i45	i45	i46	i46	i47	i47	i48	i48	i49	i49	i50	i50	i51	i51	i52	i52
p5	p5	s9	s9	s10	s10	i53	i53	i54	i54	i55	i55	i56	i56	i57	i57
i58	i58	i59	i59	i60	i60	i61	i61	i62	i62	i63	i63	i64	i64	i65	i65
p6	p6	s11	s11	s12	s12	i66	i66	i67	i67	i68	i68	i69	i69	i70	i70
i71	i71	i72	i72	i73	i73	i74	i74	i75	i75	i76	i76	i77	i77	i78	i78
p7	p7	s13	s13	s14	s14	i79	i79	i80	i80	i81	i81	i82	i82	i83	i83
i84	i84	i85	i85	i86	i86	i87	i87	i88	i88	i89	i89	i90	i90	i91	i91
p8	p8	s15	s15	s16	s16	i92	i92	i93	i93	i94	i94	i95	i95	i96	i96
i97	i97	i98	i98	i99	i99	i100	i100	i101	i101	i102	i102	i103	i103	i104	i104
p9	p9	s17	s17	s18	s18	i105	i105	i106	i106	i107	i107	i108	i108	i109	i109
i110	i110	i111	i111	i112	i112	i113	i113	i114	i114	i115	i115	i116	i116	i117	i117
p10	p10	s19	s19	s20	s20	i118	i118	i119	i119	i120	i120	i121	i121	i122	i122
i123	i123	i124	i124	i125	i125	i126	i126	i127	i127	i128	i128	i129	i129	i130	i130

(After JTC (AIR) / 95.02.02 – 037R1)

Table 10.18   Multiplexing Matrix for Forward Traffic Channel
(Write Operation at 16 kbps)

p1	p1	p1	p1	s1	s1	s1	s1	i1	i1	i1	i1	i2	i2	i2	i2
i3	i3	i3	i3	i4	i4	i4	i4	i5	i5	i5	i5	i6	i6	i6	i6
p2	p2	p2	p2	s2	s2	s2	s2	i7	i7	i7	i7	i8	i8	i8	i8
i9	i9	i9	i9	i10	i10	i10	i10	i11	i11	i11	i11	i12	i12	i12	i12
p3	p3	p3	p3	s3	s3	s3	s3	i13	i13	i13	i13	i14	i14	i14	i14
i15	i15	i15	i15	i16	i16	i16	i16	i17	i17	i17	i17	i18	i18	i18	i18
p4	p4	p4	p4	s4	s4	s4	s4	i19	i19	i19	i19	i20	i20	i20	i20
i21	i21	i21	i21	i22	i22	i22	i22	i23	i23	i23	i23	i24	i24	i24	i24
p5	p5	p5	p5	s5	s5	s5	s5	i25	i25	i25	i25	i26	i26	i26	i26
i27	i27	i27	i27	i28	i28	i28	i28	i29	i29	i29	i29	i30	i30	i30	i30
p6	p6	p6	p6	s6	s6	s6	s6	i31	i31	i31	i31	i32	i32	i32	i32
i33	i33	i33	i33	i34	i34	i34	i34	i35	i35	i35	i35	i36	i36	i36	i36
p7	p7	p7	p7	s7	s7	s7	s7	i37	i37	i37	i37	i38	i38	i38	i38
i39	i39	i39	i39	i40	i40	i40	i40	i41	i41	i41	i41	i42	i42	i42	i42
p8	p8	p8	p8	s8	s8	s8	s8	i43	i43	i43	i43	i44	i44	i44	i44
i45	i45	i45	i45	i46	i46	i46	i46	i47	i47	i47	i47	i48	i48	i48	i48
p9	p9	p9	p9	s9	s9	s9	s9	i49	i49	i49	i49	i50	i50	i50	i50
i51	i51	i51	i51	i52	i52	i52	i52	i53	i53	i53	i53	i54	i54	i54	i54
p10	p10	p10	p10	s10	s10	s10	s10	i55	i55	i55	i55	i56	i56	i56	i56
i57	i57	i57	i57	i58	i58	i58	i58	i59	i59	i59	i59	i60	i60	i60	i60

(After JTC (AIR) / 95.02.02 – 037R1)

p: Power control bit s: Signaling symbol i: Information symbol

Example 10.34   For the forward traffic I channel, consider multiplexing the power control bits, signaling bits, and information bits after symbol interleaving as shown in Table 10.18 with the 16 kbps data rate.

- Assumed power control bits: 1000010011 which corresponds to p1=1, p2=0, p3=0, p4=0, p5=0, p6=1, p7=0, p8=0, p9=1, p10=1

• The forward signaling I-channel interleaver output: 1 0 1 1 1 1 0 1 0 0 which corresponds to s1=1, s2=0, s3=1, s4=1, s5=1, s6=1, s7=0, s8=1, s9=0, s10=0

• The forward information I-channel interleaver output:

i1	i5	i10	i15	i20	i25	i30

1 1 1 1 1 0 0 1 0 1   1 1 1 0 1 1 0 0 1 1   0 0 0 1 0 0 1 0 0 1

i35	i40	i45	i50	i55	i60

0 0 0 0 0 1 1 0 1 1   1 0 1 0 1 0 1 0 0 0   1 1 0 0 1 0 0 0 1 0

Substituting these power control bits, signaling symbols (I-channel), and informaton symbols (I-channel) into the multiplexing matrix for the forward traffic either I or Q channel of Table 10.18 generates the multiplexed forward traffic I channel as shown below:

I-channel Multiplexed Symbol

1111 1111 1111 1111
1111 1111 1111 0000
0000 0000 0000 1111
0000 1111 1111 1111
0000 1111 1111 0000
1111 1111 0000 0000
0000 1111 1111 1111
0000 0000 0000 1111
0000 1111 0000 0000
1111 0000 0000 1111
1111 1111 0000 0000
0000 0000 0000 1111
0000 0000 1111 0000
1111 1111 1111 0000
0000 1111 1111 0000
1111 0000 1111 0000
1111 0000 0000 0000
1111 1111 0000 0000
1111 0000 1111 0000
0000 0000 1111 0000

This result indicates the multiplexed data symbols for the forward traffic I channel.

Example 10.35 Consider the multiplexing process for the forward traffic Q channel. Multiplexing the power control bits, signaling bits (Q-channel) and information bits (Q-channel) after symbol interleaving according to Table 10.18 generates the multiplex symbol sequence for the forward traffic Q channel.

• Assumed power control bits:
1000010011 ↔ p1, p2, p3, p4, p5, p6, p7, p8, p9, p10

- The forward signaling Q-channel interleaver output:

    $1000110100 \leftrightarrow s1, s2, s3, s4, s5, s6, s7, s8, s9, s10$

- The forward information Q-channel interleaver output:

i1	i5	i10	i15	i20	i25	i30

    0 1 0 1 1 1 0 0 1 0   1 0 1 1 0 1 0 1 0 1   1 0 0 0 1 0 0 1 0 0

i35	i40	i45	i50	i55	i60

    1 1 0 1 0 1 1 0 0 0   0 0 0 0 0 0 0 0 1   0 1 1 0 1 0 0 0 1 0

Substituting these power control bits, signaling symbols (Q-channel), and information symbols (Q-channel) to the multiplexing matrix of Table 10.18 result as shown below:

Q-channel Multiplexed Symbols

1111 1111 0000 1111

0000 1111 1111 1111

0000 0000 0000 0000

1111 0000 1111 0000

0000 0000 1111 1111

0000 1111 0000 1111

0000 0000 0000 1111

1111 0000 0000 0000

0000 1111 1111 0000

0000 1111 0000 0000

1111 1111 1111 1111

0000 1111 0000 1111

0000 0000 1111 0000

0000 0000 0000 0000

0000 1111 0000 0000

0000 0000 0000 0000

1111 0000 0000 1111

0000 1111 1111 0000

1111 0000 1111 0000

0000 0000 1111 0000

10.4.10 Discontinuous Transmission

When discontinuous transmission is enabled, prior to transmission, the forward traffic channel is gated by a time filter that allows transmission of certain punctured, multiplexed output symbols and deletion of others according to the voice activity indication. The voice activity detector (VAD) is to recognize an available traffic channel for transmission of a valid signal. The purpose of the voice activity detector is to reject high-level noise on an input channel, avoid introducing front end clipping on signals, minimize false operation on impulse noise, avoid clipping during a signaling sequence, etc. The signaling and power control bits are transmitted at all times. This process is illustrated in Fig. 10.9. As shown in Fig. 10.9, the cycle of the signaling and power control bits is gate-on (i.e., transmitted) for 46.875, 93.75, and 125μ s for transmission rates of 64, 32, and 16 kbps, respectively; and gate-off (i.e., not transmitted) for 453.125, 406.25, and 375μ s for transmission rates of 64, 32, and 16 kbps, respectively. This cycle provides ten repetitions per frame.

Each forward traffic channel has a frame offset of either 0, 1, 2 or 3 units of 1024 chips (125μ s) to shift the timing of gate-on during the discontinuous transmission of each forward traffic channel. The base station selects the appropriate frame offset for each forward traffic channel. The base station transmits this frame offset information to each personal station during call processing.

When the frame is gated-off (i.e., not transmitted) according to the voice activity detection indicator, the input data of the next frame is encoded by '0'. This is called a tail frame and it provides tail bits to the convolutional code decoder.

When the personal station detects that the voice activity indication transitions from the OFF state (i.e., not transmitted) to the ON state (i.e., transmitted), the convolutional encoder is set to the all-zero state at the start of the frame.

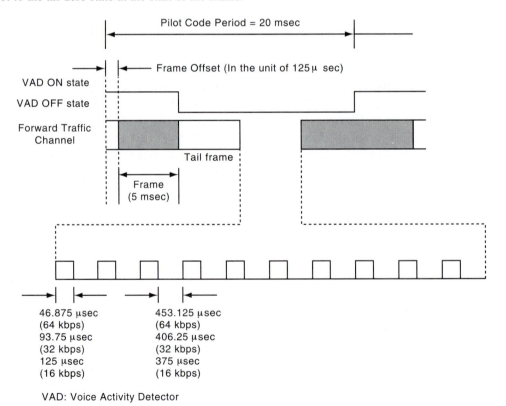

Figure 10.9 Dicontinuous transmission example

10.4.11 Forward Traffic Channel Data Scrambling

Data scrambling applies to the forward traffic channels. The forward traffic channel (FTC) is scrambled after multiplexing, that is, data scrambling is performed on the modulation of each symbol output from the multiplexor. Data scrambling is accomplished by performing the modulo-2 addition of the multiplexed output symbol with the binary value ('0' or '1') of the long code PN chip as shown in Fig. 10.10 (data rate of 64 ksps). The decimated output rates of the long code PN sequence are 64, 32, and 16 kbps corresponding to the data rates of 64, 32, and 16 ksps, respectively. This PN sequence is equivalent to the long code operation at 4.096 MHz clock rate, where only the first output of every 64 chips is used for the data scrambling. The long code may be generated as described in the following.

Example 10.36   The long code seed is sent in the message of the paging channel from the base station.   The   initial   seed   vector   consisting   of   32   bits   is   assumed   to   be 10010011001010011011010111100011.

Referring to Fig. 10.2, the register tap polynomial of the PN code generator is represented by $g(x) = 1 + x + x^2 + x^{22} + x^{32}$ and its vector form is 111000000000 00000000010000000001. The chip rate for the long code sequence is 4.096 Mcps. Using these two vectors, the long code sequence can be generated by the PN code generator as shown below:

10111101 01110100 00101110 10110010 10011110 11101011 01110011 11000001
00101000 11011010 01101101 11011100 10101101 00101100 01001010 00000101
11000010 10110101 01011011 01001111 10011110 11101101 00111011 10100000
00110000 10001111 10000001 10000000 11110011 10010010 01100000 00011010
11011100 01001000 00110010 01001000 10000111 11011001 01101001 10100001
11111111 01110001 11000000 00010101 11111001 00101011 00001010 10101011
01000101 00000000 01110101 01001101 10001000 01100011 01010010 00110100
00110100 00100101 10110111 11100001 01110100 10100101 00101110 11101101
11110100 00000110 11001001 00110011 10000000 11110000 01000011 10001010
01100000 01001001 1101100

This long code sequence is the equivalent of the PN sequence generating at 4.096 MCps.

Example 10.37   The decimated output is compiled with the first value of every 64 chips of the long code sequence. With the limited sample of the long code sequence computed in Example 10.36, the decimated output 1010110010 corresponds to the leftmost column of the long code sequence. Therefore, the extended decimated output corresponding to the sufficiently large size of the long code is given in the following.

1010110010 0101111011 0110100011 1011100100 0101111100
0101100110 1101001001 1011010100 1101011000 1101100111
1101101100 0101001110 1101000011 0001000010 0111111000
0111001010 0010111001 1111000011 1010011110 0001010101
0110000111 1111110001 0110011011 1110111110 0001001011
1001111011 0111110101 1110001001 0101010001 0100100011
0010001100 1000100001

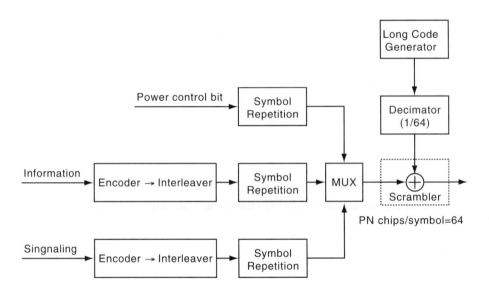

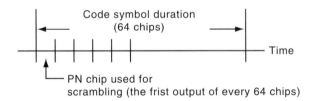

Figure 10.10 Data scrambler function

Example 10.38 Data scrambling is accomplished by performing the modulo-2 addition of each multiplexed symbol from the MUX operation in Fig. 10.10 with the long code PN chip at the 64 kbps decimated output rate. Only the first output of every 64 chips of the long code sequence are used for data scrambling. Utilizing the format of Data scrambling = Multiplexed I-channel symbol ⊕ Decimated symbol, forward traffic I channel data scrambling can be found as shown in the following:

I-channel Data Scrambling

0101001101	1010000100	1001011111	1011100100	0101000000
0110011001	0010001010	0100100100	0010100100	1101100111
0010010011	1001001110	1101111111	0010110010	0111000100
0111000101	1101000101	1111000011	1010011101	1101010101
1001000100	0000001110	0110011000	0001001110	1110001000
0101110100	0111110101	1101110110	0101010010	1000101100
0010001100	1011010001			

Example 10.39 Consider data scrambling for the forward traffic Q channel. The multiplexed symbol sequence for the forward traffic Q channel was already computed in Example 10.35. Using the same format of Data scrambling = Multiplexed Q-channel symbol ⊕ Decimated symbol, data scrambling is obtained by performing the modulo-2 addition according to the above the scrambling formula. The result of data scrambling is listed in the following table.

Q-channel Data Scrambling

0101001110	0110001011	1001011100	0111100100	0101111111
1001101001	1101001001	1000101011	1101100100	1110010111
1101101111	1010111110	1101000011	0010111101	0111111011
1011001010	1101000110	0000110011	0101011101	1101010101
1001000111	1111110001	0110011000	0010111110	0001001011
1001110100	0111110110	0010000110	1010010010	1000101100
0010001100	1011010001			

10.4.12 Forward Traffic Channel Orthogonal Spreading

Prior to transmission, the forward traffic channel after the data scrambling is orthogonally spread with a Walsh function as described below. Each forward traffic I or Q channel is spread with a Walsh code at a fixed chip rate of 4.096 Mcps to provide orthogonal channelization between these two I and Q subchannels. A forward traffic channel that is spread using Walsh code n is assigned to Walsh code index number n, $0 \le n \le 255$, for 16 kbps. All of the Walsh code tables are based on Walsh 64 and the inversion (or complement) of Walsh 64, i.e., $\overline{\text{Walsh}}$ $\overline{64}$. There can be at most 256 channels for 16 kbps. However, our analysis for orthogonal spreading here is based on Walsh code index number $n = 72$.

Example 10.40   Suppose Walsh index number $n = 72$ is assigned to a Walsh code as shown in the following:

0   Walsh chip within a Walsh code   255

Code index 72	Walsh 64	Walsh 64	Walsh 64	Walsh 64

Walsh chip within Walsh code 72 is
00000000 11111111 00000000 11111111 00000000 11111111 00000000 11111111
11111111 00000000 11111111 00000000 11111111 00000000 11111111 00000000
00000000 11111111 00000000 11111111 00000000 11111111 00000000 11111111
11111111 00000000 11111111 00000000 11111111 00000000 11111111 00000000

Example 10.41 The I-channel scrambled data is orthogonally spread with a Walsh code corresponding to the index number 72. Using Walsh code chips of Walsh 64 and $\overline{\text{Walsh 64}}$ listed in Example 10.38, the I-channel scrambled data (a gray block only, i.e., 1010000100) is orthogonally spread as follows:

I-channel Orthogonal Spreading

Scrambled data	Scrambled data $\oplus$ Walsh code 72
1 →	11111111 00000000 11111111 00000000 11111111 00000000 11111111 00000000
0 →	11111111 00000000 11111111 00000000 11111111 00000000 11111111 00000000
1 →	11111111 00000000 11111111 00000000 11111111 00000000 11111111 00000000
0 →	11111111 00000000 11111111 00000000 11111111 00000000 11111111 00000000
0 →	00000000 11111111 00000000 11111111 00000000 11111111 00000000 11111111
0 →	11111111 00000000 11111111 00000000 11111111 00000000 11111111 00000000
0 →	00000000 11111111 00000000 11111111 00000000 11111111 00000000 11111111
1 →	00000000 11111111 00000000 11111111 00000000 11111111 00000000 11111111
0 →	00000000 11111111 00000000 11111111 00000000 11111111 00000000 11111111
0 →	11111111 00000000 11111111 00000000 11111111 00000000 11111111 00000000

Example 10.42 Consider the Q-channel orthogonal spreading with Walsh code 72. Using the 256-chip within the Walsh code 72, the Q-channel orthogonal spreading is achieved by performing the modulo-2 addition of the Q-channel scrambled data sequence (a gray block only, i.e., 0110001011) with Walsh code chips.

Q-channel Orthogonal Spreading

Scrambled data	Scrambled data $\oplus$ Walsh code 72
0 →	00000000 11111111 00000000 11111111 00000000 11111111 00000000 11111111
1 →	00000000 11111111 00000000 11111111 00000000 11111111 00000000 11111111
1 →	11111111 00000000 11111111 00000000 11111111 00000000 11111111 00000000
0 →	11111111 00000000 11111111 00000000 11111111 00000000 11111111 00000000
0 →	00000000 11111111 00000000 11111111 00000000 11111111 00000000 11111111
0 →	11111111 00000000 11111111 00000000 11111111 00000000 11111111 00000000
1 →	11111111 00000000 11111111 00000000 11111111 00000000 11111111 00000000
0 →	11111111 00000000 11111111 00000000 11111111 00000000 11111111 00000000
1 →	11111111 00000000 11111111 00000000 11111111 00000000 11111111 00000000
1 →	00000000 11111111 00000000 11111111 00000000 11111111 00000000 11111111

10.4.13 Forward Traffic Channel Quadrature Modulation

The forward traffic channel after the orthogonal spreading is quadrature modulated by a pilot PN sequence at a fixed chip rate of 4.096 Mcps. Like the pilot, sync, and paging channel, the forward traffic channel also is spread in quadrature as shown in Fig. 10.8. A pilot PN sequence is generated from the long code generator as shown in Fig. 10.2. The long code satisfies the linear recursion specified by the generator tap polynomial $g(x) = 1 + x + x^2 + x^{22} + x^{32}$.

The quadrature modulation is accomplished by performing the modulo-2 addition of orthogonal spread data with the pilot code sequence.

Example 10.43 Using the same generator tap vector and the identical initial seed vector as specified in Examples 10.12 (sync channel) and 10.22 (paging channel), the pilot code sequence is generated as shown below:

10111101 01110100 00101110 10110010 10011110 11101011 01110011 11000001
00101000 11011010 01101101 11011100 10101101 00101100 01001010 00000101
11000010 10110101 01011011 01001111 10011110 11101101 00111011 10100000
00110000 10001111 10000001 10000000 11110011 10010010 01100000 00011010
11011100 01001000 00110010 01001000 10000111 11011001 01101001 10100001
11111111 01110001 11000000 00010101 11111001 00101011 00001010 10101011
01000101 00000000 01110101 01001101 10001000 01100011 01010010 00110100
00110100 00100101 10110111 11100001 01110100 10100101 00101110 11101101

Example 10.44 Consider the EX-ORing process for the forward traffic I channel. The quadrature modulation of the I-channel spread data by the pilot code sequence is performed by the following formula:

Quadrature modulation for I channel =
Orthogonally spread I data $\oplus$ Pilot code sequence
(see Example 10.41)   (see Example 10.43)
I-channel Quadrature Modulation

01000010 01110100 11010001 10110010 01100001 11101011 10001100 11000001
11010111 11011010 10010010 11011100 01010010 00101100 10110101 00000101
00111101 10110101 10100100 01001111 01100001 11101101 11000100 10100000
11001111 10001111 01111110 10000000 00001100 10010010 10011111 00011010
11011100 10110111 00110010 10110111 10000111 00100110 01101001 01011110
00000000 01110001 00111111 00010101 00000110 00101011 11110101 10101011
01000101 11111111 01110101 10110010 10001000 10011100 01010010 11001011
00110100 11011010 10110111 00011110 01110100 01011010 00101110 00010010

Example 10.45 Consider the EX-ORing process for the forward traffic Q channel. The quadrature modulation of the Q-channel spread data by the pilot code sequence is achieved by the following formula:

Quadrature modulation for Q channel =
Orthogonally spread Q data $\oplus$ Pilot code sequence
(see Example 10.42)   (see Example 10.43)
Q-channel Quadrature Modulation

10111101 10001011 00101110 01001101 10011110 00010100 01110011 00111110
00101000 00100101 01101101 00100011 10101101 11010011 01001010 11111010
00111101 10110101 10100100 01001111 01100001 11101101 11000100 10100000
11001111 10001111 01111110 10000000 00001100 10010010 10011111 00011010
11011100 10110111 00110010 10110111 10000111 00100110 01101001 01011110
00000000 01110001 00111111 00010101 00000110 00101011 11110101 10101011
10111010 00000000 10001010 01001101 01110111 01100011 10101101 00110100
11001011 00100101 01001000 11100001 10001011 10100101 11010001 11101101

10.4.14 Forward Traffic Channel Filtering

The forward traffic channel after the quadrature modulation is filtered in such a way that the I and Q impulses are applied to the inputs of the I and Q baseband filters as shown in Fig. 10.8. The baseband filters have a frequency response $S(f)$ that satisfies the limits given in Fig. 10.3.

Specifically, the normalized frequency response of the filter is contained within $\pm\delta_1$, in the passband $0 \leq f \leq f_P$ and is less than or equal to $-\delta_2$ in the stopband $f \geq f_S$. The numerical values for these parameters are $\delta_1 = 1.5$ dB, $\delta_2 = 40$ dB, $f_P = 1.96$ MHz, and $f_S = 2.47$ MHZ.

10.4.15 FTC Quadrature Phase Shift Keying

After baseband filtering, the binary I and Q data output from the baseband filters are mapped into phase according to Table 10.19, as shown in Fig. 10.6.

Table 10.19   Forward Traffic Channel I and Q Mapping

Phase (θ)	I		Q	
	NRZ	Binary	NRZ	Binary
$\pi/4$	1	0	1	0
$3\pi/4$	-1	1	1	0
$-3\pi/4$	-1	1	-1	1
$-\pi/4$	1	0	-1	1

In non-offset QPSK, the two baseband outputs coincide in time so that the carrier phase can be changed once every T_b seconds. An orthogonal QPSK waveform $s(t)$ is obtained by amplitude modulation of I and Q each onto the cosine and sine functions of a carrier wave as shown in Fig. 10.8. The in-phase impulse I is amplitude-modulated by the cosine function with an amplitude of $+1$ or -1, which produces a BPSK waveform. The quadrature-phase impulse Q modulates the sine function, resulting in a BPSK waveform orthogonal to the cosine function. Thus, the sum of the two orthogonal BPSK waveforms yields the QPSK waveform.

Let $s(t)$ denote the QPSK waveform as shown by

$$s(t) = I(t)\cos\omega_0 t + Q(t)\sin\omega_0 t$$
$$= \sqrt{2}\cos(\omega_0 t + \theta(t))$$

where $I(t) = \sqrt{2}\cos\theta(t)$ and $Q(t) = \sqrt{2}\sin\theta(t)$.

The QPSK stream $s(t)$ in accordance with the specific values of $I(t)$ and $Q(t)$ can be determined according to the chosen values of $\theta(t)$ as follows:

1. Since $I(t) = 1$ and $Q(t) = 1$ for $\theta(t) = \dfrac{\pi}{4}$, $s(t) = \sqrt{2}\cos\left(\omega_0 t - \dfrac{\pi}{4}\right)$

2. Since $I(t) = -1$ and $Q(t) = 1$ for $\theta(t) = \dfrac{3\pi}{4}$, $s(t) = \sqrt{2}\cos\left(\omega_0 t - \dfrac{3\pi}{4}\right)$

3. Since $I(t) = -1$ and $Q(t) = -1$ for $\theta(t) = -\dfrac{3\pi}{4}$, $s(t) = \sqrt{2}\cos\left(\omega_0 t + \dfrac{3\pi}{4}\right)$

4. Since $I(t) = 1$ and $Q(t) = -1$ for $\theta(t) = -\dfrac{\pi}{4}$, $s(t) = \sqrt{2}\cos\left(\omega_0 t + \dfrac{\pi}{4}\right)$

Thus, I(t) and Q(t) phase mapping for the forward W-CDMA channel (Pilot, Sync, Paging, and Forward traffic channels) are summarized in Table 10.19. Consequently, the resulting signal constellation and phase transition is as shown in Fig. 10.6.

Data for the QPSK waveform plot is tabulated in Table 10.20.

Table 10.20   Data Tabulation for the Plot of QPSK Waveform $s(t)$

$\omega_0 t$	$s(t) = \sqrt{2} \cos(\omega_0 t - \theta(t))$			
	(1,1), $\theta(t) = \pi/4$	(−1,1), $\theta(t) = 3\pi/4$	(−1,−1), $\theta(t) = -3\pi/4$	(1,−1), $\theta(t) = -\pi/4$
0	1	−1	−1	1
$\pi/4$	$\sqrt{2}$	0	$-\sqrt{2}$	0
$\pi/2$	1	1	−1	−1
$3\pi/4$	0	$\sqrt{2}$	0	$-\sqrt{2}$
π	−1	1	1	−1
$-3\pi/4$	$-\sqrt{2}$	0	$\sqrt{2}$	0
$-\pi/2$	−1	−1	1	1
$-\pi/4$	0	$-\sqrt{2}$	0	$\sqrt{2}$
2π	1	−1	1	1

Using Table 10.20, the QPSK pulse train $s(t)$ based on I(t) and Q(t) is drawn as shown in the following example.

Example 10.46   The I-channel quadrature modulation I(t) and the Q-channel quadrature modulation Q(t) discussed in Section 10.4.13 were computed in Examples 10.44 and 10.45, respectively.

Quadrature Modulation

I-channel : 01000010   01110100   11010001   10110010   . . .

Q-channel : 10111101   10001011   00101110   01001101   . . .

Considering only the gray pair of (I, Q) pulses, the QPSK waveform $s(t)$ is shown as illustrated in Fig. 10.11.

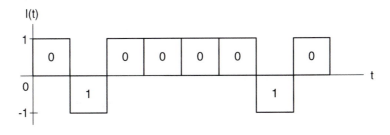

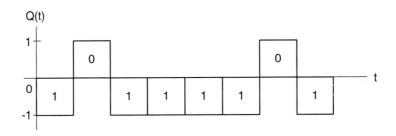

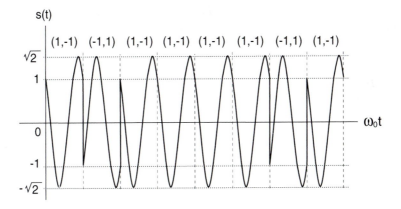

Figure 10.11 The QPSK waveform s(t) corresponding to the

given I(t)=(01000010) and Q(t)=(10111101)

References

1. ANSI X 3.92, "American National Standard for Data Encryption Algorithm (DEA)," American National Standard Institute, 1981.
2. CT-2, "Second Generation Cordless Telephone (CT-2), Common Air Interface Specifications," Department of Trade and Industry, London, May 1989.
3. Digital European Cordless Telecommunications, Part I, "Overview," DE/RES 3001-1, Common Interface, Radio Equipment and Systems, ETS 300 175-1, ETSI, B. P. 152, F-06561 Valbonne Cedex, France, August 1991.
4. Digital European Cordless Telecommunications, "Radio Equipment and Systems" European Telecommunications Standards Institute, ETSI, Valbonne, Cedex, France, May 1991.
5. EIA, "Dual-Mode Subscriber Equipment Compatibility Specification," Electronic Industries Association Specification IS-54, EIA Project Number 2215, Washington, D. C., May 1990.
6. EIA/TIA/IS-54-B, "Cellular System Dual Mode Mobile Station—Base Station Compatibility Standard," April 1992.
7. EIA/TIA, "Mobile Station—Base Station Compatibility Standard for Dual-Mode Wideband Spread Spectrum Cellular System," TIA/EIA/IS-95, July, 1993.
8. EIA/TIA-Qualcomm, Inc, "Spread Spectrum Digital Cellular System Dual-Mode Mobile Station—Base Station Compatibility Standard," Proposed EIA/TIA Interim Standard, April 21, 1992; TIA Distribution TR 45.5, April 1992.
9. "European Digital Cellular Telecommunications System (Phase 2): General Description of a GSM Public Land Mobile Network," ETSI, 06921 Sophia Antipolis Cedex, France, GSM 01-12, October 1993.
10. Feher, K., *Wireless Digital Communications*, Prentice Hall, New Jersey, 1995.
11. Guest Editors (Elkaheem, Shilling, Baier, and Nakagawa), "Code Division Multiple Access Network I," IEEE Journal on Selected Areas in Communications, Vol. 12, No. 5, pp. 560–762, May, 1994.
12. Guest Editors (Elkaheem, Shilling, Baier, and Nakagawa), "Code Division Multiple Access Network II," IEEE Journal on Selected Areas in Communications, Vol. 12, No. 5, pp. 774–983, June, 1994.
13. ISO/IEC 9594-8, "Information Proceeding Systems—Open Systems Interconnection—The Directory" Part 8: Authentication Framework, 1990.
14. ISO/IEC 9796, "Information Technology—Security Techniques—Digital Signature Giving Message Recovery," 1991.

15. ISO/IEC 9797, "Information Technology—Security Techniques—Data Integrity Mechanism Using a Crypto-graphic Check Function Employing a Block Cipher Algorithm," 1994.

16. ISO/IEC 9979, "Data Cryptographic Techniques—Procedures for the Registration of Cryptographic Algorithms," 1991.

17. ISO/IEC 10116, "Information Technology—Security Techniques—Modes of Operation for an n-bit Block Cipher Algorithm," 1991.

18. ISO/IEC 10118-1, "Information Technology—Security Techniques—Hash Functions," Part 1: General, October 15, 1994.

19. ISO/IEC 10118-2, "Information Technology—Security Techniques—Hash Functions," Part 2: Hash Functions using an n-bit Block cipher Algorithm, 1994 (E).

20. ISO/IEC 10181-2, "Information Technology—Security Frameworks for Open System," Part 2: Authentication Framework, 1995.

21. ISO/IEC 2nd CD 9798-5, "Information Technology—Security Techniques—Entity Authentication" PART 5: Mechanisms Using Zero Knowledge Techniques, 1996.

22. ITU-R Recommendation, 687, "Future Public Land Mobile Telecommunication System," Question 39/8, 1990.

23. ITU-T Recommendation E.212, "Identification Plan for Land Mobile Stations."

24. ITU-T Recommendation, G.721, "32 kbps Adaptive Differential Pulse Code Modulation (ADPCM)," Melbourne, 1988.

25. ITU-T Recommendation V.22, "1200 bits per second Duplex Modem Standardized for Use in the General Switched Telephone Network and on Point-to-Point 2-wire Leased Telephone-Type Circuits," Melbourne, 1988.

26. Kohl, J. T., "The Evolution of the Kerberos Authentication Service," European Conference Proceedings, pp. 295–313, May 1991.

27. Lai X. and J. Massey, "Hash Functions Based on Block Ciphers," Advances in Cryptology—EUROCRYPT'92 Proceedings, Berlin: Springer-Verlag, pp. 55–70, 1992.

28. Lee, W.C.Y., *Mobile Cellular Telecommunications*, McGraw-Hill, New York, 1995.

29. Merkle, R.C., "A Fast Software One-Way Hash Function," Journal of Cryptology, Vol. 3, No. 1, pp. 43–58, 1990.

30. Miller, S.P., B.C. Newman, J. I. Schiwer, and J. H. Salttzer, "Section E.2.1: Kerberos Authentication and Authorization System," MIT Project Athena, Dec. 1987.

31. Motorola, Inc. (Author & Source), "Personal Access Communications System-Unlicensed Version B (PACS-UB) Air Interface," PN 3520-Ballot text for TAG3, JTC(AIR)/95.03.13-066R2, March 13, 1995.

32. Natyas, S. M., C. H. Meyer, and J. Oseas, "Generating Strong One-Way Functions with Cryptographic Algorithms" IBM Technical Disclosure Bulletin, Vol. 27, No. 10A, pp. 5658–5659, March 1985.

33. NBS FIPS PUB 46, "Data Encryption Standard," National Bureau of Standards, U.S. Department of Commerce, January 1977.

34. NIST FIPS PUB 180, "Secure Hash Standard" National Institute of Standards and Technology, U.S. Department of Commerce, DRAFT, April 1993.

35. PCS 2000, "A Composite CDMA/TDMA Air Interface Compatibility Standard for Personal Communications in 1.8-2.2 GHz for Licensed Applications," Version 2.

36. "PDC-Digital Cellular Telecommunication System," RCR STF-27A Version, Research & Development Center for Radio System (RCR), Nippon Ericsson K. K., January 1992.

37. PHS-Personal Handy Phone Standard, Research Development Center for Radio System (RCR), "Personal Handy Phone Standard (PHS)," CRC STD-82, December 20, 1993.

38. QUALCOMM Incorporated, "CDMA System Engineering Training Handbook," Volume 2, Document Number: 80-12015, Rev. X1, April 1993.

39. QUALCOMM Incorporated, "CDMA System Engineering Training Handbook," Preliminary, Document Number: 80-12015, Rev. X1, April 1993.

40. QUALCOMM Incorporated, "The CDMA Network Engineering Handbook," Volume 1: Concepts in CDMA, Document Number: AT80-10497, Rev. X1, March 1, 1993.

41. Rhee, M. Y., *Cryptography and Secure Communications*, McGraw-Hill, New York, 1994.

42. Rhee, M. Y., "CDMA Digital Mobile Communications and Message Security," Keynote Speech at 1996 International Computer Symposium, Kaohsiung, Taiwan, December 1996.

43. Rivest R., A. Shamir, and L. Adelman, "A Method for Obtaining Digital Signatures and Public-Key Cryptosystems," Communications of the ACM, Vol. 21, No. 2, 1978.

44. Rivest, R., "The MD4 Message Digest Algorithm," RFC 1186, MIT, October 1990.

45. Rivest, R., "The MD5 Message Digest Algorithm," RFC 1321, MIT, April 1992.

46. Schneier, B., *Applied Cryptography: Protocols, Algorithms, and Source Code in C*, Wiley, New York, 1994.

47. Simmons, G. J. (Edited), *Contemporary Cryptology*: *The Science of Information Integrity*, IEEE Press, New York, 1992.

48. Stallings, W., *Network and Internetwork Security—Principles and Practice*, Prentice Hall, Englewood Cliffs, New Jersey, 1995.

49. TIA/EIA/IS-95, "Mobile Station-Base Station Compatibility Standard for Dual-Mode Wideband Spread Spectrum Cellular System," July 1993.

50. TR 45, "Recommended Minimum Performance Standards for Base Stations Supporting Dual-Mode Wideband Spread Spectrum Cellular Mobile Stations," PN-3120 (to be published as IS-97), February 18, 1994.

51. TR 45, "Recommended Minimum Performance Standards for Dual-Mode Wideband Spread Spectrum Cellular Mobile Stations," PN-3121 (to be published as IS-98), February 18, 1994.

52. Viterbi, A. J., CDMA, *Principles of Spread Spectrum Communication*, Addison-Wesley, 1995.

53. Winternity, R. S., "Producing One-Way Hash Functions from DES," Advance in Cryptology: Proceeding of Crypto 83, Plenum Press, pp. 203–207, 1984.

Index

–R–

–S–

–T–